Lecture Notes in Physics

Springer
Berlin
Heidelberg
New York
Barcelona
Hong Kong
London
Milan
Paris
Tokyo

Physics and Astronomy ONLINE LIBRARY

http://www.springer.de/phys/

Editorial Policy

The series *Lecture Notes in Physics* (LNP), founded in 1969, reports new developments in physics research and teaching -- quickly, informally but with a high quality. Manuscripts to be considered for publication are topical volumes consisting of a limited number of contributions, carefully edited and closely related to each other. Each contribution should contain at least partly original and previously unpublished material, be written in a clear, pedagogical style and aimed at a broader readership, especially graduate students and nonspecialist researchers wishing to familiarize themselves with the topic concerned. For this reason, traditional proceedings cannot be considered for this series though volumes to appear in this series are often based on material presented at conferences, workshops and schools (in exceptional cases the original papers and/or those not included in the printed book may be added on an accompanying CD ROM, together with the abstracts of posters and other material suitable for publication, e.g. large tables, colour pictures, program codes, etc.).

Acceptance

A project can only be accepted tentatively for publication, by both the editorial board and the publisher, following thorough examination of the material submitted. The book proposal sent to the publisher should consist at least of a preliminary table of contents outlining the structure of the book together with abstracts of all contributions to be included.
Final acceptance is issued by the series editor in charge, in consultation with the publisher, only after receiving the complete manuscript. Final acceptance, possibly requiring minor corrections, usually follows the tentative acceptance unless the final manuscript differs significantly from expectations (project outline). In particular, the series editors are entitled to reject individual contributions if they do not meet the high quality standards of this series. The final manuscript must be camera-ready, and should include both an informative introduction and a sufficiently detailed subject index.

Contractual Aspects

Publication in LNP is free of charge. There is no formal contract, no royalties are paid, and no bulk orders are required, although special discounts are offered in this case. The volume editors receive jointly 30 free copies for their personal use and are entitled, as are the contributing authors, to purchase Springer books at a reduced rate. The publisher secures the copyright for each volume. As a rule, no reprints of individual contributions can be supplied.

Manuscript Submission

The manuscript in its final and approved version must be submitted in camera-ready form. The corresponding electronic source files are also required for the production process, in particular the online version. Technical assistance in compiling the final manuscript can be provided by the publisher's production editor(s), especially with regard to the publisher's own Latex macro package which has been specially designed for this series.

Online Version/ LNP Homepage

LNP homepage (list of available titles, aims and scope, editorial contacts etc.):
http://www.springer.de/phys/books/lnpp/

LNP online (abstracts, full-texts, subscriptions etc.):
http://link.springer.de/series/lnpp/

M. Lemoine G. Sigl (Eds.)

Physics and Astrophysics of Ultra-High-Energy Cosmic Rays

Springer

Editors

Martin Lemoine
Günter Sigl
Centre National de la Recherche Scientifique
Institut de Astrophysique de Paris
98bis, Boulevard Arago
75014 Paris, France

Cover picture: Hadronic, electromagnetic, and muonic (in order of increasing range) components of a giant air shower (adapted from P. Billoir's contribution).

Library of Congress Cataloging-in-Publication Data applied for.

Die Deutsche Bibliothek - CIP-Einheitsaufnahme

Physics and astrophysics of ultra high energy cosmic rays / M. Lemoine ; G. Sigl (ed.). - Berlin ; Heidelberg ; New York ; Barcelona ; Hong Kong ; London ; Milan ; Paris ; Tokyo : Springer, 2001
(Lecture notes in physics ; Vol. 576)
(Physics and astronomy online library)
ISBN 3-540-42899-2

ISSN 0075-8450
ISBN 3-540-42899-2 Springer-Verlag Berlin Heidelberg New York

Springer-Verlag Berlin Heidelberg New York
a member of BertelsmannSpringer Science+Business Media GmbH

http://www.springer.de

Typesetting: Camera-ready by the authors/editors
Camera-data conversion by Steingraeber Satztechnik GmbH Heidelberg
Cover design: *design & production*, Heidelberg

Printed on acid-free paper
SPIN: 10857548 54/3141/du - 5 4 3 2 1 0

Preface

The International School on Physics and Astrophysics of Ultra High Energy Cosmic Rays (UHECR2000) was held at the Observatoire de Paris–Meudon on June 26-29, 2000. This was the first international school specifically dedicated to ultra high energy cosmic rays. Its aim was to familiarize with and attract students, physicists and astronomers into this quickly developing new research field.

The mysterious and currently unknown origin of the most energetic particles observed in Nature has triggered in recent years theoretical speculations ranging from electromagnetic acceleration to as yet undiscovered physics beyond the Standard Model. It has also lead to the development of several new detection concepts and experimental projects, some of which are currently under construction. By its nature, the field of ultra high energy cosmic rays is therefore highly interdisciplinary and borrows from astrophysics and cosmology, via particle physics, to experimental physics and observational astronomy. One main aspect of the school was to emphasize and take advantage of this interdisciplinarity. The lectures were grouped into subtopics and are reproduced in this volume in the following order: After a general introductory lecture on cosmic rays follow two contributions on experimental detection techniques, followed by three lectures on acceleration in astrophysical objects. The next four contributions cover all major aspects of propagation and interactions of ultra high energy radiation, including speculative issues such as new interactions. The last lecture discusses "top-down" scenarios where cosmic rays are produced by decay from higher energies close to Grand Unification scale instead of being accelerated. The volume is rounded off by a critical summary of the topics covered. We hope that this topical book will be useful to a broad range of people interested in ultra high energy cosmic rays, from beginning students looking for interesting research projects to senior researchers.

The school was financially supported by the Observatoire de Paris–Meudon, by the Centre National de la Recherche Scientifique (CNRS), and by the Programme National de Cosmologie (PNC). We warmly thank our co-organizers Murat Boratav, Antoine Letessier-Selvon and Patrick Peter.

Paris,
September 2001

Martin Lemoine
Günter Sigl

Table of Contents

List of Contributors

Eun-Joo Ahn
Physics Department, Seoul National University, Seoul, Korea

Pijushpani Bhattacharjee
Indian Institute of Astrophysics
Koramangala
Bangalore 560 034, India
pijush@iiap.ernet.in

Peter L. Biermann
Max-Planck Institute for Radioastronomy
Auf dem Hügel 69
53121 Bonn, Germany
plbiermann@mpifr-bonn.mpg.de

Pierre Billoir
CNRS, Universités PARIS VI
et PARIS VII
4 place Jussieu
75252 Paris Cedex 05, France
billoir@in2p3.fr

Murat Boratav
Université Paris 6, LPNHE
4, place Jussieu
75252 Paris Cedex 5, France
boratav@in2p3.fr

Philipp P. Kronberg
Physics Department, University of Toronto
60 St. George Street
Toronto, Ontario, M5S 1A7, Canada
kronberg@astro.utoronto.ca

Antoine Letessier-Selvon
Université Paris 6, LPNHE
4, place Jussieu
75252 Paris Cedex 5, France
Antoine.Letessier-Selvon@in2p3.fr

Gustavo Medina-Tanco
Instituto Astronômico e Geofisico,
Universidado de São Paulo
Av. Miguel Stefano 4200
CEP 04301-904, São Paulo, Brazil
gustavo@iagusp.usp.br

Guy Pelletier
Laboratoire d'Astrophysique de Grenoble
BP 53
38041 Grenoble Cedex 9, France
Guy.Pelletier@obs.ujf-grenoble.fr

Bronislaw Rudak
Copernikus Astronomical Center
Rabiańska 8
87100 Toruń, Poland
bronek@ncac.torun.pl

Günter H.W. Sigl
CNRS, Institut d'Astrophysique de Paris
98bis, boulevard Arago
75014 Paris, France
sigl@iap.fr

Todor Stanev
Bartol Research Institute, University of Delaware
Newark, DE 19716, USA
stanev@bartol.udel.edu

Eli Waxman
Condensed Matter Physics Department, Weizman Institute
Rehovot 76100, Israel
waxman@wicc.weizmann.ac.il

Shigeru Yoshida
Institute for Cosmic Ray Research, University of Tokyo
5-1-5 Kashiwa-no-Ha
Kashiwa City, Chiba 277-8582, Japan
syoshida@icrr.u-tokyo.ac.jp

Introduction to Cosmic Rays

Peter L. Biermann[1] and Günter Sigl[2]

[1] Max-Planck Institute for Radioastronomy, and
Department for Physics and Astronomy, University of Bonn, Bonn, Germany
[2] Institut d'Astrophysique de Paris, Paris, France

Abstract. Energetic particles, traditionally called *Cosmic Rays*, were discovered nearly a hundred years ago, and their origin is still uncertain. Their main constituents are the normal nuclei as in the standard cosmic abundances of matter, with some enhancements for the heavier elements; there are also electrons, positrons and antiprotons. Today we also have information on isotopic abundances, which show some anomalies, as compared with the interstellar medium. And there is antimatter, but no anti-nuclei. The known spectrum extends over energies from a few hundred MeV to 300 EeV ($= 3 \times 10^{20}$ eV), and shows few clear spectral signatures: There is a small spectral break near 5×10^{15} eV, commonly referred to as the *knee*, where the spectrum turns down; there is another spectral break near 3×10^{18} eV, usually called the *ankle*, where the spectrum turns up again. Up to the ankle the cosmic rays are usually interpreted as originating from supernova explosions, i.e. those cosmic ray particles are thought to be Galactic in origin; however, the details are not clear. We do not know what the origin of the knee is, and what physical processes can give rise to particle energies in the energy range from the knee to the ankle. The particles beyond the ankle have to be extragalactic, it is usually assumed, because the Larmor radii in the Galactic magnetic field are too large; this argument could be overcome if those particles were very heavy nuclei as Fe, an idea which appears to be inconsistent, however, with the airshower data immediately above the energy of the ankle. Due to interaction with the cosmic microwave background (CMB), a relic of the Big Bang, there is a strong cut-off expected near 50 EeV ($=5 \times 10^{19}$ eV), which is, however, not seen; this expected cutoff is called the GZK-cutoff after its discoverers, Greisen, Zatsepin and Kuzmin. The spectral index α is near 2.7 below the knee, near 3.1 above the knee, and again near 2.7 above the ankle, where this refers to a differential spectrum of the form $E^{-\alpha}$ in numbers. The high energy cosmic rays beyond the GZK-cutoff are the challenge to interpret. We will describe the various approaches to understand the origin and physics of cosmic rays.

1 Introduction and History

Cosmic Rays were discovered by Hess [1] and Kohlhörster [2] in the beginning of the twentieth century through their ionizing effect on airtight vessels of glas enclosing two electrodes with a high voltage between them. This ionizing effect increased with altitude during balloon flights, and therefore the effect must come from outside the Earth. So the term *Cosmic Rays* was coined. The Earth's magnetic field acts on energetic particles according to their charge, they are differently affected coming from East and West, and so their charge was detected, proving once and for all that they are charged particles. At the energies near 10^{18} eV there is observational evidence, that a small fraction of the particles are

Martin Lemoine, Günter Sigl (Eds.): LNP 576, pp. 1–26, 2001.

neutral, and in fact neutrons; these events correlate on the sky with the regions of highest expected cosmic ray interactions, the Cygnus region and the Galactic center region. From around 1960 onwards particles were detected at or above 10^{20} eV, with today about two dozen such events known. It took almost forty years for the community to be convinced that these energies are real, and this success is due to the combination of air fluorescence data with ground-based observations of secondary electrons/positrons and muons, as well as Čerenkov light; the Fly's Eye [3], Haverah Park [4] and AGASA [5] arrays are those with the most extensive discussion of their data out and published; other arrays have also contributed a great deal, like Yakutsk [8], Volcano Ranch [9] and SUGAR [10]. Already in the fifties it was noted that protons with energies above 3×10^{18} eV have Larmor radii in the Galactic magnetic field which are too large to be contained, and so such particles must come from outside [11]. After the CMB was discovered, in the early 1960s, it was noted only a little later by Greisen [12], and Zatsepin and Kuzmin [13], in two papers, that near and above an energy of 5×10^{19} eV (called the GZK-cutoff) the interaction with the CMB would lead to strong losses, if these particles were protons, as is now believed on the basis of detailed airshower data. In such an interaction, protons see the photon as having an energy of above the pion mass, and so pions can be produced in the reference frame of the collision, leading to about a 20 % energy loss of the proton about every $\simeq 6\,$Mpc in the observer frame. Therefore for an assumed cosmologically homogeneous distribution of sources for protons at extreme energies, a spectrum at Earth is predicted which shows a strong cutoff at 5×10^{19} eV, the GZK-cutoff. This cutoff is not seen, leading to many speculations as to what the nature of the particles beyond the GZK-energy, and their origin might be.

Cosmic rays are measured with balloon flights, satellites, now with instruments such as AMS [14] on the Space Shuttle, and soon also with instruments on the International Space Station [15], and with Ground Arrays. The instrument chosen depends strongly on what is being looked for, and the energy of the primary particle. One of the most successful campaigns has been with balloon flights in Antarctica, where the balloon can float at about 40 km altitude and circumnavigate the South Pole once, and possibly even several times during one Antarctic summer. For very high precision measurements very large instruments on the Space Shuttle or soon the International Space Station have been or will be used, such as for the search for antimatter. The presently developed new experiments such as the fluorescence detector array HiRes [17] and the hybrid array Auger [18], are expected to contribute decisively to the next generation of data sets for the highest energies.

Critical measurements are today the exact spectrum of the most common elements, Hydrogen and Helium, the energy dependence of the fraction of anti-particles (anti-protons and positrons), isotopic ratios of elements such as Neon and Iron, the ratio of spallation products such as Boron to the primary nuclei such as Carbon as a function of energy, the chemical composition near and beyond the knee, at about 5×10^{15} eV, and the spectrum and nature of the particles beyond the ankle, at 3×10^{18} eV, with special emphasis on the particles beyond the expected GZK-cutoff, at $\simeq 5 \times 10^{19}$ eV. The detection of anti-nuclei would constitute a rather extreme challenge. One of the most decisive points is the quest for the highest energy events and the high energy cutoff in the spectrum. This is also the main topic of the present volume. The data situation and experimental issues involved at the highest energies have been reviewed in Refs. [19,20].

Relevant reviews and important original papers have been published over many years, e.g., [21,22,23,24,25,26,27,28].

2 Physical Concepts

2.1 Cosmic Ray Spectrum and Isotropy

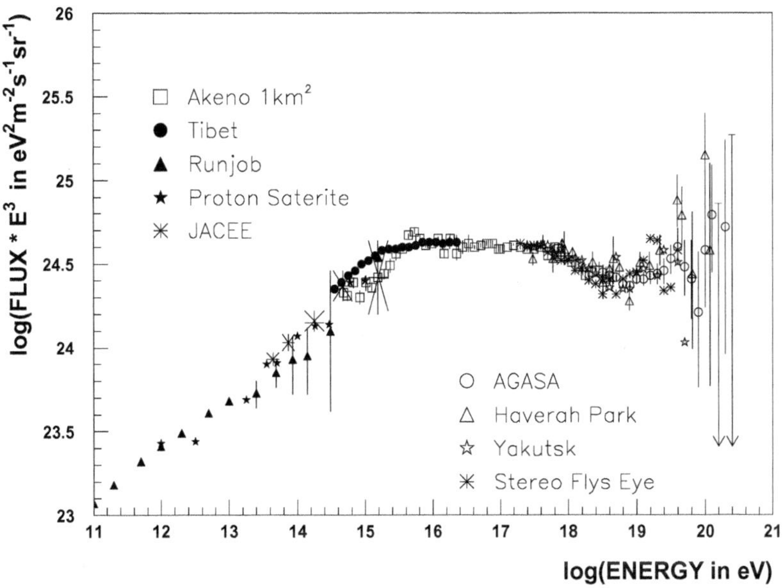

Fig. 1. The CR all-particle spectrum observed by different experiments above 10^{11} eV (from Ref. [20]). The differential flux in units of events per area, time, energy, and solid angle was multiplied with E^3 to project out the steeply falling character. The "knee" can be seen at $E \simeq 4 \times 10^{15}$ eV, the "second knee" at $\simeq 3 \times 10^{17}$ eV, and the "ankle" at $E \simeq 5 \times 10^{18}$ eV

The number of particles at a certain energy E within a certain small energy interval dE is called the spectrum. Cosmic rays have usually a powerlaw spectrum, which is referred to as a non-thermal behaviour, since non-thermal processes are thought to be producing such spectra. Flux is usually expressed as the number of particles, coming in per area, per second, per solid angle in steradians (all sky is 4 π), and per energy interval. Cosmic rays have a spectrum near $E^{-2.7}$ up the the knee, at about 5×10^{15} eV, and then about $E^{-3.1}$ beyond, up the ankle, at about 3×10^{18} eV, beyond which the spectrum becomes hard to quantify, but can very approximately again be described by $E^{-2.7}$. There is no other strong feature in the spectrum, especially no cutoff at the upper end. There is some limited evidence from the newest experiments (AGASA [5] and HiRes [17,29]) for another feature, at about 3×10^{17} eV, called *the second knee*, where the spectrum appears to dip. Both the first and the second knee may be at an energy which is proportional to charge [30], i.e. at a constant Larmor radius, and therefore may imply a range in energies per particle. Figure 1 shows the overall cosmic ray spectrum.

There is no anisotropy except for a weak hint near 10^{18} eV [31,32,33], and the suggestive signal for pairing at energies near and beyond the GZK-cutoff [34].

2.2 Fermi Acceleration

In a compressing system the particles gain energy; the walls can be magnetic irregularities which reflect charged particles through magnetic resonance between the gyromotion and waves in the ionized magnetic gas, the plasma. Such magnetic irregularities usually exist everywhere in a plasma that gets stirred by, e.g., stellar ultraviolet radiation and their ionization fronts, by stellar winds, supernova explosions, and by the energetic particles moving through. Considering now the two sides of a shock, one realizes that this is a permanently compressing system for charged particles which move much faster than the flow in the shock frame. Therefore particles gain energy, going back and forth. In one cycle they normally gain a fraction of U_{sh}/c in momentum (adopting relativistic particles here), and the population loses a fraction of also U_{sh}/c. Here U_{sh} is the shock velocity. For the original articles by E. Fermi see Ref. [21], Ref. [35] for a recent review, and see also the contribution by G. Pelletier in this volume.

The density jump r in an adiabatic shockfront is given by the adiabatic index of the gas γ and the upstream Mach number of the shock M_1

$$r = \frac{\gamma + 1}{\gamma - 1 + 2/M_1^2} \tag{1}$$

The general expression for the spectral index of the particle momentum distribution p^{-a} is

$$a = \frac{3r}{r - 1} \tag{2}$$

This is in three-dimensional phase space; the energy distribution then is given by E^{2-a}, for relativistic particles. This means, for instance, that for a very large Machnumber and the standard case of $\gamma = 5/3$ the density jump is 4, and the spectral index is $a - 2 = 2$. For $\gamma = 4/3$, as would be the case in a gas with a relativistic equation of state (like a radiation dominated gas) the density jump is 7, and the spectral energy index of the particles is $a - 2 = 3/2$. The time scale for acceleration is given in, e.g., [36,37].

In a relativistic shock wave the derivation no longer holds so simply for the spectrum; however, it is worth noting that the density jump can go to infinity both in the case of a relativistic shockwave as in the case of a strong cooling shock. Then the spectral index in energy approaches $a-2=1$. However, detailed Monte-Carlo simulations for relativistic shocks, taking into account the highly anisotropic nature of the scattering as well as the particle distribution, again find a spectrum near 2 [38]. For more details on Fermi acceleration see also the contribution by G. Pelletier in this volume.

2.3 Spallation

Spallation is the destruction of atomic nuclei in a collision with another energetic particle, such as another nucleus, commonly a proton [39,40,41]. In this destruction many pieces of debris can be formed, with one common result the stripping of just one proton or neutron, and another common result a distribution of lighter nuclei. Since the proton number determines the chemical element, these debris are usually other nuclei, such as Boron, from the destruction of a Carbon nucleus. It is an interesting question, whether these collisions lead to a new state of matter, the quark-gluon plasma; the Relativistic Heavy Ion Collider (RHIC) experiment [42] performed at Brookhaven collides heavy nuclei with each other, in order to find evidence for this new state. Both in our upper atmosphere and out in the Galaxy such collisions happen all the time, at very much higher energy than possible in the laboratory, and may well be visible in the data. Conversely, the existing data could be used perhaps to derive limits on what happens when a quark-gluon plasma is formed.

As a curiosity we mention that collisions of energetic cosmic rays with each other and with large objects such as the moon have been used to constrain the risk that high energy collisons in terrestrial accelerators could produce particles or new vacuum states that would trigger a phase transition to a lower energy state such as strange quark matter which would destroy the Earth [43]. This risk can be determined by calculating how much more often such processes occurred naturally involving cosmic rays since the birth of our Universe.

2.4 Chemical Abundances

The chemical abundances in cosmic rays are rather similar to first approximation to those in the interstellar medium [44]. We consider them in the following framework: We plot the number of particles per energy interval as a function of energy per particle, and normalize at 1 TeV energy per particle, so as to be free of any solar modulation effect [45]. And we refer to Silicon for the comparison, so by definition the abundance for Silicon is adopted to be equal for cosmic rays and for the so-called cosmic abundances in the interstellar medium. In this well defined frame-work we then note the following differences:

- The abundance of Hydrogen is very much less for cosmic rays, as is the ratio of Hydrogen to Helium.

- The abundances of the elements Lithum, Beryllium and Boron are very much larger in cosmic rays than in the interstellar medium, by several powers of ten.
- The abundances of the sub-Iron elements are also larger than relative to Iron for cosmic rays.
- The abundances of odd-Z elements are larger.
- And, finally, those elements with a low first ionization potential are systematically more abundant.

These tendencies can be seen in Fig. 2 which compares solar System abundances with abundances in cosmic rays at 1 TeV.

In addition, the isotopic ratios among a given element are sometimes very similar to those in the interstellar medium, and for other cases, very different, indicating rather specific source contributors.

In all versions of theories it is acknowledged that spallation of abundant elements plays a major role, especially for the light elements, where spallation and subsequent ionization loss can even explain the abundances of the light elements in the interstellar medium. This is an especially interesting test using the light element abundances in stars formed in the young years of our Galaxy [46].

2.5 Cosmic Ray Airshower

When a primary particle at high energy, either a photon, or a nucleus, comes into the upper atmosphere, the sequence of interactions and cascades form an airshower. This airshower can be dominated by Čerenkov light, a bluish light, produced when particles travel at a speed higher than the speed of light c divided by the local index of refraction (which is 4/3 in water, for instance, and about 1.0003 in air). Observing this bluish light allows observations of high GeV to TeV photon sources in the sky. For particles, such as protons, or atomic nuclei, the resulting airshower is dominated by air fluorescence, when normal emission lines of air molecules are excited, and by a pancake of secondary electrons and positrons as well as muons. Most modern observations of very high energy cosmic rays are done either by observing the air fluorescence, (arrays such as Flys's Eye [3], HiRes [17], or Auger [18]), or by observing the secondary electrons and positrons (in arrays such as Haverah Park [4], AGASA [5], Yakutsk [8], or also Auger [18]). In the further future such observations may be possible from space, by observing the air fluorescence, or also the reflected Čerenkov light, from either the International Space Station, or from dedicated satellites. Fly's Eye was and HiRes is in Utah, USA, Auger is in Argentina, AGASA is in Japan, Yakutsk is in Russia, and Haverah Park was in the United Kingdom. Future planned experiments are EUSO [47] on the space station, built by the ESA, and, even later, a satellite mission, OWL [49], discussed by NASA. For reviews of experimental techniques to detect giant airshowers see Refs. [19,20].

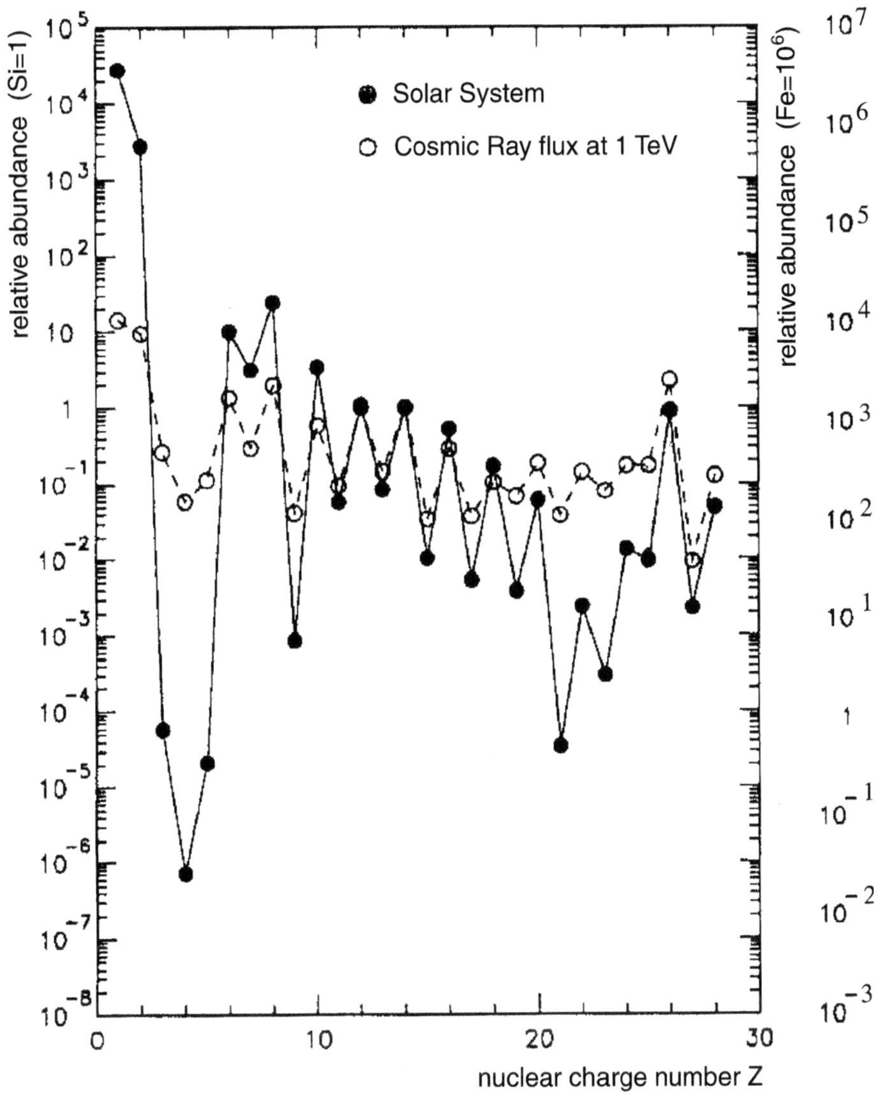

Fig. 2. The chemical composition of cosmic rays relative to Silicon and iron at 1 TeV, and in the solar System, as a function of nuclear charge Z, from Ref. [45]

2.6 Cosmic Ray GZK-Cutoff

The interactions with the CMB should produce a strong cutoff in the observed spectrum, at 5×10^{19} eV, called the GZK-cutoff [12,13,50]. This is expected provided that a) these particles are protons (or neutrons), and b) the source distribution is homogeneous in the universe. This cutoff is not seen; in fact, no cutoff is seen at any energy, up to the limit of data, at $\simeq 3 \times 10^{20}$ eV, or 300 EeV. This is one of the most serious problems facing cosmic ray physics today. Assuming a source distribution just as the observed galaxy distribution alleviates the problem, but does not solve it [51,52] (see also the contribution by G. Medina Tanco in this volume).

2.7 Black Holes

It is now believed that almost all galaxies have a massive black hole at their center, with masses sometimes ranging up 10^{10} solar masses, but usually much less. There are also stellar mass black holes, but their number is not well known, probably many thousands in each galaxy. The growth of these black holes has almost certainly put an enormous amount of energy into the universe, possibly commensurate with other forms of baryonic energy. The ratio of the masses of the black holes and the stellar spheroidal component of older stars has a narrow distribution which is limited from above by about 1:300. There is a near perfect correlation between black hole mass and the velocity dispersion of the inner stars of the central cusp around the black hole [53,54,55,56]. These black holes can be expected to interact strongly with their environment, both in stars and in gas [57].

2.8 Our Galaxy

Our galaxy is a flat distribution of stars and gas, mixed with interstellar dust, and embedded in a spheroidal distribution of old stars. The age of this system is about 15 billion years; its size is about 30 kpc across, and its inner region is about 6 kpc across. At its very center there is a black hole with 2.6×10^6 solar masses [58]. The gravitational field is dominated in the outer parts of the Galaxy by an unknown component, called dark matter, which we deduce only through its gravitational force. In the innermost part of the galaxy normal matter dominates. The mass ratio of dark matter to stars to interstellar matter in our Galaxy is about 100:10:1. Averaged over the nearby universe these ratios are shifted in favor of gas, with gas dominating over stars probably, but with dark matter still dominating over stars and gas by a large factor. The universal ratios of baryonic matter, dark matter and the Λ-term have been tightly constrained by the observation of the first three waves in the fluctuation spectrum of the CMB by the balloon experiments BOOMERanG [59], MAXIMA [60], and the ground detector DASI [61], as well as by measurements of the relation between apparent magnitude and redshift of certain type Ia supernovae which serve as "standard candles" of known absolute luminosity [62]. All experiments agree rather well in these conclusions [63]. In very small galaxies the dark matter component dominates over baryonic matter even at the center [64,65].

2.9 Interstellar Matter

The gas in between the stars in our Galaxy is composed of very hot gas (order 4×10^6 K), various stages of cooler gas, down to about 20 K, dust, cosmic rays, and magnetic fields [66,67,68,69]. All three components, gas, cosmic rays, and magnetic fields, have approximately the same energy density, which happens to be also close to the energy density of the CMB, about 1 eV per cm^3. The average density of the neutral hydrogen gas, of temperature a few 10^3 K, is about 1 particle per cm^3, in a disk of thickness about 100 pc ($= 3 \times 10^{20}$ cm). The very hot gas extends much farther from the symmetry plane, about 2 kpc on either side.

2.10 Magnetic Fields

Magnetic fields are ubiquitous in the Universe [70,71,72] (see also the contribution by G. Medina Tanco in this volume). In our Galaxy they have a total strength of about 6 - 7 microGauss (μG) in the solar neighborhood, and about 10 μG further in, at around 3 kpc from the center. The magnetic field is partially irregular, partially regular, with roughly 1/2 to 2/3 of it in a circular ring-like pattern; other galaxies demonstrate that the underlying symmetry is dominated by a spiral structure with the overall magnetic field pointing inwards along the spiral. One level down in scale, the fine structure is then of occasional reversals, but still mostly parallel to a circle around the center. At the small scales, less than the thickness of the hot disk, it appears that the magnetic field can be described as a Kolmogorov turbulence spectrum [73], all the way down to dissipation scales.

The origin of the magnetic field is not understood [74,75,76]. Comparing our Galaxy with others, in the starburst phase, and also at high redshift makes it obvious that the magnetic field is regenerated at time scales which are less or at most equal to the rotation time scale, with circumstantial evidence suggesting that this happens at a few times 10^7 years. Interestingly, this is the same time scale at which convection losses transport energy from the disk of the Galaxy, and on which cosmic ray energy is lost. We do not have a real understanding of what drives the energy balance of the interstellar medium.

2.11 Transport of Cosmic Rays

From the ratio of radioactive isotopes resulting from spallation to stable isotopes we can deduce the time of transport of cosmic rays near 1 GeV: It is about 3×10^7 years. This is very similar to the sound crossing time scale across the hot thick disk of the interstellar medium, and also to the Alfvénic time scale across the same thick disk. It is unlikely that these numerical coincidences are chance.

The transport of cosmic rays is dominated by a variety of effects [39,40,41]:

- Ionization losses, mosty relevant for protons and nuclei. This limits the lower energy of protons to about 50 MeV *after* traversing most of the interstellar medium path, as derived from the ionizing effect [77].
- Spallation - discussed separately above. For any given isotope, spallation is a loss and a gain-process in the equation of balance.
- Radioactive decay. For any specific isotope this can be a loss and a gain-process. The resulting observed ratios provide a clock for cosmic ray transport.
- Synchrotron and Inverse Compton losses, only relevant for electrons and positrons. Above about 10 GeV these losses dominate over diffusive losses, and so the spectrum is steepened by unity. Then one deduces from the observed spectrum of $E^{-3.3}$, that injection must have happened with about $E^{-2.3}$.
- Diffusive loss from the disk. This is almost certainly governed by the spectrum of turbulence, in an isotropic approximation best described by a Kolmogorov spectrum [73]. This entails that the time scale of loss is proportional to $E^{-1/3}$. In an equilibrium situation this steepens the observed spectrum by $1/3$ over the injection spectrum. Along this line of reasoning one deduces, that without re-acceleration the injection spectrum ought to be $E^{-2.35}$ approximately, as noted immediately above providing a very important consistency check.
- Convective loss from the disk. This is likely to dominate at energies below about 1 GeV for protons, or the corresponding energy of other nuclei with the same Larmor radius.
- Magnetic field irregularities; in analogy with the Sun, it is conceivable that the magnetic field is very inhomogeneous, contains flux tubes of much higher than average field, and then the transport of cosmic ray particles is governed by a mixture of streaming, convection, and diffusion by pitch angle scattering on these magnetic irregularities.
- Some cosmic rays almost certainly come from outside the Galaxy, coming down the galactic wind - of which the existence is very likely, but not certain. Using then the analogy with the solar wind, we need to again ask the question what the most likely turbulence spectrum is in the wind, and that may be quite different from a Kolmogorov spectrum [73], $k^{-5/3}$, where k is the wavenumber, and the spectrum denotes the energy per volume per wavenumber in isotropic phase space. Such a Kolmogorov spectrum is observed in the solar wind over some part of the wavenumber spectrum. Some have argued that it could be governed by the repeated injection of supernova shockwaves, and so best be described by a k^{-2} spectrum. Interestingly, for just such a spectrum the scattering in the irregularities of cosmic ray particles becomes independent of energy, and so there would be no critical energy, below which the cosmic ray spectrum coming in from the outside is cut off. This situation would then be quite different from the solar wind, where all cosmic rays below about 500 MeV/nucleon (measured on the outside) are cut off altogether.

The transport of ultra-high energy cosmic rays, photons, and neutrinos in extragalactic space is dominated by various processes: Pion production (leading to the GZK effect) for nucleons above $\simeq 5 \times 10^{19}$ eV, electromagnetic cascades for $\gamma-$rays and, under certain circumstances, weak interactions with, for example, production and decay of Z-bosons for ultra-high energy neutrinos propagating from large redshifts. Furthermore, protons and nuclei are significantly deflected by or even diffuse in large scale extragalactic magnetic fields [78]. For a detailed discussion of these effects see the contribution by G. Sigl in this volume.

2.12 Supernovae

All stars above an original mass of more than 8 solar masses are expected to explode at the end of their life-time, after they have exhausted nuclear burning; the observable effect of such an explosion is called a supernova. When they explode, they emit about 3×10^{53} erg in neutrinos, and also about 10^{51} erg in visible energy, such as in shock waves in ordinary matter, the former stellar envelope and interstellar gas. These neutrinos have an energy in the range of a few MeV to about 20 MeV. When stars are in stellar binary systems, they can also explode at low mass, but this process is believed to give only 10 % or less of all stellar explosions. There appears to be a connection to Gamma Ray Bursts (GRBs), but the physical details are far from clear at present; some suggest a highly anisotropic explosion, others an explosion running along a pre-existing channel. It is noteworthy that above an original stellar mass of about 15 solar masses, stars also have a strong stellar wind, which for original masses above 25 solar masses becomes so strong, that it can blow out most of the original stellar mass, even before the star explodes as a supernova. The energy in this wind, integrated over the lifetime of the star, can attain the energy of the subsequent supernova, as seen in the shockwave of the explosion.

2.13 Gamma Ray Bursts

Bursts of gamma ray emission [79] come from the far reaches of the universe, and are almost certainly the result of the creation of a stellar mass black hole. The duration of these bursts ranges from a fraction of a second to usually a few seconds, and sometimes hundreds of seconds. Some such GRBs have afterglows in other wavelengths like radio, optical and X-rays, with an optical brightness which very rarely comes close to being detectable with standard binoculars. The emission peaks near 100 keV in observable photon energy, and appears to have an underlying powerlaw character, suggesting non-thermal emission processes. See the contribution by E. Waxman in this volume for a detailed discussion.

2.14 Active Galactic Nuclei

When massive black holes accrete, then their immediate environment, usually thought to be an accretion disk and a powerful relativistic jet (i.e. where the material is ejected with a speed very close to the speed of light) emits a luminosity often far in excess of the emission of all stars in the host galaxy put together. There is the proposal of a "unified scheme", which contains the elements of a black hole, an accretion disk, a jet and a torus of surrounding molecular material. The mass range of these black holes appears to extend to 3×10^9 solar masses. As an example such black holes of a mass near 10^8 solar masses have a size of order the diameter of the Earth orbit around the Sun, and their accretion can produce a total emission of 1000 times that of all stars in our Galaxy. When the emission of the jets gets very strong, and the jet very powerful, then the radio image of such a galaxy can extend to 300 kpc, or more, dissipating the jet in radio hot spots embedded in giant radio lobes, very rarely to several Mpc. The space density of such radio galaxies, with powerful jets, hot spots and lobes, is low, less than 1/1000 of all galaxies, but on the radio sky they dominate due to their extreme emission. The activity is thought to be fed by inflow of gas and/or stars into the black hole, maybe usually fuelled by galaxy-galaxy interaction [80]. High energy particle interactions in active galactic nuclei and their surroundings may be detectable through the neutrino emission, even at cosmological distances [81]. See also the contribution by G. Pelletier in this volume.

2.15 Topological Defects and Supermassive Particles

Particle accelerator experiments and the mathematical structure of the Standard Model of the weak, electromagnetic and strong interactions suggest that these forces should be unified at energies of about 2×10^{16} GeV (1 GeV$= 10^9$ eV) [82], 4-5 orders of magnitude above the highest energies observed in cosmic rays. The relevant "Grand Unified Theories" (GUTs) predict the existence of X particles with mass m_X around the GUT scale of $\simeq 2 \times 10^{16}\,\mathrm{GeV}/c^2$. If their lifetime is comparable or larger than the age of the Universe, they would be dark matter candidates and their decays could contribute to cosmic ray fluxes at the highest energies today, with an anisotropy pattern that reflects the expected dark matter distribution [83]. However, in many GUTs supermassive particles are expected to have lifetimes not much longer than their inverse mass, $\sim 6.6 \times 10^{-41} (10^{16}\,\mathrm{GeV}/m_X c^2)$ sec, and thus have to be produced continuously if their decays are to give rise to ultra-high energy cosmic rays. This can only occur by emission from topological defects which are relics of cosmological phase transitions that could have occurred in the early Universe at temperatures close to the GUT scale. Phase transitions in general are associated with a breakdown of a group of symmetries down to a subgroup which is indicated by an order parameter taking on a non-vanishing value. Topological defects occur between regions that are causally disconnected, such that the orientation of the order parameter cannot be communicated between these regions and thus will adopt different values. Examples are cosmic strings [1], magnetic monopoles [2], and do-

[1] Strings correspond to the breakdown of rotational symmetry $U(1)$ around a certain direction; a laboratory example are vortices in superfluid helium.

[2] Magnetic monopoles correspond to the breakdown of arbitrary 3-dimensional rotations $SO(3)$ to rotations $U(1)$ around a specific direction.

main walls [3]. The Kibble mechanism states [84] that about one defect forms per maximal volume over which the order parameter can be communicated by physical processes. The defects are topologically stable, but in the case of GUTs time dependent motion can lead to the emission of GUT scale X particles.

One of the prime cosmological motivations to postulate inflation, a phase of exponential expansion in the early Universe [85], was to dilute excessive production of "dangerous relics" such as topological defects and superheavy stable particles. However, right after inflation, when the Universe reheats, phase transitions can occur and such relics can be produced in cosmologically interesting abundances where they contribute to the dark matter, and with a mass scale roughly given by the inflationary scale. The mass scale is fixed by the CMB anisotropies to $\sim 10^{13}\,\mathrm{GeV}/c^2$ [86], and it is not far above the highest energies observed in cosmic rays, thus motivating a connection between these primordial relics and ultra-high energy cosmic rays which in turn may provide a probe of the early Universe.

Within GUTs the X particles typically decay into jets of particles whose spectra can be estimated within the Standard Model. Very roughly, one expects a few percent nucleons and the rest in neutrinos and photons [87]; these neutrinos and photons then cascade in the big bang relic neutrinos and photons, and so produce a universal photon and neutrino background (see the contribution by G. Sigl in this volume). It is not finally settled at which level we need to observe a background to confirm or refute this expected background. The resulting hadron spectrum can be a fair bit flatter than any background resulting from cosmic accelerators such as radio galaxies. Therefore any background from the decay of topological defects or other relics should produce observable signatures in neutrinos, photons and hadrons with characteristic properties. For more details on the top-down scenario see the contribution by P. Bhattacharjee and G. Sigl in this volume.

2.16 Magnetic Monopoles

The physics of electric and magnetic fields contains electric charges but no magnetic charges. In the context of particle physics it is likely that monopoles, basic magnetically charged particles, also exist. Such monopoles are a special kind of topological defects. The basic property of monopoles can be described as follows: a) Just as electrically charged particles shortcircuit electric fields, monopoles shortcircuit magnetic fields. The observation of very large scale and permeating magnetic fields in the cosmos shows that the universal flux of monopoles must be very low; the implied upper limit from this argument is called the *Parker limit.* b) Monopoles are accelerated in magnetic fields, just as electrically charged particles are accelerated in electric fields. In cosmic magnetic fields, the

[3] Domain walls correspond to the breakdown of a discrete symmetry where the order parameter is only allowed to take several discrete values; a laboratory example are the Bloch walls separating regions of different magnetization along the principal axis of a ferromagnet.

energies which can be attained are of 10^{21} eV, or even more. Any relation to the observed high energy cosmic rays is uncertain at present [88].

2.17 Primordial Black Holes and Z-bursts

In the early universe it is possible, that very small black holes were also formed. At sufficiently small mass, they can decay, and produce a characteristic spectrum of particles rather similar to topological defects [89].

Another way to obtain very energetic hadrons is to start with a neutrino at very high energy and at distances possibly much larger than the energy loss lengths $\sim 50\,\mathrm{Mpc}$ for photons, nucleons, and nuclei and have it interact with the relic neutrino background, the neutrino analogue of the CMB [90], within $\sim 50\,\mathrm{Mpc}$. Such neutrino-neutrino interactions produce a Z boson, a carrier of the electroweak interactions, which immediately decays into hadrons and other particles, thus producing a proton possibly quite near to us in the Universe. For more details on this "Z-burst" mechanism see the contributions by G. Sigl and by S. Yoshida on neutrino cascades in this volume.

3 Energies, Spectra, and Composition

The solar wind prevents low energy charged particles to come into the inner solar system, due to interaction with the magnetic field in the solar wind, a steady stream of gas going out from the Sun into all directions, originally discovered in 1950 from the effect on cometary tails: they all point outwards, at all latitudes of the Sun, and independent on whether the comet actually comes into the inner solar system, or goes outwards, in which case the tail actually precedes the head of the comet. This prevents us from knowing anything about the energies lower than about 300 MeV of interstellar energetic particles. From about 10 GeV per charge unit Z of the particle, the effect of the solar wind becomes negligible. Since cosmic ray particles are mostly fully ionized nuclei (i.e. with the exception of electrons and positrons), this is a strong effect.

Our Galaxy has a magnetic field of about 6×10^{-6} Gauss in the solar neighbourhood; the energy of such a field corresponds approximately to 1 eV per cm^3, just like the other components of the interstellar medium. In such a magnetic field charged energetic particles gyrate, with a radius of gyration, called the Larmor radius, which is proportional to the momentum of the particle perpendicular to the magnetic field direction. For highly relativistic particles this entails, that around 3×10^{18} eV protons - or other nuclei of the same energy to charge ratio - no longer gyrate in the disk of the Galaxy, i.e. their radius of gyration is larger than the thickness of the disk. So they cannot possibly originate in the Galaxy, they must come from outside; and indeed, at that energy there is evidence for a change both in chemical composition, and in the slope of the spectrum.

The energies of these cosmic ray particles, that we observe, range from a few hundred MeV to $\simeq 300\,\mathrm{EeV}$. The integral flux ranges from about 10^{-5} per cm^2, per s, per steradian, at 1 TeV per nucleus for Hydrogen, or protons, to 1 particle per steradian per km^2 and per century around 10^{20} eV, a decrease by a factor of 3×10^{19} in integral flux, and a corresponding decrease by a factor of 3×10^{27} in differential flux, i.e. per energy interval (see also Fig. 1). Electrons have only been measured to a few TeV.

As already discussed in Sect. 2.1, the total particle spectrum is about $E^{-2.7}$ below the knee, and about $E^{-3.1}$ above the knee, at 5 PeV, and flattens again to about $E^{-2.7}$ beyond the ankle, at about 3 EeV. Electrons have a spectrum, which is similar to that of protons below about 10 GeV, and steeper, near $E^{-3.3}$ above this energy. The lower spectrum of electrons is inferred from radio emission, while the steeper spectrum at the higher energies is measured directly.

The chemical composition is rather close to that of the interstellar medium, with a few strong peculiarities relative to that of the interstellar medium, see Sect. 2.4 for a general discussion. Concerning the energy dependence towards the knee, and beyond, the fraction of heavy elements appears to continuously increase, with moderately to heavy elements almost certainly dominating beyond the knee [91], all the way to the ankle, where the composition seems to become light again [3]. This means, at that energy we observe a transition to what appears to be mostly Hydrogen and Helium nuclei. At much higher energies we can only show consistency with a continuation of these properties, we cannot prove unambiguously what the nature of these particles is.

The fraction of antiparticles is a few percent for positrons and a few 10^{-4} for anti-protons. No other anti-nuclei have been found [92].

4 Origin of Galactic Cosmic Rays

4.1 Injection

For the injection of cosmic rays the following reasons have been suggested, and we will group the answers into three segments following the very different paths of arguments.

There is first the suggestion, that low mass stars with their coronal activity provide the injection mechanism (mostly due to M. Shapiro, [93]). The main argument for this reasoning is the observation that the selection effects for the different elements among energetic particles are very similar in the solar wind and in cosmic rays. Since low mass stars are often observed to be very active, their possible contribution is expected to be substantial. In fact, in a few other stars, these selection effects have been checked [94,95].

The argument then proceeds as follows:

- Low mass stars in their coronal activity accelerate selectively certain elements to supra-thermal energies, and so inject them into the interstellar medium.
- Normal supernova explosions then accelerate them, via shock waves running through the interstellar medium.

There is second the suggestion that the injection of cosmic rays starts with ionized dust particles, and finishes by a break-up of the energetic dust. Many of the selection effects governing dust formation, and also the sites of dust formation then rule the abundances of the final cosmic ray particles.

- This model has been developed on the one hand by Luke Drury and his collaborators [96], and on the other by the group of the late Reuven Ramaty and his collaborators [46].
- One of the biggest successes of this theory is the rather good explanation for the various abundances of the chemical elements just using the known properties of dust, and the observed fact that dust is abundant everywhere.
- A challenging aspect is the possibility to explain the observational fact that the light elements such as Boron were already abundant at early times in the Galaxy, when the general abundances of all heavy elements were low; dust is formed early around the supernovae of massive stars, such as supernova 1987a, as observations clearly indicate, and so the general abundance of dust in the interstellar medium is of no significance. This aspect is one of the strengths of the approach by Ramaty. He elegantly solves the problem of the abundances of the light elements in the young Galaxy.
- The isotopic ratios of certain elements clearly suggest that at least some massive stars, such as Wolf Rayet stars, do contribute at some level. However, in this approach, they play a minor role.

There is a third, competing theory, which emphasizes the role played by the very massive stars, and their winds.

- Here the difference is noted, that massive stars come in three well-understood varieties, i) those with a zero age main sequence mass between 8 and 15 solar masses, which explode into the interstellar medium, ii) those with a mass between 15 and about 25 solar masses, which explode into their stellar wind, which is enriched mostly in Helium, and finally those with a mass above about 25 solar masses, which explode as blue supergiants, Wolf Rayet stars, for which the wind is heavily enriched in Carbon and Oxygen.
- The interstellar turbulence spectrum is taken to be of Kolmogorov type [73], as indicated by an abundance of observations and theoretical work [97].
- The injection happens from the stellar wind abundances, explaining the general features of the abundances. However, since some elements are doubly ionized, their injection is enhanced, leading to a selection effect well known from the active zones of the Sun and the solar wind, and also seen in some active stars. Therefore, this picture also uses the analogy between the solar wind, and assumes that similar selection effects play a role in the winds of massive stars.

4.2 Primary Acceleration

It has been long surmised that supernova explosions provide the bulk of the acceleration of cosmic rays in the Galaxy [98]. The acceleration is thought to be a kind of ping-pong between the two sides of the strong shock wave sent out by the explosion of the star. This ping pong is a repeated reflection via magnetic resonant interaction between the gyromotion of the energetic charged particles, and waves of the same wavelength as the Larmor motion in the magnetic thermal gas. Since the reflection is usually thought to be a gradual diffusion in direction, the process is called diffusive shock acceleration, or after its discoverer Fermi acceleration [21]; see the contribution by G. Pelletier in this volume for a detailed discussion.

For a shock wave sent out directly into the interstellar gas this kind of acceleration easily provides particle energies up to about 100 TeV. While the detailed injection mechanism is not quite clear, the very fact that we observe the emission of particles at these energies in X-rays provides a good case, and a rather direct argument for highly energetic electrons. Even though protons are by a factor of about 100 more abundant at energies near 1 GeV than electrons, we cannot prove yet directly that supernova shocks provide the acceleration; only the analogy with electrons can be demonstrated.

However, we observe what are probably Galactic cosmic rays up to energies near the knee, and beyond to the ankle, i.e. 3 EeV.

The energies can be provided by several possibilities, with the only theory worked out to a quantitative level suggesting that those particles also get accelerated in supernova shock waves, in those which run through the powerful stellar wind of the predecessor star. In this first possibility it can easily be shown, that energies up to 3 EeV per particle are possible (mostly Iron then). An alternate, second, possibility is that a ping pong between various supernova shockwaves occurs, but in this case seen from outside. In either (or any other) such theory it is a problem, that we observe a knee, i.e. a bend down of the spectrum at an energy per charge ratio which appears to be fairly sharply defined. In the concept (the first possibility) that stellar explosions are at the origin it entails that all such stars are closely similar in their properties, including their magnetic field, at the time of explosion; while this is certainly possible, we have too little information on the magnetic field of pre-supernova stars to verify or falsify this. In the case of the other concept (the second possibility) it means that the transport through the interstellar gas has a change in properties also at a fairly sharply defined energy to charge ration, indicating a special scale in the interstellar gas, for which there is no other evidence.

Galactic cosmic rays get injected from their sources with a certain spectrum. While they travel through the Galaxy, from the site of injection to escape or to the observer, they have a certain chance to leak out from the hot galactic magnetic disk of several kpc thickness. This escape becomes easier with higher energy. As a consequence their spectrum steepens, comparing source and observed spectrum. The radio observations of other galaxies show consistency with the understanding that the average spectrum of cosmic rays at least in the GeV to many GeV energy range is always the same, in various locations in a Galaxy, and also the same in different galaxies. During this travel inside a galaxy the cosmic rays interact with the interstellar gas, and in this interaction produce gamma ray emission from pion decay, positrons, and also neutrons, anti-protons, and neutrinos. The future gamma ray emission observations will certainly provide very strong constraints on this aspect of cosmic rays.

One kind of evidence where cosmic rays exactly come from, what kind of stars and stellar explosions really dominate among their sources is the isotopic ratios of various isotopes of Neon, Iron and other heavy elements; these isotope ratios suggest that at least one population is indeed the very massive stars with strong stellar winds; however, whether these stars provide most of the heavier elements, as one theory proposes, is still quite an open question.

There is some evidence now, that just near EeV energies there is one component of galactic cosmic rays, which is spatially associated in arrival direction with the two regions of highest activity in our Galaxy, at least as seen from Earth (by AGASA and SUGAR): the Galactic Center region as well as the Cygnus region show some weak enhancement [33]. Such a directional association is only possible for neutral particles, and since neutrons at that energy can just about travel from those regions to here, before they decay (only free neutrons decay, neutrons bound into a nucleus do not decay), a production of neutrons is conceivable as one explanation of these data. One major difficulty with this interpretation is the lack of discernible high energy gamma ray emission associated with the regions of presumed neutron emission; the CASA-MIA experiment only provided stringent upper limits [99], which appear on first sight to rule out the possibility that related interactions might provide the neutrons. On the other hand, these two regions are clearly those two parts of the Galaxy, where cosmic ray interactions are the strongest, as evidenced by both lower energy gamma data as well as radio data.

4.3 Beyond the Knee

There are several ideas how to get particles accelerated to energies near and beyond the knee, at about 5×10^{15} eV. The observations of air showers suggest that the knee is a feature in constant energy per charge, or rigidity, as surmised already by B. Peters [30]. The same may be true of the "second knee", near 3×10^{17} eV.

There are again several approaches conceivable, with only one quantitative theory for this energy range:

- Obviously, a new accelerator, such as pulsars, might take over; however, then the steeper spectrum with a matching flux at the knee energy is a serious problem, and so this notion is normally discounted today.
- In the context of the injection from energetic particles from low mass active stars, an additional unidentified process provides further acceleration to those energies beyond the knee.
- In the model using dust particles as primary injection mechanism there is no account of the cosmic ray spectrum beyond the knee. A development of the theory, using acceleration between the expanding shells and shocks of different supernovae might solve this problem.

- In the theory using the supernova shock racing through stellar winds, their shell, and the immediate surroundings, all particle energies up to the ankle can be explained due to shock acceleration in the wind, which is magnetized. The knee is explained as due to a diminution of the acceleration efficiency when drift acceleration is reduced due to the matching of the Larmor radius of the motion of the particle, and the spatial constraints in a shocked shell, racing through the stellar wind.

4.4 Transport in the Galaxy

Cosmic ray particles are diffusively transported through the Galaxy, interacting all the time with the matter, magnetic fields and photons. The various theories differ in which interaction site dominates.

- In the theory using dust particles the injection is with a spectrum of $E^{-2.1}$ approximately, and so an interstellar turbulence spectrum such that it would lead to a steepening in $E^{-0.6}$ is required, for which there is little convincing observational nor theoretical evidence, except indirectly through using an adopted model of a leaky box for cosmic ray transport. Again, a further development of the theory might remedy this aspect. Especially, re-acceleration in the interstellar medium might help, as argued by Seo and Ptuskin [100].
- In the theory using stellar winds the cosmic ray interaction happens in the shells around the stellar winds [101,102], and their immediate environments, explaining readily the energy dependence of the ratio of the secondary elements from spallation and the primary elements, with $E^{-5/9}$. This also explains the gamma ray spectrum, which is observed to be best approximated by an interaction spectrum of $E^{-2.3}$. And, furthermore, this approach also explains the electron spectrum, observed to be $E^{-3.3}$, and since it is dominated by losses, requires an injection close to a spectrum of $E^{-2.3}$, as noted earlier.

For an example for detailed modeling of cosmic ray progagation and secondary production in the Galaxy see, e.g., Ref. [103].

4.5 Key Tests

In all these theories, there are critical aspects which are not yet developed, and will surely determine in the future, which of these proposals, if anyone of them, does explain what Nature is doing.

- In the picture using energetic particles from low mass active stars a key test would be the isotopic abundances, comparing those in the solar wind, and those in cosmic rays.
- In the theory using dust particle injection the expected gamma ray spectrum from cosmic ray interactions has not been worked out yet, and may finally confirm this approach, or falsify it. Also, the isotopic abundances provide key tests.

- What has yet to be done, and may well finally prove or falsify the theory involving stellar winds is the very detailed accounting of all the abundances of the chemical elements and their isotopic abundances.
- And, finally, once we observe the high energy gamma ray emission spectrum, its spatial distribution, as well as the neutrino spectrum from the inner part of our Galaxy, then we can expect to finalize our physical understanding of where cosmic rays come from.

Observations such as [104] may provide key tests for progress from the knee on up.

5 The Cosmic Rays between 3 EeV and 50 EeV

The cosmic rays between the ankle and the expected GZK-cutoff are readily explained by many possible sources, almost all outside our galaxy.

Some, but not all of these proposals can also explain particles beyond the GZK-cutoff, discussed in Sect. 6 below.

Pulsars, especially those with very high magnetic fields, called magnetars, can possibly accelerate charged particles to energies of 10^{21} eV (see contribution by B. Rudak in this volume). There are several problems with such a notion, one being the adiabatic losses on the way from close to the pulsar out to the interstellar gas, and another one the sky distribution, which should be anisotropic given the distribution and strength of Galactic magnetic fields. On the other hand if this concept could be proven, it would certainly provide a very easy explanation, why there are particles beyond the GZK-cutoff: for Galactic particles the interaction with the CMB is totally irrelevant, and no GZK-cutoff is expected.

Another proposal is GRBs, and is discussed in detail in the contribution by E. Waxman in this volume. However since ultimately we do not yet know what constitutes a GRB, their contribution cannot be settled with full certainty.

Shock waves running through a magnetized and ionized gas accelerate charged particles, as we know from in situ observations in the solar wind already; and this forms the basis of almost all theories to account for Galactic Cosmic Rays. The largest shock waves in the universe have scales of many tens of Mpc, and have shock velocities of around 1000 km/s. These shock waves arise in the cosmological large scale structure formation, seen as a soap-bubble like distribution of galaxies in the universe. The accretion flow to enhance the matter density in the resulting sheets, filaments and clusters is still continuing, and causes shock waves to exist all around us. In the shock waves, which also have been shown to form around growing clusters of galaxies, particles can be accelerated, and can attain fairly high energies. However, the maximum energies can barely reach the energy of the GZK-cutoff, and so a strong contribution to the overall flux is unlikely [105].

The most conventional explanation is radio galaxies, which provide with their hot spots an obvious acceleration site: These hot spots are giant shock waves, often of a size exceeding that of our entire Galaxy. The shock speeds may approach several percent, maybe even several tens of percent of the speed of light, if sporadic. Integrating over all known radio galaxies readily explains flux and spectrum, as well as chemical composition of the cosmic rays in this energy range [24,106,107]. In this proposal it is the greatest challenge to identify the single radio galaxy dominating the highest energy; for this M87 has been proposed already some time ago (see also the contribution by P. Biermann et al. in this volume).

6 Particles beyond the GZK-cutoff

For these energies there is no argument, whether these particles are really protons, as an extrapolation from lower energies might suggest. However, everything we know is quite consistent with such an assumption [20].

Apart from the more "conservative" astrophysical mechanisms involving "bottom-up" acceleration, there are many exciting approaches to account for these particles:

- Decay of topological defects (TDs), or other relics from the big bang, the so-called "top-down" scenario. This theory can account readily for the apparent upturn in the spectrum beyond the GZK cutoff, and explains those events with a mixture of nucleons and $\gamma-$rays. These models predict significant diffuse $\gamma-$ray fluxes in the 100 MeV-GeV region and thus are strongly constrained by the observed fluxes in this energy range. There are many variants of top-down models [108], some of them with a quite predictive power.
- Decay of primordial black holes. The final particle distribution is rather similar to that expected from the decay of TDs [89].
- Violation of the Lorentz invariance [109]: At some very high energy, where the four basic forces of Nature combine, Lorentz Invariance may no longer hold, and a ripple effect of this is anticipated at lower energies. One possible result would be that protons might survive much longer in the bath of the CMB. In fact, observations of photons of energies up to $\simeq$ 20 TeV from Markarian 501, where absorption in the infrared background is expected to be strong, was considered as a possible signature of violation of Lorentz invariance [110,111]. Furthermore, photons at different energies would have divergent travel times, conceivably measurable with GRBs [111].

7 Outlook

The next few years promise to give great advances to our physical understanding of both the macro and the microcosmos. On the one side, this is due to our increased theoretical understanding on how to combine accelerator data and cosmic ray and astrophysical data to arrive at strong constraints, for example, on new physics. On the other hand, it is due to an expected enormous increase of data from new experiments, especially on the cosmic ray and astrophysics side. Ground arrays, Balloons, Space Station experiments will proliferate within the next few years and hold great promise for us. On a somewhat longer time scale, powerful new particle accelerators such as the LHC will directly test new physics in the TeV region, an energy range which is also, somewhat more indirectly, probed by cosmic ray, γ−ray and neutrino experiments.

Acknowledgement

Work on these topics with PLB has been supported by grants from NATO, the BMBF-DESY, the EU, the DFG and other sources. PLB has to thank many people, especially all those with whom he has recently published and worked in this area, or is doing it at present, especially Eun-Joo Ahn, Alina Donea, Torsten Ensslin, Heino Falcke, Glennys Farrar, Martin Harwit, Hyesung Kang, Tom Kephardt, Phil Kronberg, Norbert Langer, Sera Markoff, Gustavo Medina-Tanco, Ray Protheroe, Giovanna Pugliese, Wolfgang Rhode, Dongsu Ryu, Eun-Suk Seo, Ramin Sina, Todor Stanev, Samvel Ter-Antonyan, Amri Wandel, Yip-ing Wang, Tom Weiler, Stefan Westerhoff, and of course, the present coauthor. GS is especially grateful to the late David Schramm for his continuous encouragment to work on the exciting subject of ultrahigh energy cosmic rays. GS also wishes to thank his recent collaborators on this subject, Luis Anchordoqui, Phijushpani Bhattacharjee, Paolo Coppi, John Ellis, Christopher Hill, Claudia Isola, Karsten Jedamzik, Sangjin Lee, Martin Lemoine, Angela Olinto, Gustavo Romero, Diego Torres, Craig Tyler, Shigeru Yoshida, and, of course, the current coauthor.

References

1. V. F. Hess: Phys. Z. **13**, 1084 (1912)
2. W. Kohlhörster: Phys. Z. **14**, 1153 (1913)
3. D. J. Bird, et al.: Phys. Rev. Lett. **71**, 3401 (1993)
   D. J. Bird, et al.: Astrophys. J. **424**, 491 (1994)
   D. J. Bird, et al.: Astrophys. J. **441**, 144 (1995)
4. See, e.g., M. A. Lawrence, R. J. O. Reid, A. A. Watson: J. Phys. G Nucl. Part. Phys. **17**, 733 (1991), and references therein
   see also http://ast.leeds.ac.uk/haverah/hav-home.html
5. N. Hayashida, et al.: Phys. Rev. Lett. **73**, 3491 (1994)
   S. Yoshida, et al.: Astropart. Phys. **3**, 105 (1995)
   M. Takeda, et al.: Phys. Rev. Lett. **81**, 1163 (1998)
   see also http://icrsun.icrr.u-tokyo.ac.jp/as/project/agasa.html
6. Proc. International Symposium on *Astrophysical Aspects of the Most Energetic Cosmic Rays*, eds. M. Nagano and F. Takahara (World Scientific, Singapore, 1991)
7. Proc. of International Symposium on *Extremely High Energy Cosmic Rays: Astrophysics and Future Observatories*, ed. M. Nagano (Institute for Cosmic Ray Research, Tokyo, 1996)
8. N. N. Efimov, et al.: Ref. [6], p. 20
   B. N. Afnasiev: in Ref. [7], p. 32.
9. J. Linsley: Phys. Rev. Lett. **10**, 146 (1963)
   J. Linsley: Proc. *8th International Cosmic Ray Conference* **4**, 295 (1963)

10. R. G. Brownlee, et al.: Can. J. Phys. **46**, S259 (1968)
M. M. Winn, et al.: J. Phys. G **12**, 653 (1986)
see also http://www.physics.usyd.edu.au/hienergy/sugar.html
11. G. Cocconi: Nuovo Cimento, 10th ser. **3**, 1433 (1956)
12. K. Greisen: Phys. Rev. Lett. **16**, 748 (1966)
13. G. T. Zatsepin, V. A. Kuzmin, Pis'ma Zh. Eksp. Teor. Fiz. **4**, 114 (1966) [JETP. Lett. **4**, 78 (1966)]
14. AMS collaboration: Phys. Lett. **B490**, 27 (2000)
AMS collaboration: Phys. Lett. **B494**, 193 (2000)
15. See http://spaceflight.nasa.gov/station/science/space/index.html
16. Proc. *24th International Cosmic Ray Conference* (Istituto Nazionale Fisica Nucleare, Rome, Italy, 1995)
17. S. C. Corbató, et al.: Nucl. Phys. B (Proc. Suppl.) **28B**, 36 (1992)
D. J. Bird, et al., in Ref. [16], Vol. **2**, 504; Vol. **1**, 750
M. Al-Seady, et al., in Ref. [7], p. 191
see also http://hires.physics.utah.edu/
18. J. W. Cronin: Nucl. Phys. B (Proc. Suppl.) **28B**, 213 (1992)
The Pierre Auger Observatory Design Report (2nd edition), March 1997
see also http://www.auger.org/ and
http://www-lpnhep.in2p3.fr/auger/welcome.html
19. S. Yoshida, H. Dai: J. Phys. G **24**, 905 (1998)
X. Bertou, M. Boratav, A. Letessier-Selvon: Int. J. Mod. Phys. **A15**, 2181 (2000), and references therein.
20. M. Nagano, A. A. Watson: Rev. Mod. Phys. **72**, 689 (2000)
21. E. Fermi: Phys. Rev. 2nd ser., **75**, no. 8, 1169 (1949)
E. Fermi: Astrophys. J. **119**, 1 (1954)
22. V. L. Ginzburg: Usp. Fiz. Nauk **51**, 343 (1953)
See also *Nuovo Cimento* **3**, 38 (1956)
23. V. L. Ginzburg: Dokl. Akad. Nauk SSSR **92**, 1133 (1953) (NSF-Transl. 230)
24. V. L. Ginzburg, S. I. Syrovatskii: *The origin of cosmic rays* (Pergamon Press, Oxford 1964), Russian edition (1963)
25. T. K. Gaisser: *Cosmic Rays and Particle Physics* (Cambridge Univ. Press 1990)
26. V. L. Ginzburg: Phys. Usp. **36**, 587 (1993)
27. V. L. Ginzburg: Uspekhi Fizicheskikh Nauk **166**, 169 (1996)
28. V. S. Berezinskii, et al.: *Astrophysics of Cosmic Rays* (North-Holland, Amsterdam 1990), especially chapter IV.
29. T. Abu-Zayyad, et al.: Astrophys. J. **557**, 686 (2001)
30. B. Peters: Nuovo Cimento Suppl. **XIV**, 436 (1959)
B. Peters: Nuovo Cimento **XXII**, 800 (1961)
31. D. J. Bird, et al.: Astrophys. J. **511**, 739 (1999)
32. N. Hayashida, et al.: Astropart. Phys. **10**, 303 (1999)
33. J. A. Bellido, R. W. Clay, B. R. Dawson, M. Johnston-Hollitt: Astropart. Phys. **15**, 167 (2001)
34. N. Hayashida, et al.: Phys. Rev. Lett. **77**, 1000 (1996)
M. Takeda, et al.: Astrophys. J. **522**, 225 (1999); and astro-ph/0008102
35. R. D. Blandford, R. D. Eichler: Phys. Rep. **154**, 1 (1987)
36. L. O'C Drury: Rep. Prog. Phys. **46**, 973 (1983)
37. J. R. Jokipii: Astrophys. J. **313**, 842 (1987)
38. J. Bednarz, M. Ostrowski: Phys. Rev. Lett. **80**, 3911 (1998)
39. M. Garcia-Munoz, G. M. Mason, J. A. Simpson: Astrophys. J. **217**, 859 (1977)

40. M. Garcia-Munoz, J. A. Simpson, T. G. Guzik, J. P. Wefel, S. H. Margolis: Astrophys. J. Suppl. **64**, 269 (1987)
41. C. H. Tsao, R. Silberberg, A. F. Barghouty: Astrophys. J. **549**, 320 (2001)
42. See http://www.bnl.gov/rhic/
43. See, e.g.:
R. L. Jaffe, W. Busza, J. Sandweiss, F. Wilczek: Rev. Mod. Phys. **72**, 1125 (2000)
A. Dar, A. De Rújula, U. Heinz: Phys. Lett. **B470**, 142 (1999)
A. Kent, hep-ph/0009130
44. H. Reeves: Ann. Rev. Astron. Astroph. **12**, 437 (1974)
H. Reeves: Rev. Mod. Phys. **66**, 193 (1994)
45. B. Wiebel-Sooth, P. L. Biermann: Landolt-Börnstein, vol. **VI/3c**, Springer Verlag 1999, pp. 37 - 90
46. R. Ramaty, B. Kozlovsky, R. E. Lingenfelter, H. Reeves: Astrophys. J. **488**, 730 (1997)
47. See http://www.ifcai.pa.cnr.it/Ifcai/euso.html
48. Proc. 25th *International Cosmic Ray Conference*, eds.: M. S. Potgieter, et al. (Durban, 1997)
49. J. F. Ormes, et al.: in Ref. [48], Vol. **5**, 273
Y. Takahashi, et al.: in Ref. [7], p. 310
see also http://lheawww.gsfc.nasa.gov/docs/gamcosray/hecr/OWL/
50. J. P. Rachen, P. L. Biermann: Astron. Astrophys. **272**, 161 (1993)
J. P. Rachen, T. Stanev, P. L. Biermann, Astron. Astrophys. **273**, 377 (1993)
51. G. A. Medina Tanco: Astrophys. J. Lett. **510**, L91 (1999)
52. M. Blanton, P. Blasi, A. V. Olinto: Astropart. Phys. **15**, 275 (2001)
53. J. Kormendy, D. Richstone: Ann. Rev. Astron. Astrophys. **33**, 581 (1995)
54. S. M. Faber, et al.: Astron. J. **114**, 1771 (1997)
55. J. Magorrian,, et al.: Astron. J. **115**, 2285 (1998)
56. K. Gebhardt, et al.: Astrophys. J. Lett. **539**, L13 (2000)
57. A. P. Lightman, S. L. Shapiro: Rev. Mod. Phys. **50**, 437 (1978)
58. F. Melia, H. Falcke: Ann. Rev. Astron. Astrophys. (in press) (2001)
59. C. B. Netterfield, et al.: astro-ph/0104460
60. A. T. Lee, et al.: astro-ph/0104459
61. N. W. Halverson, et al.: astro-ph/0104488, 0104489, 0104490
62. S. Perlmutter, et al.: Astrophys. J. **517**, 565 (1999)
A. G. Riess, et al.: Astron. J. **116**, 1009 (1998)
63. see also N. A. Bahcall, J. P. Ostriker, S. Perlmutter, P. J. Steinhardt: Science **284**, 1481 (1999)
R. Stompor, et al.: astro-ph/0105062
64. J. P. Ostriker: Ann. Rev. Astron. Astrophys. **31**, 689 (1993)
65. M. Mateo: Ann. Rev. Astron. Astrophys. **36**, 435 (1998)
66. S. L. Snowden, et al.: Astrophys. J. **485**, 125 (1997)
67. H. Kaneda, K. Makishima, S. Yamauchi, K. Koyama, K. Matsuzaki, N. Y. Yamasaki: Astrophys. J. **491**, 638 (1997)
68. A. Valinia, F. E. Marshall: Astrophys. J. **505**, 134 (1998)
69. J. Pietz, J. Kerp, P. M. W. Kalberla, W. B. Burton, D. Hartmann, U. Mebold: Astron. Astrophys. **332**, 55 (1998)
70. P. P. Kronberg: Rep. Prog. Phys., **57**, 325 (1994)
71. R. Beck, A. Brandenburg, D. Moss, A. Shukurov, D. Sokoloff: Ann. Rev. Astron. Astrophys. **34**, 155 (1996)
72. J. P. Vallée: Fund. Cosm. Phys. **19**, 1 (1997)

73. A. M. Obukhov: Dokl. Akad. Nauk SSSR **32**, 22 (1941)
A. N. Kolmogorov: Dokl. Akad. Nauk SSSR **30**, 299 (1941); ibid. **31**, 538 (1941); ibid. **32**, 19 (1941)
W. Heisenberg: Z. Physik **124**, 628 (1948)
R. Z. Sagdeev: Rev. Mod. Phys. **51**, 1 (1979)
M. L. Goldstein, D. A. Roberts, W. H. Matthaeus: Ann. Rev. Astron. Astroph. **33**, 283 (1995)
74. F. Krause, R. Beck: Astron. Astrophys. **335**, 789 (1998)
75. R. M. Kulsrud: Ann. Rev. Astron. Astrophys. **37**, 37 (1999)
76. D. Grasso, H. R. Rubinstein: Phys. Rep. **348**, 163 (2001)
77. B. B. Nath, P. L. Biermann: Month. Not. R. Astron. Soc. **267**, 447 (1994)
78. D. Ryu, H. Kang, P. L. Biermann: Astron. Astrophys. **335**, 19 (1998)
79. T. Piran: Phys. Rep. **314**, 575 (1999)
80. D. B. Sanders, I. F. Mirabel: Ann. Rev. Astron. Astrophys. **34**, 749 (1996)
81. J. G. Learned, K. Mannheim: Ann. Rev. Nucl. Part. Sci. **50**, 679 (2000)
82. see, e.g., S. Weinberg: *The Quantum Theory of Fields Vol 2: Modern Applications*, Cambridge University Press (Cambridge, 1996)
83. see, e.g., W. Evans, F. Ferrer, S. Sarkar: astro-ph/0103085; to appear in Astropart. Phys.
84. T. W. B. Kibble: J. Phys. **A9**, 1387 (1976)
for a review see A. Vilenkin: Phys. Rep. **121**, 263 (1985)
85. see, e.g., E. W. Kolb, M. S. Turner: *The Early Universe* (Addison-Wesley, Redwood City, California, 1990)
86. for a brief review see V. Kuzmin, I. Tkachev: Phys. Rep. **320**, 199 (1999)
87. V. Berezinsky, M. Kachelriess: Phys. Rev. D **63** 034007 (2001)
88. S. D. Wick, T. W. Kephart, T. J. Weiler, P. L. Biermann: astro-ph/0001233
89. A. Barrau: Astropart. Phys. **12**, 269 (2000)
90. T. J. Weiler: Phys. Rev. Lett. **49**, 234 (1982)
T. J. Weiler: Astrophys. J. **285**, 495 (1984)
T. J. Weiler: Astropart. Phys. **11**, 303 (1999)
D. Fargion, B. Mele, A. Salis: Astrophys. J. **517**, 725 (1999)
G. Gelmini, A. Kusenko: Phys. Rev. Lett. **82**, 5202 (1999); ibid. **84**, 1378 (2000)
91. see, e.g., K.-H. Kampert, et al.: astro-ph/0102266
92. see, e.g., WiZard/CAPRICE Collaboration: astro-ph/0103513
T. Maeno, et al., BESS Collaboration: Astropart. Phys. **16**, 121 (2001)
M. A. Huang: astro-ph/0104229
93. M. M. Shapiro: in *LiBeB, Cosmic rays, and related X- and γ-rays*, ASP conf. No. 171, Eds. R. Ramaty et al., p. 139 - 145 (1999)
94. J. H. M. M. Schmitt, B. M. Haisch, J. J. Drake: Science **265**, 1420+ (1994)
95. J. J. Drake, et al.: Science **267**, 1470+ (1995)
96. J.-P. Meyer, L. O'C Drury, D. C. Ellison: Astrophys. J. **487**, 182 (1997)
97. B. J. Rickett: Ann. Rev. Astron. Astroph. **28**, 561 (1990)
98. W. Baade, F. Zwicky: Proc. Nat. Acad. Science **20**, no. **5**, 259 (1934)
99. A. Borione, et al.: Astrophys. J. **493**, 175 (1998)
100. E.-S. Seo, V. S. Ptuskin: Astrophys. J. **431**, 705 (1994)
101. P. L. Biermann: in *Cosmic winds and the Heliosphere*, Eds. J. R. Jokipii, et al., Univ. of Arizona press, p. 887 - 957 (1997), astro-ph/9501030
102. P. L. Biermann, N. Langer, E. Seo, T. Stanev: Astron. Astrophys. **369**, 269 (2001)

103. A. W. Strong, I. V. Moskalenko: astro-ph/0101068, invited talk at COSPAR2000 Symposium, E1.3: Origin and Acceleration of Cosmic Rays, ed. M.H. Israel , to appear in Adv. Sp. Res. Vol. **27** No. 4
I. V. Moskalenko, A. W. Strong: Astrophys. Sp. Sci. **272**, 247 (2000), and references therein
104. M. A. K. Glasmacher, et al.: Astropart. Phys. **10**, 291 (1999); ibid. **12**, 1 (1999)
105. H. Kang, D. Ryu, T. W. Jones: Astropart. Phys. **456**, 422 (1996)
H. Kang, J. P. Rachen, P. L. Biermann: Mon. Not. R. Astron. Soc. **286**, 257 (1997)
106. A. M. Hillas: Ann. Rev. Astron. Astroph. **22**, 425 (1984)
107. P. L. Biermann, P. A. Strittmatter: Astrophys. J. **322**, 643 (1987)
108. for a detailed review see P. Bhattacharjee, G. Sigl: Phys. Rep. **327**, 109 (2000)
109. S. Coleman, S. L. Glashow: Phys. Rev. D **59**, 116008 (1999)
G. Amelino-Camelia, J. Ellis, N. E. Mavromatos, D. V. Nanopoulos, S. Sarkar: Nature **393**, 763 (1998)
T. Kifune, Astrophys. J. Lett. **518**, L21 (1999)
R. Aloisio, P. Blasi, P. L. Ghia, A. F. Grillo, Phys. Rev. D **62**, 053010 (2000)
110. H. Meyer, R. Protheroe: Phys. Lett. **B493**, 1 (2000)
111. G. Amelino-Camelia, T. Piran: Phys. Lett. **B497**, 265 (2001)

Phenomenology of Ultra-High-Energy Atmospheric Showers

Pierre Billoir

LPNHE (CNRS, Universités PARIS VI et PARIS VII)
4 place Jussieu
F-75252 Paris Cedex 05, France

Abstract. We describe the cascade of hadronic and electromagnetic interactions which gives rise to an extensive atmospheric shower. Its development depends on the nature and the energy of the primary particle; it may be described by a longitudinal profile and a lateral extension. We explained how remote observations allow to reconstruct the characteristics of the primary object, with a special emphasis on neutrinos and photons, and we discuss the uncertainties related to the simulation of interaction modeling.

1 Introduction

We focus here on showers (cascade of particles) induced in the atmosphere by ultra high energy cosmic rays (UHECR, beyond 1 EeV $= 10^{18}$ eV). These rays are generally believed to be protons or nuclei accelerated in some astrophysical sites (whatever their origin, electrons encounter magnetic fields and lose rapidly their energy), see the introduction by P. Biermann and G. Sigl and the contribution by G. Pelletier in this volume. Protons and nuclei may also interact in an intermediate medium and give stable subproducts: for example, photoproduction of mesons on the cosmic microwave background (the so-called "GZK effect" $p + \gamma_{\mathrm{CMB}} \to p + \pi$) is a source of photons and neutrinos (for more details on UHECR propagation see the contribution by G. Sigl in this volume). On the other hand, ultra-high energy particles may be created in a process starting at even higher energy, for example, decay of an object with a mass at the level of the Grand Unification ($\sim 10^{25}$ eV): in such scenarios, particles arriving at the Earth are mainly neutrinos, and possibly photons (above 10^{20} eV, their mean free path is at least a few Mpc). These so called "top-down models" are discussed in more detail in the contribution by P. Bhattacharjee and G. Sigl in this volume. We may also consider other possibilities: neutrons (of galactic origin), which give atmospheric showers similar to those from protons, and secondary electrons (produced in the geomagnetic field as explained below).

The physical processes at such energies (more than 100 TeV in the center-of-mass frame) cannot be studied with present accelerators, even in colliders. The theory of electromagnetic interactions is assumed to be still valid, and exact computations of individual processes are possible. The hadronic interactions are more problematic, because they require far extrapolations of more or less empirical models, tuned on experimental data at lower energy. They induce uncertainties on the first steps of the cascade, which cannot be directly observed.

Martin Lemoine, Günter Sigl (Eds.): LNP 576, pp. 27–44, 2001.

We describe the general features of the shower development, and we state some approximate scaling laws. Then we show how some quantities related to the longitudinal and transverse development may be measured from optical observations (telescopes) and/or from ground particle detectors, allowing a precise determination of the direction of arrival on Earth, and an estimation of the energy, without too much dependence on the high energy interaction model. Moreover, to some extent, the nature of the primary particle may be identified on a statistical basis (light/heavy nuclei discrimination) in relation to the speed of the development of the cascade; this identification is sensitive to modelling uncertainties.

We also want to point out that primary photons or neutrinos (signature of "top-down" models for the origin of ultra-high energy cosmic rays) may give showers clearly different from nucleic ones, independent of the high energy model.

2 Shower Development

2.1 Main Features

Let us first assume that the incoming projectile is a nucleon or a nucleus with atomic number A (in practice $A \leq 56$, because nuclei heavier than iron are not abundant). The primary interaction is hadronic, it occurs in the upper atmosphere (generally more than 20 km of altitude above sea level); as a first approximation, if collective effects are neglected, a nucleus A of energy E is

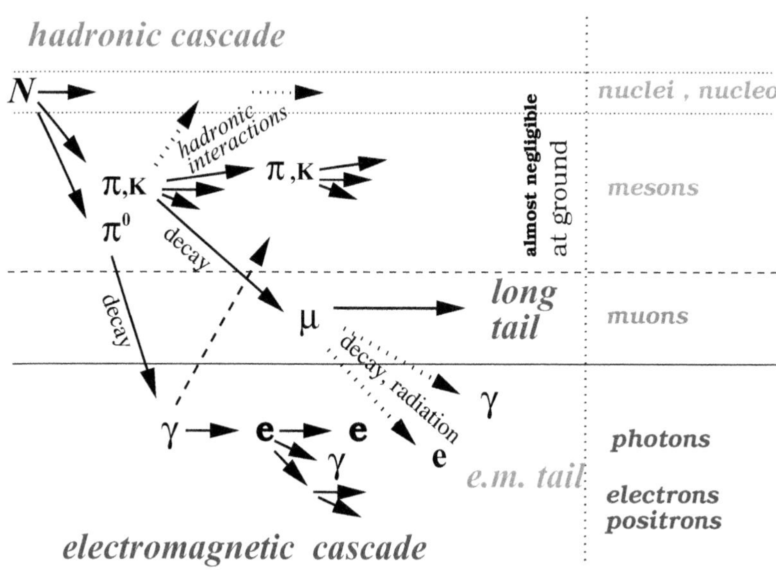

Fig. 1. General structure of an air shower: hadronic cascade, electromagnetic cascade, muonic tail

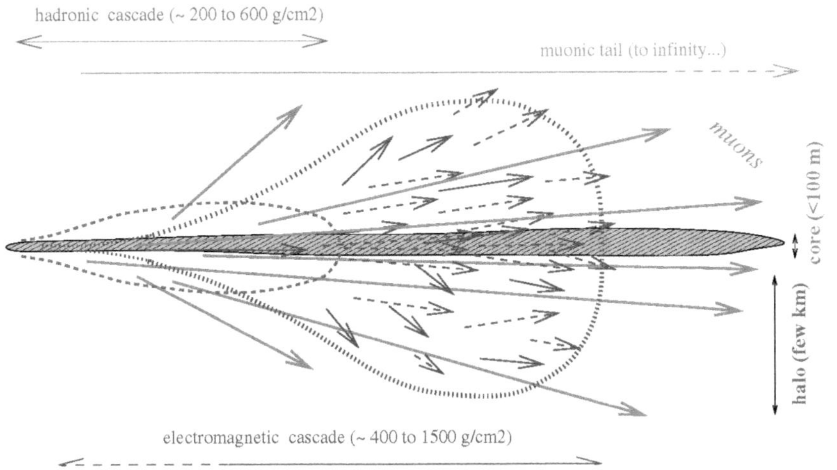

Fig. 2. Spatial extension of the shower components

equivalent to the superposition of A independent nucleons, each with energy E/A. The primary interaction, in addition to the nucleons of the projectile and the target, produces a large number of secondary particles (mainly pions), which give rise to further hadronic interactions, and so on: this is the *hadronic cascade.* Neutral pions decay into photons before reinteracting (except at Lorentz factors above 10^{11}, that is at energy above 10 EeV), so that at each step of the hadronic cascade, about 1/3 of the energy is transferred to photons, giving rise to the *electromagnetic cascade*: photons produce e^+e^- pairs and Compton electrons, and electrons/positrons radiate photons through bremsstrahlung on atmospheric nuclei. A small fraction of the electromagnetic component is re-injected in the hadronic cascade because of hadronic interactions of the photons; as a consequence, a shower induced by a primary photon (or an electron) would also develop an hadronic cascade. Figure 1 summarizes the processes generating the cascades, and Fig. 2 gives an idea of the spatial extension of the different components.

The hadronic cascade (except a few nucleons) ends up with the decay of charged pions into muons, at intermediate altitudes (around 6 km, with a large spread), when their Lorentz factor γ is such that the decay length $\beta\gamma c\tau$ becomes comparable to the interaction length, of the order of $50\,\mathrm{g\,cm^{-2}}$ (400 m at sea level, more at high altitude). With $c\tau \simeq 8$ m, this corresponds to an energy around 7 GeV, then the muons are produced with a typical energy of a few GeV, increasing with the altitude of production. On the other hand they inherit the transverse momentum of their parents (a few hundred MeV): their divergence (angle with the shower axis) is relatively small, and strongly anticorrelated to their energy. Figure 3 gives the characteristics of the muons in a vertical shower.

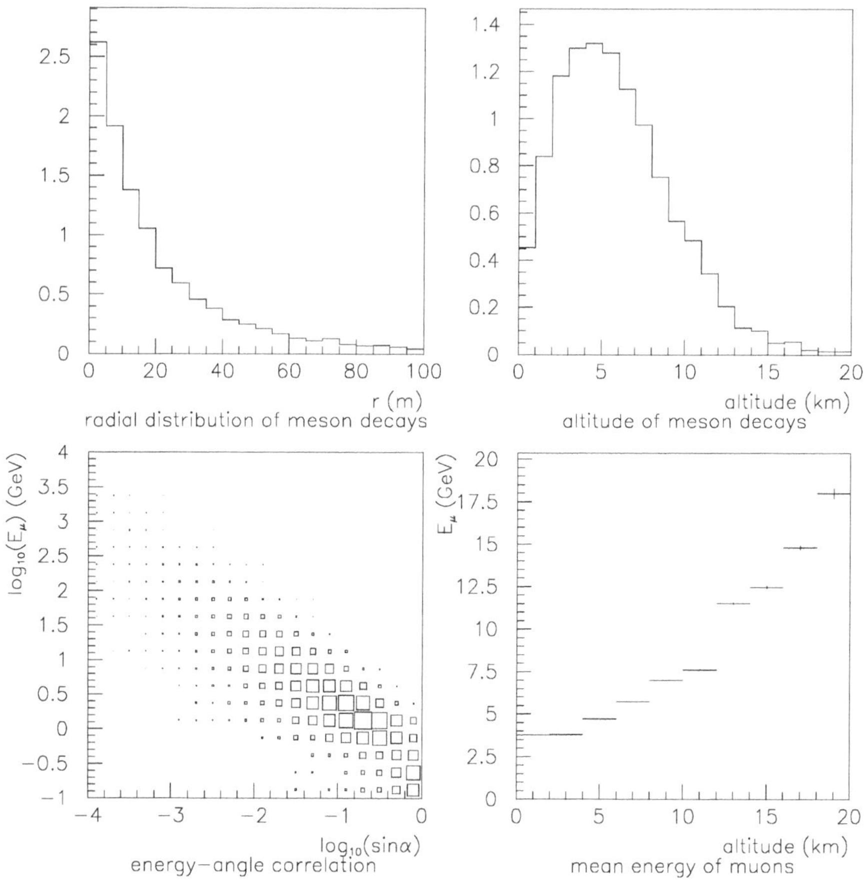

Fig. 3. Characteristics of muons produced in meson decays, in a typical shower (primary energy 100 EeV; zenith angle 40°). The distributions do not depend strongly on the initial conditions. Top: spatial extension of the meson decays. Bottom: correlation of the muon energy to the divergence α from axis (left), and to the altitude (right)

A large fraction of them reach the ground before decaying, with a non-negligible energy loss, $\simeq 2\,\mathrm{MeV}/(\mathrm{g\,cm}^{-2})$.

The electromagnetic cascade continues down to energies below 1 MeV, until electrons slow down through ionization without further radiating, and finally stop. Except for very inclined showers, this process is not achieved at ground level.

It is important to notice that the energy lost along the cascade is essentially deposited through low energy charged particles, and that only a few percent of the initial energy goes into neutrinos (mainly produced in meson and muon decays). From this point of view, the atmosphere behaves as a giant calorimeter with a good linearity (but the shower is not fully contained)

2.2 Longitudinal Development

As long as meson decays and density effects (for example the LPM effect, see below) are negligible, the longitudinal development (for a given primary particle at a given energy $E_{\rm prim}$) depends only on the *cumulated slant depth* X (the thickness of air already crossed), usually expressed in g cm^{-2}: $X = X_{\rm vert}/\cos\theta$, where θ is the zenith angle of the shower axis (the direction of the primary particle) and $X_{\rm vert}$ is the vertical thickness of the atmosphere above this point (1000 g cm^{-2} at sea level).

As a first approximation, we can consider the development of the cascade as a succession of steps (slices in X) where the number of particles is multiplied by a constant factor, and their energy divided by this factor, until the particles decay or stop, at an energy level which do not depend on the initial state. More precisely, after a few initial steps, there is a quasi-continuous spectrum in energy, at a scale decreasing exponentially with X, with a reproducible succession of macroscopic states, down to the extinction; in other terms, all showers have the same shape, except for a translation and a global factor of intensity, proportional to $E_{\rm prim}$. In the scaling approximation mentioned above, this translation is a logarithmic function of $E_{\rm prim}$. Actually, this is essentially true for the electromagnetic cascade, which is the dominating process in the range where the longitudinal profile can be observed.

For example, the number of charged particles (mainly e^- and e^+) may be adequately parametrized by the Gaisser-Hillas function [1], which fits well to the observations:

$$N_e = N_{\rm max}\left(\frac{X - X_0}{X_{\rm max} - X_0}\right)^{\frac{X_{\rm max}-X_0}{70}} \exp\left(\frac{X_{\rm max} - X}{70}\right)$$

where X_0 is the depth of the first interaction, and $X_{\rm max} - X_0$ depends on the energy and the nature of the primary; $X_{\rm max}$ is the position of the maximum of N_e. The quantity $X - X_{\rm max}$ is an indicator of the stage of evolution (the "age") of the shower. $X_{\rm max}$ increases logarithmically with the energy: $X_{\rm max} \simeq X_i + 55\log_{10} E_{\rm prim}$ (in g cm^{-2}). The value of X_i depends on the nature of the primary; in the model of independent nucleon superposition, starting from $E_{\rm prim}/A$ is equivalent to a shift in $X_{\rm max}$ proportional to $\log A$; in practice, at a given energy, $X_{\rm max}(p) - X_{\rm max}(\rm Fe) \simeq 100$ g cm^{-2}. This is the essential feature allowing a discrimination between protons, light and heavy nuclei. More generally, such a dependence is expected within any reasonable model, where the cross sections and/or the multiplicities are increasing with A, giving a faster development at the beginning of the hadronic cascade.

In our energy range, $X_{\rm max}$ is of the order of 700 to 800 g cm^{-2}, that is slighty less than the total vertical thickness of the atmosphere: then at any zenith angle θ, the maximum of the shower is above the ground.

Figure 4 gives the longitudinal profile of the different particles in a typical shower, induced by a proton of 10 EeV at $\theta = 40°$.

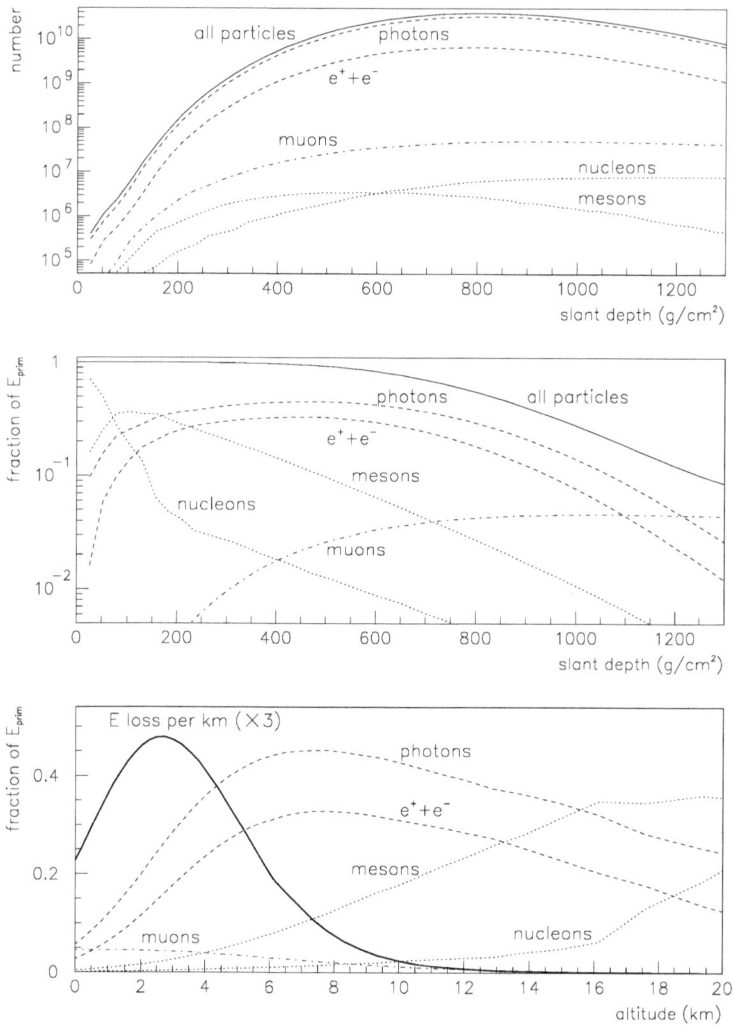

Fig. 4. Longitudinal profile of a typical shower. Top: number of particles, as a function of X. Medium: fraction of primary energy carried, as a function of X. Bottom: the same, as a function of the altitude (Note that, in comparison with the other panels the horizontal scale is reversed and distorted). In addition (thick line): rate of energy loss, related to the profile of Čerenkov light emission and fluorescence

If the primary particle is a photon or an electron, the first steps are purely electromagnetic, and the initial descent in energy is slower than in a hadronic cascade. As a result, $X_{\max}$ is expected to be larger than in a proton/nucleus shower.

2.3 Lateral Extension

Physical processes in the cascade give generally products with a moderate transverse momentum, whatever the energy; then most high energy particles are collimated along the initial axis: they constitute the "core" of the shower. The lateral extension of the core depends on the mean free path, then it is proportional to the inverse of the density. It may be expressed in terms of the *Moliere radius* R_{M}, such that 90 % of the energy is contained within a distance from axis $r < R_{\mathrm{M}}$; in atmospheric showers $R_{\mathrm{M}} \simeq 60$ m. However, low energy photons and electrons, as well as muons, extend far away from the core: this "halo" has a detectable density up to a few kilometers from the axis (depending on the primary energy). The electromagnetic part of the halo increases with the depth, as long as the core remains active, reaches its maximum around $X_{\mathrm{max}} + 100$ g cm^{-2} and then decreases rapidly: it is completely extinguished at $X_{\mathrm{max}} + 1000$ g cm^{-2}. On the other hand, the steepness of the lateral distribution decreases with X (the distribution gets flatter).

Most muons travel beyond the electromagnetic cascade, giving a "muonic tail", with an increasing spread in r, due to a simple straightforward propagation; far from the core, the muon density is approximately an exponential function of the distance (reflecting the distribution of the transverse momentum at the end of the hadronic cascade). After a long range lower energy muons (with larger divergence) decay, and the angular distribution of the muonic tail shrinks: the spread in r does not increase linearly with the distance. On the other hand, independently of the electromagnetic cascade, the muons generate an "electromagnetic tail" through their decay, δ-ray production and radiative processes (bremsstrahlung and pair production), important above a few 10 GeV.

Figure 5 shows the lateral distribution of the different species, at ground level, and their characteristics (energy, divergence).

2.4 Muon Fraction

The muon content is an important feature depending on the nature of the primary. If the primary particle is a nucleus, the fraction of energy remaining in the hadronic cascade decreases exponentially with the number of steps; then the energy carried by the muons depends essentially on the number of steps needed to reach the energy level where the charged pions decay. As a consequence, a heavy nucleus gives more muons than a light one, or a proton; the exact ratio depends on the model of hadronic interactions.

Of course, within a given shower, the muon fraction depends on the longitudinal and transverse position (it increases with X and with the distance from the core). To be complete, one has to account also for the variation of the density, and then consider a dependence on θ: at large zenith angle, the hadronic cascade develops at a higher altitude, then the pions decay earlier (in terms of the slant depth), giving less muons, with a larger mean energy, carrying globally a larger fraction of E_{prim}.

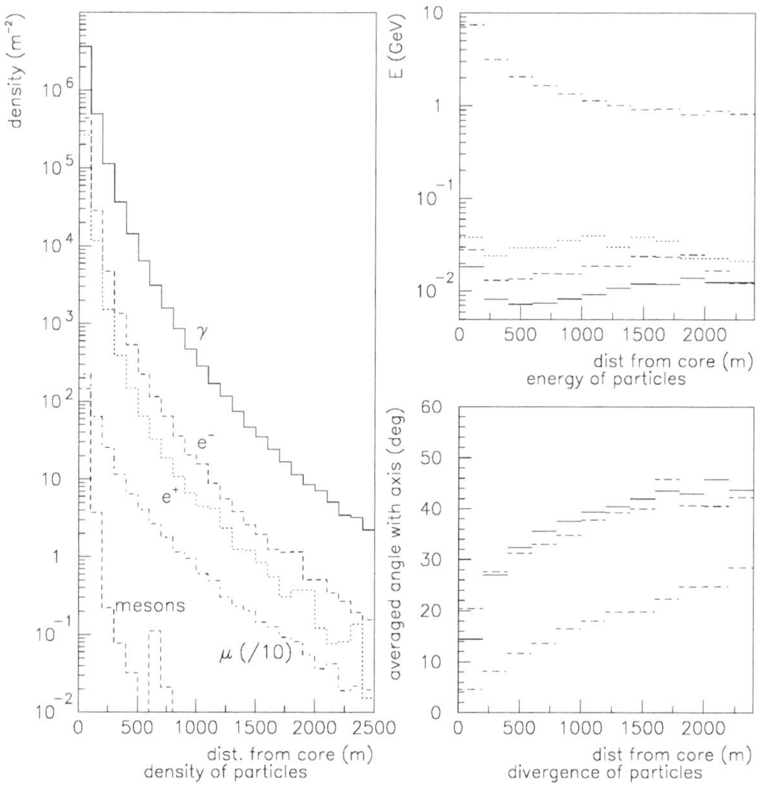

Fig. 5. Lateral distributions in a typical shower

If the primary is a photon or an electron, the hadronic cascade is initiated at a lower energy, then is has less steps, but it remains globally weak and the muon content is much reduced compared to a proton shower.

2.5 Time Structure

Let us consider now the time structure of the "signal" seen at a given position (distribution in time of the various components). As long as the particles are ultrarelativistic (at energies well above 1 GeV), the core is a kind of "fire ball" with a slowly increasing radius, moving at the speed of light. Let us call "front plane" the plane perpendicular to the shower axis, moving at speed c. The structure of the halo may be described in terms of the delay with respect to the front plane. Its components have different behaviours:

- the nucleons survive down to lowest energies: their arrival is spread over a long time (typically tens of microseconds at ground level). However this component is almost negligible, and does not extend far away from the shower axis.

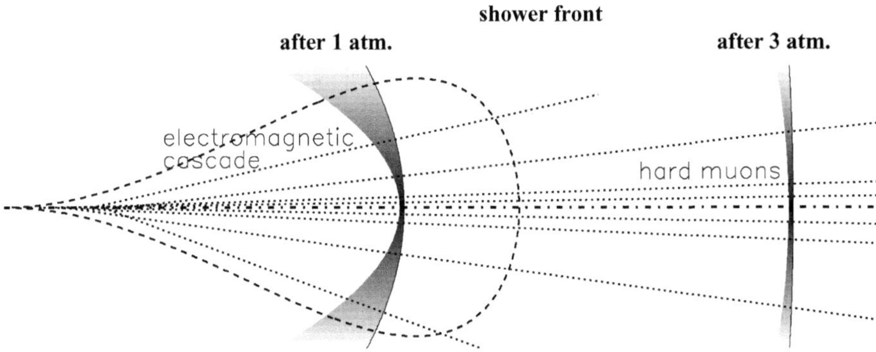

Fig. 6. Space-time structure (evolution of the shape of the front)

- the muons (and surviving pions) are generally highly relativistic ($1-\beta$ is small, but not negligible), and come almost in straight line from the hadronic core: then their delay increases with the distance r to the core, because of a simple geometric effect, and also because the velocity is anticorrelated to the angle of emission, thus to r. When going forward, both the longitudinal distance from the source and the mean energy increase, then the muonic front gets flatter and thinner. In practice, for nearly vertical showers, at ground level, the mean delay increases quadratically up to $r = 2$ km (~ 1.5 μs), then less rapidly; the time dispersion is also increasing with r.
- the electromagnetic halo may be considered as the result of a *diffusive* process, continuously generated from the core, with a relatively stable structure: a mean delay and a dispersion roughly proportional to r (typically 2.5 ± 1 μs at 2 km). There is no significant difference between the repartitions of electrons, positrons and photons.

Considered as a whole, the shower halo may be seen as a largely opened cone (a delay of 1 μs correspond to a longitudinal distance of 300 m), with a clear forward front of rounded shape, and a more diffuse backward boundary. The muons are more concentrated in the forward part. The curvature of the front and the thickness decrease as the shower propagates. After 2000 g cm^{-2}, the muonic tail (including its electromagnetic byproducts) is a very thin and almost flat disk. Figure 6 illustrates this evolution.

2.6 Fluctuations

Fluctuations, that is differences between showers produced from the same initial conditions, originate mainly from the very first steps, and especially from the position X_0 of the first interaction (whose uncertainty is of the order of the interaction length). As a result, the position of $X_{\max}$ has an intrinsic fluctuation of the order of a few 10 g cm^{-2}, decreasing with A.

The lateral distribution depends on X and is also affected by this initial fluctuation, but the evolution of the shape is such that, at a given X, there is a distance from core where the dependence on X_0 is stationary, then the fluctuation is reduced (less than 10%). At ground level, at 10 EeV, this distance is about 1000 m (it increases with energy). This feature is exploited in energy measurement by ground arrays.

There may be also small fluctuations in the atmosphere itself (temperature and pressure), but they just change the density, thus the scale of X versus the altitude, and not the total thickness; the transverse scale and the muon fraction may be slighly affected.

2.7 LPM Effect in the Electromagnetic Cascade

The Landau-Pomeranchuk-Migdal (LPM) effect [2] is a suppression of the electromagnetic processes at very high energy. These processes are characterized by a "formation length" inversely proportional to the longitudinal momentum transfer, which decreases with increasing energy (it may be interpreted as the wavelength of the exchanged virtual photon, or the distance required for the final particles to separate at least by a Compton wavelength) . If it is long, the formation becomes sensitive to other interactions in the medium, and the coherence is broken, for example by multiple scattering in a relatively dense medium. A detailed presentation of environmental factors may be found in [3]. These processes begins to be reduced above a characteristic energy×density product: $E_{\mathrm{LPM}} = m^2c^3\alpha X_0/4\pi\hbar = 7.7\,\mathrm{TeV/cm}\ X_0$. At the density of the upper atmosphere, E_{LPM} is of the order of 10 EeV (10^{19} eV). The density increases gradually when the cascade goes down; however, the descent in energy is faster, and only the first steps are above LPM threshold. In a hadronically induced shower, the electromagnetic cascade is a secondary process: it begins generally at an energy scale well below E_{prim}, then it is not strongly affected.

The situation is quite different if the primary particle is a photon (or an electron) well above 10 EeV. Then the first interaction is delayed, and several of the following steps may also be affected. Moreover, a large delay on the first step pushes the next one towards higher densities, thus it increases its interaction length, and so on as long as the energy is above E_{LPM}). As a consequence, there is a positive correlation between the step lengths at the beginning of the cascade: the initial fluctuations are "self-amplifying", and for given initial conditions, the fluctuations on X_{max}, and other features related to the "age" of the shower, are enhanced. Figure 7 illustrates both the lenghtening of the longitudinal profile and the increase of fluctuations.

3 Possible Observations

The rate of UHECR is so low that a direct detection of the shower core is practically excluded. In the same way, the Čerenkov light emitted along the core is strongly collimated and cannot be exploited in the same way as in high energy gamma detection, for example. The showers may be observed either through the fluorescence of nitrogen (isotropic emission, detectable up to tens of kilometers),

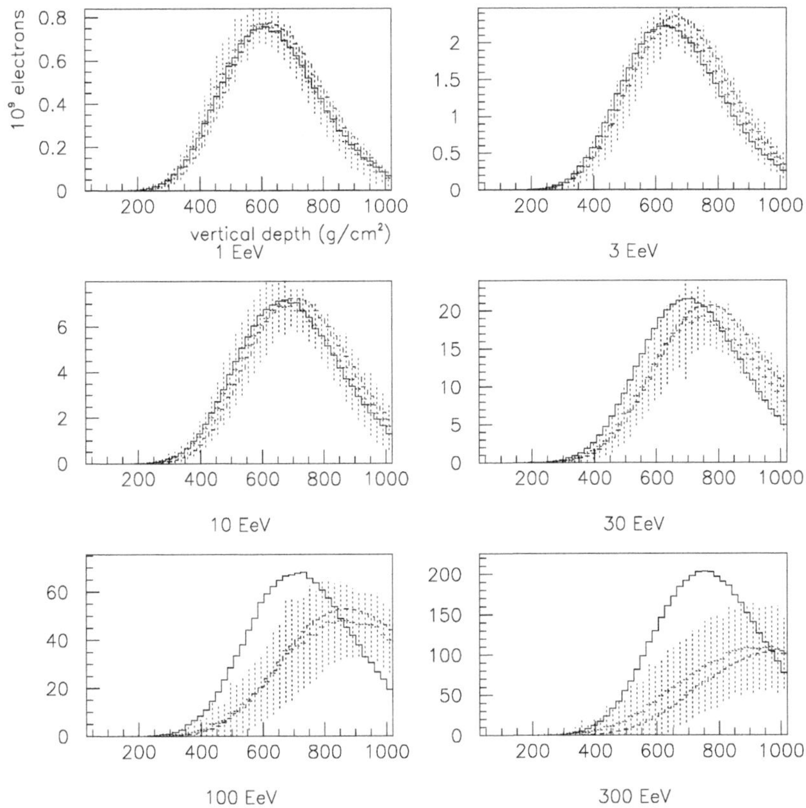

Fig. 7. Delay of electromagnetically induced showers due to LPM effect. Solid: shower simulation without LPM effect. Dotted: simulation inluding LPM effect, and semi-analytical evaluation. The curves are an average over different showers with the same initial conditions, and the vertical bars indicate the shower-to-shower fluctuations

or through a sampling of the halo with a ground array of particle detectors (within 2 to 3 km from the core). Another technique was also proposed, based on the radio emission due to the magnetic separation of charges [4].

3.1 Fluorescence Profile

The nitrogen molecules, excited by charged particles, may emit near-UV radiation. The probability of collisional de-excitation increases with the density, so that the rate of fluorescence per unit length is roughly independent of the altitude up to 20 km ($\simeq$ 5 photons per meter per charged particle). The typical detector is a set of telescopes covering a large part of the sky (the so-called "fly's eye" configuration), with a fast acquisition able to record rapidly varying pulses (over 10 to 30 μs, with a resolution around 100 ns). A good geometrical reconstruction of the shower axis ($\sim 0.5°$) is obtained with a stereoscopic view (two distant telescopes observing the same shower). After various corrections for attenuation (Rayleigh scattering on molecules and Mie scattering on aerosols), subtraction of the scattered Čerenkov light, and other observational biases, the longitudinal profile may be reconstructed. Of course, such telescopes can only work during dark nights, with a cloudless sky (in practice, 10 to 15 % of the time in semi-desertic sites). For more details the reader is referred to the contribution by S. Yoshida on the air fluorescence method in this volume.

3.2 Particle Detectors at Ground

To be efficient a ground observatory should cover a very large area. In practice it is an array of detectors giving a local sampling of the halo; at ultra-high energy the spacing of the detectors may be at the kilometer scale. The signal seen by individual detectors may be simply the density of charged particles (e.g. thin scintillators), or have a special sensitivity to penetrating particles, mainly muons (Čerenkov light in water tanks). With enough sampling points, it is possible to reconstruct different quantities:

- the *direction* of the shower axis, using the differences of time between the detectors, provided there are at least 3 points in the region where the shower front is close to a plane (precision $1-2°$). With more points, it is possible to measure the *curvature* of the front, related to the age of the shower.
- the *primary energy*, from the amplitudes of the signals: for example an interpolation is performed to estimate the density at a distance from the core where the fluctuations are minimal (typically 1 km); if no external information is available to determine the position of the core, it should be fitted together with a normalization factor, using a parametrization of the lateral density function, for example:

  $\rho(r) \propto \left(\frac{r}{r_\mathrm{M}}\right)^{s-2}\left(1+\frac{r}{r_\mathrm{M}}\right)^{s-4.5}$ (Nishimura-Kamata-Greisen function) [5]

  $\rho(r) \propto r^{-(\eta+r/4000)}$ (Haverah Park parametrization) [6].

  These formulae need corrective terms at large distances r. The parameter s or η (indicating the steepness of the distribution) is related to the age of the shower.
- the *rise time* of the signal (defined e.g. as the time where 50 % of the integrated amplitude is reached) which increases roughly linearly with r, and is also sensitive to the muonic fraction (if the detector has a large sensitivity to muons), thus to the nature of the primary.

4 Characterization of "Exotic" Primaries (Neutrinos and Photons)

4.1 Neutrino Interactions in the Deep Atmosphere

UHE neutrinos are usually produced as secondaries of hadronic interactions in the decay of a light meson into $\mu\nu_\mu$, followed by a muon decay into $e\nu_\mu\nu_e$ (we do not distinguish antileptons from leptons), thus a ratio 2:1 between ν_μ and ν_e. The rate of ν_τ, from hadrons with a heavy quark (b or c) is expected to be much lower. However a large mixing between ν_τ and ν_μ (if it is confirmed experimentally, see, e.g., the Super-Kamiokande results [7]) could change the proportions at arrival, and give equal fluxes of the three species. We do not consider this possibility here. For neutrino production in $\gamma-$ray bursts see the contribution by E. Waxman and see the contribution by P. Bhattacharjee and G. Sigl in this volume for "top-down" scenarios.

The main process in the atmosphere is the interaction on nuclei:

$$\begin{array}{ll} \nu + N \to \nu + \text{hadrons} & n.c. \text{ (neutral current)} \\ \nu + N \to l + \text{hadrons} & c.c \text{ (charged current)} \end{array}$$

where l is the corresponding charged lepton. It may be expressed in terms of the exchange of a Z^0 (n. c.) or a $W^\pm$ (c.c.) between the neutrino and a quark of the nucleus. The cross section of such processes increases with the energy E_ν of the neutrino, linearly up to $\simeq$ 10 TeV (as long as the the center-of-mass energy is below the W mass), and then more slowly. In our energy range [8] it is still much smaller than the hadronic and electromagnetic cross sections, but not completely negligible; the cross section of the c. c. reaction is roughly the same for ν and $\bar{\nu}$ and may be parametrized as:

$$\sigma_{\rm cc}(E_\nu) \simeq 10^{-32} \left(\frac{E_\nu}{10^{18}\ \text{EeV}} \right)^{0.4} \quad \text{cm}^2 \qquad \text{(per nucleon)}$$

The n.c. cross section is lower by a factor $\simeq 2.5$. The precise value of these cross sections, as well as the fraction of the energy transferred to hadrons (of the order of 0.2 at our energies), depend on the structure functions of the target nuclei; to evaluate them, an extrapolation is needed from collider data, and the result has a theoretical uncertainty (probably less than 50 %).

After the primary interaction, an hadronic shower develops normally, and also, in the case of a c.c. of ν_e or $\bar{\nu_e}$, an electromagnetic shower induced by the electron/positron; of course, due to the Lorentz boost, these two components are aligned with the initial direction and are seen as a unique "mixed" shower. A ν_μ or $\bar{\nu_\mu}$ is seen only through its hadronic products (a highly energetic muon radiates strongly, but not enough to lose a large part of its energy within the atmosphere, and thus to produce a detectable shower); this is an important difference with underwater neutrino detection, which is mainly sensitive to muons, because of their long range.

Around 10 EeV, the probability of a neutrino interaction through the whole atmosphere is typically $10^{-5}/\cos\theta$. It is distributed proportionally to the density, that is mainly in the lower 10 km. With reasonable expectations on the fluxes, such events are very rare compared to the total rate of UHECR. Then they need to be strongly discriminated from "normal" showers, induced in the upper atmosphere. In practice, they may be searched for at large zenith angles ($\theta >$

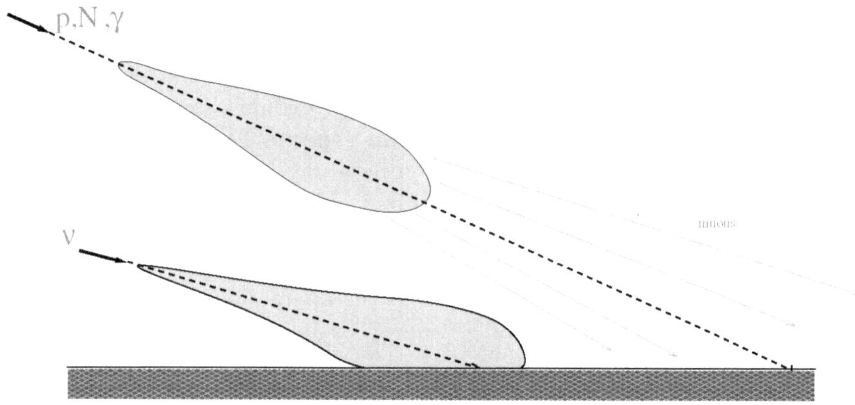

Fig. 8. Principle of the discrimination of neutrino induced showers

70°): at these angles, "normal" showers develop and decay at high altitude, and only their muonic tail hits the ground, while "deep" showers may arrive at ground at any stage of their development (see Fig. 8). Then the discriminating criteria may be:

- A deep quasi-horizontal longitudinal profile, as seen e.g. with a fluorescence detector; unfortunately these detectors are not optimized for such conditions.
- An electromagnetic component at ground level; as seen above, the signatures are the spread of arrival times of particles at a given point, the curvature of the shower front and the steepness of the lateral shape.

It should be pointed out that very inclined showers give a longitudinally extended "ground spot" (region where particles are detectable), so that they may be seen in several detectors of a ground array, even if their lateral extension is less than the array spacing, that is at energies well below the threshold for nearly vertical showers. A few signals can suffice to give a signature of the electromagnetic component, and a rough estimation of the direction; however, evaluating the primary energy is difficult if no other information is available to determine the position of the core.

4.2 Geomagnetic Photon Cascade

It was pointed out by McBreen and Lambert [9], using a theoretical review of electromagnetic interactions in extreme conditions by T. Erber [10], that γ-rays with energy above 10^{19} eV have a large probability to convert into an e^+e^- pair in the magnetic field of the Earth before entering the atmosphere. Then the electrons radiate strongly and produce a large number of photons; some of them may also give secondary pairs. As a result, instead of a unique photon, there is a electromagnetic "preshower" entering the upper atmosphere. Figure 9 gives the probability of pair conversion at typical values of the field around the Earth, and Fig. 10 shows that, whatever their initial energy and their injection

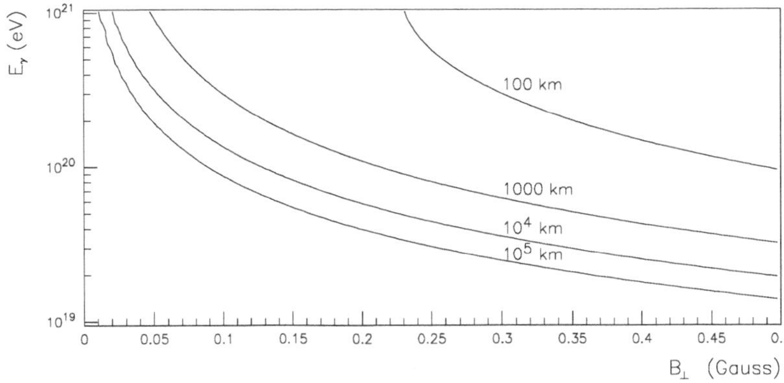

Fig. 9. Probability of geomagnetic conversion of a photon into a e^+e^- pair: the curves give the interaction length as a function of the energy and the transverse magnetic field

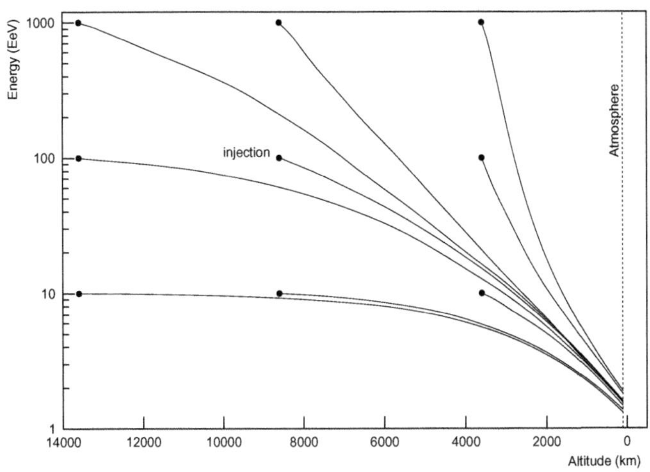

Fig. 10. Energy loss of e^+ and e^- through synchrotron radiation in the geomagnetic field

altitude, electrons and positrons radiate in such a way that they arrive in the atmosphere with an energy well below 10 EeV. The probability of conversion depends on the parameter $E_\gamma/2m_ec^2 \times B_\perp/B_{\rm cr}$, where $B_\perp$ is the field transverse to the direction of the photon, and $B_{\rm cr} = m_e^2\,c^2/e\,\hbar \simeq 4 \times 10^9\,{\rm T}$ is the "critical field"; then, this effect is expected to depend on the direction of observation with respect to the Earth frame (see for example Fig. 11). Such a dependence is a very strong signature of primary photons.

As seen above, the electromagnetic interactions are reduced at highest energies. In the atmosphere, showers induced by photons above a few 10^{19} eV will develop slowly. On the contrary, if the photon was converted, the atmospheric

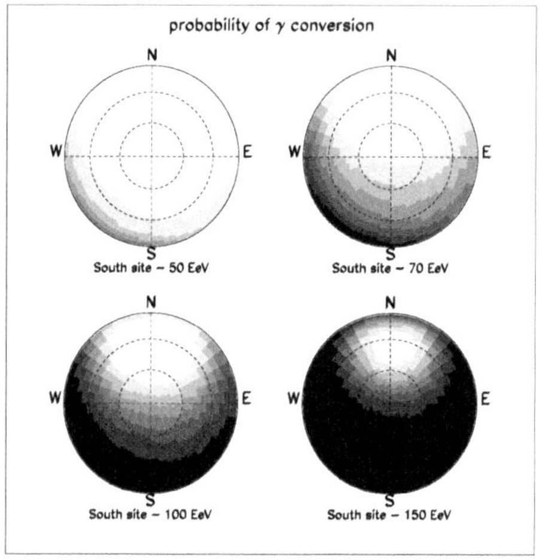

Fig. 11. Anisotropy of photon conversion probability: the level of grey indicate the probability (integrated over the photon trajectory) from 0 (white) to 1 (black), as a function from the direction of origin; the computation was done for the southern AUGER observatory (35S,69W), where the field is 25 μT, inclined by 35° over the horizon. The radial coordinate of this plot represents the zenith angle from 0 (center) to 90°

shower begins with photons, electrons and positrons below the LPM energy, and undergoes a "normal" development. The discrimination may be done either directly from the fluorescence profile, or indirectly from the ground observables (lateral steepness, curvature of the front, rise time of the signal); the difference between a converted photon and an unconverted one is generally larger than the shower-to-shower fluctuation, and also larger than the modelling uncertainties in the shower development. Then even a few unconverted photons could appear as clear anomalies compared to a background of hadronic showers.

5 Summary and Discussion

The atmospheric showers induced by ultra-high energy cosmic rays develop over more than one atmosphere depth ($1000\,\mathrm{g\,cm^{-2}}$), through a hadronic and an electromagnetic cascade. They consist of a narrow core with high density of particles (mainly photons and electrons), and a halo extending up to a few kilometers from the core. The core carries most of the energy, and produces detectable light (collimated Čerenkov emission, and isotropic nitrogen fluorescence); the halo is observable with particle detectors at ground. The decay of mesons at the end of the hadronic cascade produces a penetrating muonic component.

To answer the yet unsolved question of the origin of these rays, one needs to know their energy spectrum, their angular distribution (possible point sources), and their nature (light or heavy nuclei, "exotic" particles). The properties of the primary object may be reconstructed from the optical and ground observations:

- the direction, from geometrical and timing data, with a precision of the order of 1 degree or better.
- the energy, from an integration over the longitudinal profile, or a normalization of the lateral profile. The precision on individual events is limited to $\simeq 10\ \%$ by shower-to-shower fluctuations; the dependence on physical modelling is moderate.
- the nature, from the stage of development, e.g. the position $X_{\max}$ of the maximum of the longitudinal profile, or related properties of the shower front. Here the modeling uncertainties are important, because they are of the same order as the differences to be observed (for example between a proton and an iron nucleus). However, photons or neutrinos could be distinguished clearly, in certain conditions, from the nucleic background.

The properties of hadronic interactions at ultra-high energies need to be extrapolated from measurements of accelerator experiments, because QCD calculations cannot yet be performed exactly. Different models have been proposed, for example HPDM [11], VENUS [12], SIBYLL [13], QGSJET [14]; a review may be found in [15]. To some extent, these models may be tested if enough information is available. For example, the KASCADE detector [16] is able to measure the hadron and the muon rate in a lower energy range (below 10^{16} eV) and to compare the observations with the predictions of the models [17]. For the moment, there is no fully satisfactory model; QGSJET seems to be the preferred one. On the other hand, external constraints may be set (for example, a nucleus cannot be lighter than a proton and is unlikely heavier than iron). Assuming that a scaling law is valid (e.g. $X_{\max}$ increases logarithmically with energy) allows to see an evolution of the composition with energy (see for example [18]).

Of course, one can imagine that new physics plays an important role in this energy domain, for example collective effects in nuclear collisions [19], or more *ad hoc* hypotheses: symmetry breaking at energies below the Grand Unification scale, allowing for example neutrinos to undergo hadronic interactions, or Lorentz invariance violation. Unless specific and precise signatures are exhibited, it is difficult to take all possibilities into account.

References

1. T.K. Gaisser, A.M. Hillas: proc. 15th Int. Cosmic Ray Conf. (plovdiv), **8**, 353 (1977)
2. L.D. Landau, I. Ya. Pomeranchuk: Dokl. Akad. Nauk SSSR, **92**, 535, 735 (1953) A.B. Migdal: Phys. Rev. **103**, 1811 (1956)
3. Spencer Klein: Rev. Mod. Phys. **71(5)** 1501 (1999)
4. see, e.g., First International Workshop on Radio Detection of High-Energy Particles, University of California, Los Angeles, November 16-18 2000, to appear in the AIP conference proceedings.
5. J. Nishimura: Handbuch der Physik, vol. **46/2**, 1-114, Springer Verlag, 1967
6. R.N. Coy, G. Cunningham, C.L. Pryke, A.A. Watson: Astrop. Phys. **6**, 263 (1997)

7. Y. Fukuda, et al. (Super-Kamiokande collaboration): Phys. Rev. Lett. **81**, 1562 (1998)
8. R. Gandhi, C. Quigg, M.H. Reno, I. Sarcevic: Astropart. Phys. **5**, 81 (1996)
9. B. McBreen, C. J. Lambert: Phys. Rev. D **24**, 2536 (1981)
10. T. Erber: Rev. Mod. Phys. **38**, 626 (1966)
11. J. N. Capdevielle, et al.: Report KfK 4998 (1992)
12. K. Werner: Physics Reports **232** 87 (1993)
13. R. S. Fletcher, T. K. Gaisser, P. Lipari, T. Stanev: Phys. Rev. D **50**, 5710 (1994)
14. N. N. Kalmykov, S. S. Ostapchenko: Yad. Fiz. **56**, 105 (1993)
N. N. Kalmykov, S. S. Ostapchenko: Phys. At. Nucl. **56(3)**, 346 (1993)
15. J. Knapp, D. Heck, G. Schatz: Forschungszentrum Karlsruhe Report FZKA 5828 (1996)
16. H.O. Klages, et al. (KASCADE Collaboration): Nucl. Phys. B, Proc. Suppl. **52B**, 92 (1997)
17. T. Antoni, et al., J. Phys. G: Nucl. Part. Phys. **25**, 2161 (1999)
18. J. A. Hinton, et al.: Proc. *26*th *International Cosmic Ray Conference*, Salt Lake City, vol. **3**, 288 (1999)
19. C. Pajares, D. Sousa, R. A. Vázquez: Astropart. Phys. **12**, 291 (2000)

The Air Fluorescence Method for Measuring Extremely-High-Energy Cosmic Rays

Shigeru Yoshida

Institute for Cosmic Ray Research, University of Tokyo, 5-1-5 Kashiwa-no-Ha, Kashiwa City, Chiba 277-8582, Japan

Abstract. The air fluorescence technique for detection of Extremely High Energy Cosmic Rays (EHECRs) is reviewed. A general overview is given based on a series of simple equations to approximate the detection mechanism. The calorimetric energy measurement, which is the heart of the fluorescence technique is discussed in detail. Some technical issues and possible systematic uncertainties associated with the fluorescence measurement are briefly summarized.

1 Introduction

The purpose of the air fluorescence technique is to detect Extensive Air Showers (EAS) by the measurement of the ultraviolet fluorescence of molecular nitrogen generated by the air shower particles. Unlike Čerenkov radiation, this fluorescence is isotropic and hence it can be seen from any angle by appropriate detectors. This fluorescence yields about four photons per meter of ionizing trajectory along the EAS axis, which are collected by a light collector system such as reflection mirrors and recorded with an ultraviolet-sensitive camera like a mosaic of photomultiplier tubes. As an air shower cascade develops, emitted ultraviolet photons translate into time-dependent signals passing through the field of view of the optical detectors. This defines a moving track through the atmosphere as shown in Fig. 1, which is able to reconstruct the longitudinal shower profile. The integral of the reconstructed profile is directly proportional to the primary energy of a EHECR initiating the EAS. This method is essentially calorimetric, measuring the total energy deposition in the atmosphere by means of its fluorescence. It does not need a complex Monte Carlo simulation to determine the energy scale, as is required in the ground array technique. The identification of the primary particle is made by examining the shape of the longitudinal profile of the shower. The atmospheric depth of the shower maximum ($X_{\rm max}$) is a good parameter for the identification. This technique has a great potential to discriminate gamma rays and neutrinos from cosmic ray hadrons, which is the essential feature of the experiment to identify the origin of EHECRs.

2 Overview

General features of the fluorescence method using an optical detector which consists of a reflection mirror and a mosaic of phototubes can be easily obtained by

Martin Lemoine, Günter Sigl (Eds.): LNP 576, pp. 45–57, 2001.

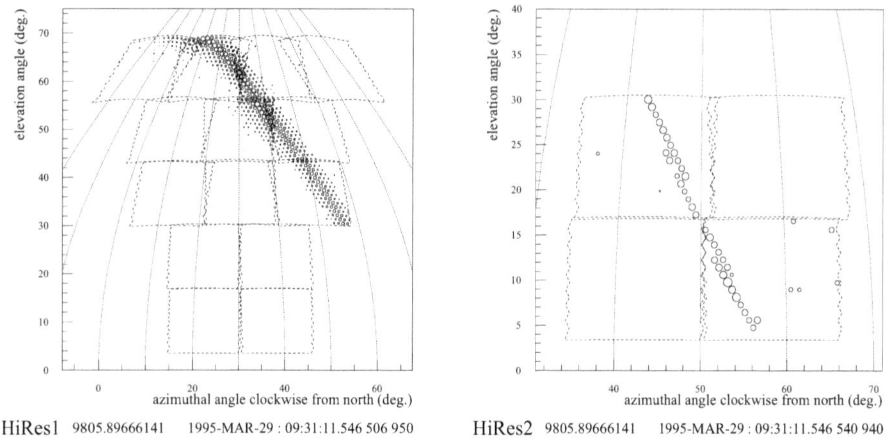

Fig. 1. An example of the event tracks detected by the HiRes prototype [1]. The tracks passing through fields of views of several mirrors are projected on the plane of mirror mosaics for plotting. This shower was viewed simultaneously by both HiRes1 and HiRes2 which are located at separation of 12.6 km. It is called a "stereo" event which provides significant redundancy

the following arguments [2]: A phototube in the camera is useful in an event only if it collects more air fluorescence light emitted from the shower track than the fluctuation of night sky background light during its integration time $t_{\rm gate}$. Some of the fluorescence light is scattered out during its propagation to the detector due to collisions with air molecules (the Rayleigh scattering) or with dust, pollution, fog, and clouds (the Mie scattering). The expected air fluorescence signal is thus given by

$$N_{\rm ph} = \frac{A_{\rm mir} N_e Q}{4\pi r_p^2} e^{-\frac{r_p}{r_0}} e_{\rm eff} r_p \Delta\theta \tag{1}$$

where r_p is distance to a fluorescence emission point along the shower axis from the detector, $A_{\rm mir}$ is the area of a mirror in the detector, N_e is the number of electrons in the shower cascade viewed by that phototube, r_0 is the extinction length of light due to the atmospheric scattering, $e_{\rm eff}$ is the fluorescence light yield from one electron (photons per meter), $\Delta\theta$ is the phototube pixel size and Q is the quantum efficiency of the phototube. The background light is given by

$$N_{\rm BG} = n_{\rm NB} t_{\rm gate} A_{\rm mir} Q (\Delta\theta)^2 \tag{2}$$

where $n_{\rm NB}$ is the night sky photon intensity and $t_{\rm gate}$ is the gate time for signal collection. Then the signal to noise ratio $n_{\rm th}$ gives the threshold shower electron size for triggering a channel as a function of r_p as follows:

$$N_{e,{\rm th}} = n_{\rm th} 4\pi r_p \exp\left(\frac{r_p}{r_0}\right) e_{\rm eff}^{-1} \left(\frac{n_{\rm NB} t_{\rm gate}}{A_{\rm mir} Q}\right)^{1/2} . \tag{3}$$

When $r_p \gg r_0$ which is likely for detection of EHECRs, to an accuracy of 35 %, this equation can be written as

$$\begin{aligned} \log N_{e,\mathrm{th}} &= 7.54 + \left(\frac{r_0}{8\mathrm{km}}\right)^{-\frac{4}{5}} 8.23 \times 10^{-2} \left(\frac{r_p}{\mathrm{km}}\right) \\ &+ \log\left[n_{\mathrm{th}}\left(\frac{r_0}{8\mathrm{km}}\right)\left(\frac{4\mathrm{m}^{-1}}{e_{\mathrm{eff}}}\right)\left(\frac{1\mathrm{m}}{R_{\mathrm{mir}}}\right)\right] \\ &+ \frac{1}{2}\log\left[\left(\frac{n_{\mathrm{NB}}}{10^6\mathrm{m}^{-2}sr^{-1}\mu\mathrm{s}^{-1}}\right)\left(\frac{t_{\mathrm{gate}}}{5\mu\mathrm{s}}\right)\right] \end{aligned} \tag{4}$$

where log is the logarithm to base 10, R_{mir} is the radius of the mirror of the optical detector and Q is assumed to be 30 %.

The atmospheric slant width during which the shower cascade contains more electrons than this threshold size $N_{e,\mathrm{th}}$ can be numerically obtained as the following expression:

$$X_t^{100\%} \equiv X_t(N_e \geq N_{e,\mathrm{th}}) = 100(-\eta^2 - 8\eta + 2) \qquad \mathrm{g/cm}^2 \tag{5}$$

$$\eta = \log(N_{e,\mathrm{th}}) - \log\left(\frac{E_e}{\mathrm{GeV}}\right) \tag{6}$$

Using Eq. (4), η can be written as a function of r_p and thus $X_t^{100\%}$ is a function of E_e and r_p. To trigger showers with a given geometry and energy, $X_t^{100\%} \geq 0$ must be required, which leads to a maximum shower distance at which the optical detector will trigger:

$$r_p^{\max} = 12.15\left(\frac{r_0}{8\mathrm{km}}\right)^{\frac{4}{5}} f \qquad \mathrm{km}, \tag{7}$$

where

$$f = 2.7 + \log\left(\frac{E}{10^{19}\mathrm{eV}}\right) \tag{8}$$

$$\begin{aligned} &+ \log\left[n_{\mathrm{th}}\left(\frac{r_0}{8\mathrm{km}}\right)\left(\frac{4\mathrm{m}^{-1}}{e_{\mathrm{eff}}}\right)\left(\frac{1\mathrm{m}}{R_{\mathrm{mir}}}\right)\right] \\ &+ \frac{1}{2}\log\left[\left(\frac{n_{\mathrm{NB}}}{10^6\mathrm{m}^{-2}sr^{-1}\mu\mathrm{s}^{-1}}\right)\left(\frac{t_{\mathrm{gate}}}{5\mu\mathrm{s}}\right)\right] \end{aligned} \tag{9}$$

At 10^{19} eV for $n_{\mathrm{th}} = 2$ (i.e., 2σ significance) $r_p^{\max} \sim 29$ km for the detectors operating in a desert atmosphere.

From the arguments above, general consequences on this detection method can be obtained. First, the typical distance scale to observable EAS from the optical detector is 30–40 km as expressed in Eqs. (7) and (9) which only weakly depend on the detector specification such as the mirror area and the pixel size of a phototube. This is because most of the light from the showers is scattered out by the Rayleigh and Mie scattering and reduced significantly. The exponential

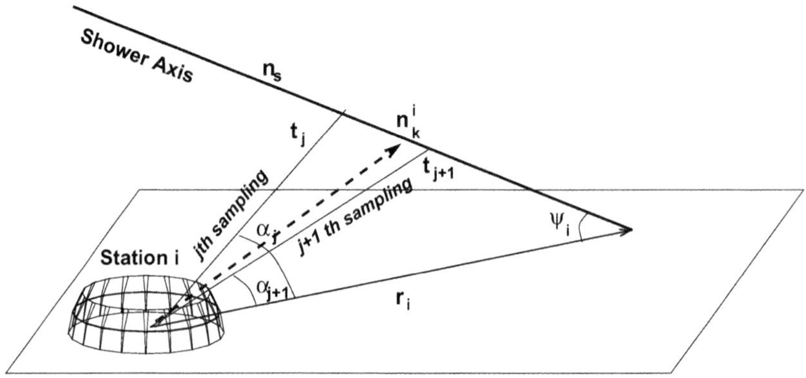

Fig. 2. Geometrical relations between the event track and the optical detectors

term in Eq. (1) dominates in the overall contribution. This consideration leads to the second consequence: The atmospheric monitoring to measure the extinction length r_0 is crucial. The primary energy of an EHECR particle is approximately proportional to the signal from the initiated air shower and Eq. (1) shows that the uncertainty in the energy determination is related to the extinction length as

$$\frac{\Delta E}{E} \simeq \frac{r_p}{r_0}\frac{\Delta r_0}{r_0}. \tag{10}$$

This means that we must determine r_0 with an accuracy of 5 % for estimating the energy of events at $\sim$ 30 km from the detector within systematic errors of 10 %. This is challenging, but not impossible because contribution of the Rayleigh scattering dominates over the Mie scattering process in the fluorescence light propagation and effects of the Rayleigh scattering can be accurately predicted because it is a rather simple electromagnetic process. Many methods to measure the extinction length have been proposed and performed [3,4]. They are based on measurements of laser and "flasher" shots fired through the detector aperture [5]. Two approaches are possible. One uses select geometries to deconvolve the effects of the Mie scattering and extract the transmission. The second fits an aerosol model to the observed data to determine the model parameters. These parameters include a horizontal attenuation length, an aerosol scale height, and a scattering dependence or phase function. Details of the technical issues are found in [5].

The third consequence from these arguments is that the estimation of primary energies relies on the geometrical reconstruction of observed events because the signal strength depends heavily on r_p as expressed in Eq. (1). A superhigh energy EAS could only produce very weak signals if it is far from the detector, in contrast to the ground array technique where the higher energy event puts the larger and denser footprint at the array surface. Therefore the accuracy of the geometrical reconstruction is important not only for studies of the arrival directions, but also for a reliable energy determination. Fortunately, the achievable resolution of the geometrical reconstruction is good enough since the event geometry can be deduced not only by the geometrical information of the recorded event track in the camera but also by its timing information. In other words, the signal profile such as strength and duration strongly depends on the geometrical relation between the detector and the shower axis. Figure 2 illustrates how the event geometry determines the signal profile. Provided the data recording system is capable of sampling the signal strength from the shower with constant frequency, the longitudinal direction α_j of the light spot at the station i along the shower track is related to a given event geometry and relative timing at the j-th sampling as follows.

$$\alpha_j = \pi - \psi_i - 2\tan^{-1}\left[\frac{c}{R_p^i}\left(t_j - t_0 - \frac{\mathbf{n_s r_i}}{c}\right)\right], \qquad (11)$$

where R_p^i is the impact parameter from station i, $\mathbf{n_s}$ is the direction of the shower axis, $\mathbf{r_i}$ is the vector from the station to the core location, ψ_i is the angle between the shower axis and the vector from the location of the station to the shower core location, and t_0 is the absolute origin of the timing, see also Fig. 2. Consequently the signal profile at every sampling time, i.e., how the light spot crosses the PMT is a function of the geometrical parameters via Eq. (11). Fine geometrical resolution is hence obtained by minimizing the χ^2 built by comparison of the prediction of the signal profile by the Monte Carlo simulation with the profile recorded at every sampling frequency. Another key to the fine resolution is to view the events in stereoscopic manner. Having two or more stations of the optical detectors at $10 \sim 40$ km separation makes stereoscopic measurement of the EAS profile possible, which provides a model independent way of checking the energy and depth of shower maximum ($X_{\max}$) resolution. By comparing the energy and $X_{\max}$ values independently reconstructed from each station using the stereo geometry (assuming the stereo geometry is well determined), the resolution of the deduced parameters like the energy can be experimentally examined without relying on a complex Monte Carlo simulation. In today's fluorescence detectors, the stereoscopic capability is considered as a mandatory function.

3 Calorimetric Energy Measurement

The calorimetric energy measurement is at the heart of the air fluorescence method. The total energy deposition in the atmosphere is

$$E^{\text{deposit}} = \int dX N_e(X)\alpha(\epsilon, S(X)) \qquad (12)$$

where $N_e(X)$ is number of electrons in the shower as a function of depth in the atmosphere where X is measured in unit of g/cm^2. The longitudinal shower profile $N_e(X)$ can be analytically well described by the Gaisser-Hillas formula (see also contribution by P. Billoir in this volume):

$$N_e(X, X_{\max}, X_1, N_{\max}) = N_{\max} \left(\frac{X - X_1}{X_{\max} - X_1} \right)^{\frac{X_{\max} - X_1}{\lambda}} \exp \left(\frac{X_{\max} - X}{\lambda} \right) . \tag{13}$$

Here, $X_{\max}$ is the shower maximum, X_1 is the atmospheric depth at the first interaction point of the shower, and λ is the attenuation length which is approximately 70 g/cm^2. Further, α in Eq. (12) is the energy loss per slant depth per charged particle in air via ionization, which is given by [6]

$$\alpha(\epsilon, S(X)) = \int_{\epsilon} dE_e \frac{dE_e}{dX} n_e(E_e, S) , \tag{14}$$

where n_e is the differential energy spectrum of electrons normalized to 1, dE_e/dX is the ionization loss rate for an electron as a function of its kinematic energy E_e. Also, ϵ is the threshold energy in this integral, which is presumably zero in real events, and $S(X)$ is the "age" parameter, which is defined as

$$S(X) = \frac{3(X - X_1)}{(X - X_1) + 2(X_{\max} - X1)} . \tag{15}$$

The number of the fluorescence photons is given by

$$N_\gamma^{\text{fl}} = \int dX N_e(X) \frac{dL}{dX} \int_{\epsilon} dE_e \frac{dY^{\text{fl}}}{dL} n_e(E_e, S) , \tag{16}$$

where dY^{fl}/dL is the air fluorescence yield in units of photons per unit length which can be formalized as [7]

$$\frac{dY^{\text{fl}}}{dL} = \kappa^{-1} \frac{dE_e}{dX} f(\rho, T) = \kappa^{-1} \frac{dE_e}{dX} \sum_i \rho \frac{A_i}{1 + B_i \rho \sqrt{T}} . \tag{17}$$

Here, ρ is the air density in g/cm^3, T is the temperature in Kelvin, and A_i and B_i are constant coefficients that take into account its various wavelength dependences. Further, κ is for normalization and chosen as

$$\kappa = \left(\frac{dE_e}{dX} \right) \bigg|_{E_e = 1.4\text{MeV}} \simeq 1.668 \frac{\text{MeV}}{\text{gcm}^{-2}} \tag{18}$$

for practical reasons. The fact that the fluorescence yield is proportional to dE_e/dX as expressed in Eq. (17) enables the fluorescence detectors to measure E^{deposit} in a calorimetric way. Comparing Eqs. (12), and (14) with Eq. (16), we get

$$\frac{dE^{\text{deposit}}}{dX} = \frac{dN_\gamma^{\text{fl}}}{dL} \kappa f^{-1}(\rho, T) \tag{19}$$

which is independent of $\alpha(\epsilon, S)$ and directly shows the calorimetric energy measurement.

It should be remarked, however, that it is necessary to calculate $\alpha(\epsilon, S)$ to obtain the longitudinal shower profile $N_e(X)$ from the measured number of fluorescence photons $N_\gamma^{\rm fl}$. From Eqs. (12), (16) and (17), we obtain that the reconstructed shower profile $N_e^{\rm rec}$ is given by

$$N_e^{\rm rec}(X, \epsilon) = \frac{dN_\gamma^{\rm fl}}{dL} \kappa f^{-1}(\rho, T) \alpha^{-1}(\epsilon, S) \,. \tag{20}$$

Calculation of $\alpha(\epsilon, S)$ relies on Monte Carlo simulations, and for technical reasons the threshold energy ϵ can not be set to zero. In the actual simulations, $\epsilon = 0.1\,\mathrm{MeV}$ has often been used, and the reconstructed shower profile $N_e^{\rm rec}(X)$ is a function of ϵ in this sense that causes simulation-dependences and weakly depends on the mass of the primary particles. However, the most important parameter $X_{\rm max}$ to deduce the mass composition of the primary EHECR particles is mainly determined by $\alpha(\epsilon, S)$ at around $S = 1$ which is very stable and almost independent of the primary mass and energy, $\alpha(0.1\,\mathrm{MeV}, 1) \simeq 2.19\,\mathrm{MeV/(g/cm^2)}$ [6]. Hence there is no major systematics in the $X_{\rm max}$ measurement caused from the simulation-dependent uncertainties. Note that Eq. (20) also gives

$$E^{\rm deposit} = \int dX N_e^{\rm rec}(X, \epsilon) \alpha(\epsilon, S(X)) = \int dX N_e(X) \alpha(0, S(X)) \tag{21}$$

indicating again that the energy deposition measurement is independent of ϵ and free from assumptions in the shower Monte Carlo simulations.

For most of the observed events, however, the primary energy measurement partly relies on the reconstructed shower profile $N_e^{\rm rec}(X, \epsilon)$ derived by the simulation-dependent procedure via $\alpha(\epsilon, S)$ because only a part of the longitudinal profile is usually within a field of view of the optical detectors and we must extrapolate the invisible part of the profile by the Gaisser-Hillas formula with $X_{\rm max}$ and X_1 determined by the detected part of the profile. The detected signal strength per unit length in units of number of photoelectrons per meter is related to $N_e^{\rm rec}(X, \epsilon)$ as follows:

$$\begin{aligned} \frac{dN_{pe}}{dL} &= \frac{N_e^{\rm rec}(X, \epsilon)}{4\pi r_p^2} \kappa^{-1} f(\rho, T) \alpha(\epsilon, S) A_{\rm mir} \\ &\times \int d\lambda T_{\rm M}(\lambda) \exp\left[-\frac{\Delta X_{\rm det}}{\lambda_{\rm R}} \left(\frac{400\mathrm{nm}}{\lambda} \right)^4 \right] f_{\rm fl}(\lambda) \epsilon_{\rm det}(\lambda) \,, \end{aligned} \tag{22}$$

where $\Delta X_{\rm det}$ is the atmospheric slant depth between the location of a mirror and the light-emission point along the shower axis, $f_{\rm fl}(\lambda)$ is the fluorescence flux at wavelength λ, $T_{\rm M}$ is the transmission factor of propagation of light taking into account Mie scattering, $\lambda_{\rm R}$ is the extinction length for Rayleigh scattering, and $\epsilon_{\rm det}(\lambda)$ is the overall detection efficiency determined by the quantum efficiency of a PMT tube, transmission factor of an optical filter, reflectivity of the mirrors, dead space of the camera and so on. This equation is essentially equivalent to Eq. (1) but takes into account the wavelength-dependence. Then the reconstructed $N_e^{\rm rec}(X,\epsilon)$ deduces $E^{\rm deposit}$ by the energy integral of Eq. (12). It should be noted that this integral is mostly determined by the shower profile at around $X_{\rm max}$. Therefore, as long as the profile around the shower maximum is directly viewed by the detectors, the dominant contribution in the energy integral is given by the calorimetric manner as expressed by Eq. (19) and the resultant energy deposition measurement is almost independent of $\alpha(\epsilon,S)$.

It is true that $E^{\rm deposit}$ is calorimetrically measured, but the primary energy of an EHECR particle E would not be the same as $E^{\rm deposit}$ because a part of the primary energy is channeled into neutrinos, high energy muons and nuclear excitation. Even for γ-ray induced showers there is a tiny "missing energy" because of the photo-nuclear interactions and the $\mu^+\mu^-$ pair production. The missing energy for γ-ray induced showers is only $\sim 1\%$ of the primary energy while that of hadronic showers is not negligible and some corrections are necessary. However, the correction factor decreases with increasing primary energy because charged pions produced in more energetic showers are more likely to interact than decay into muons and neutrinos. The simulation study [6] using CORSIKA [8] shows

$$\frac{E^{\rm deposit}}{E} = 0.959 - 0.082\left(\frac{E^{\rm deposit}}{10^{18}{\rm eV}}\right)^{-0.15} \quad . \tag{23}$$

This factor depends on primary mass and there is about 5 % difference between proton and iron-induced showers. The above function is for their average behavior.

4 Photomultiplier and Electronics

In a typical design, the camera equiped in the optical detectors basically consists of photomultiplier tubes (PMTs), optical UV filters passing only photons with wavelengths in the UV air fluorescence emission (300$\sim$ 400 nm), and the front-end electronics for reading out the signals.

The main requirements for the PMT specifications for the air fluorescence measurement are the gain stability under the night sky background photon intensities and its fluctuation, a sufficiently good linearity to record EHECR events passing near the detector, and uniformity of the PMT cathode. The expected night sky background intensity collected by a reflection mirror of 3m diameter is $\sim$ 25 photoelectrons in 200 ns for a PMT with field of view (FOV) of 1 degree. Thus the anode current is

$$I_{\rm BG} = eN_{\rm BG}G = 1.6\mu{\rm A}\left(\frac{N_{\rm BG}}{130{\rm MHz}}\right)\left(\frac{G}{8\times 10^4}\right) \tag{24}$$

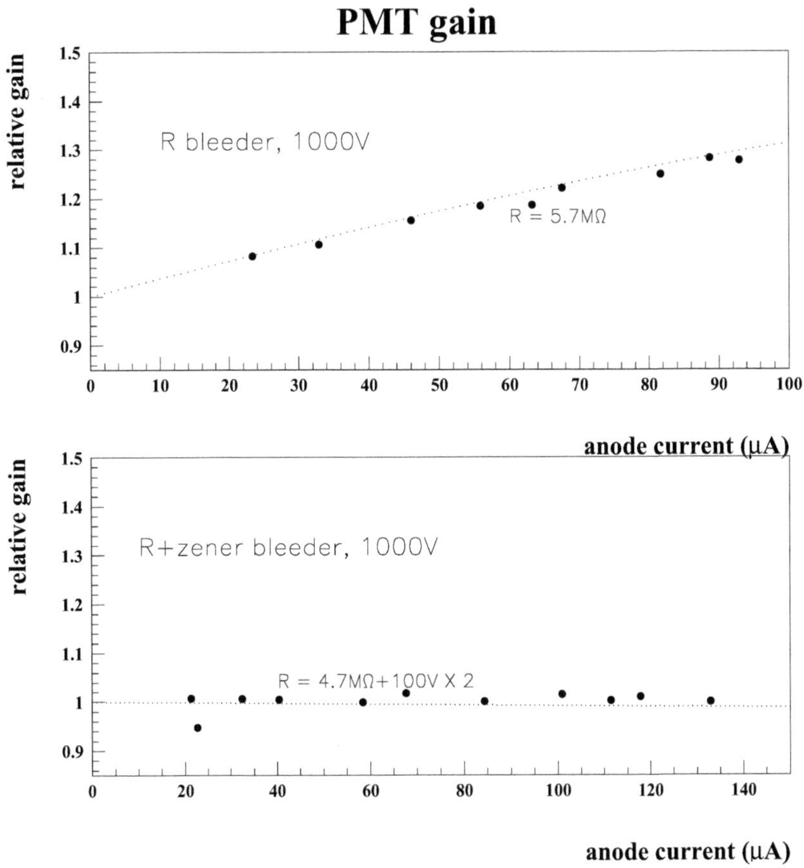

Fig. 3. Pulse gain variation against background anode current, with the ordinary bleeder circuit (a) and with the hybrid bleeder circuit (b)

when a PMT is operated with gain G. This is not completely negligible compared with the maximum bleeder current of ~ 0.2 mA. Moreover, the anode current fluctuates continuously during the observations. This fluctuation changes the PMT gain significantly when we use an ordinary bleeder circuit by the register chain. One of the possible solution is a hybrid bleeder circuit using Zener diodes at the last 2 stages of the dynode chain. The measured variation of the pulse gain against the background anode current is shown in Fig. 3. As shown in this figure, the pulse gain of the hybrid bleeder is almost constant in the wide range of anode currents. As with the linearity, the possible maximum strength of the signals is

$$I_{\text{sig}} = eG(N_{\text{ph}}/\Delta t) = 0.16\text{mA} \left(\frac{N_{pe}}{2 \times 10^3} \right) \left(\frac{\Delta t}{200\,\text{nsec}} \right)^{-1} \left(\frac{G}{8 \times 10^4} \right) \quad (25)$$

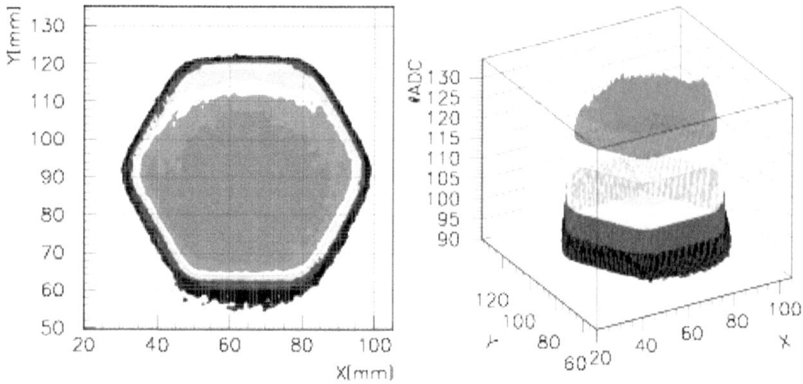

Fig. 4. Contour map of the measured anode gain (relative)

when the detector receives $\sim 2 \times 10^3$ photoelectrons at gate width of 200 nsec, which could happen if a shower with primary energy of 10^{20} eV passes at a distance of 10 km from the detector. Since the linearity of PMT depends on the relative voltages applied to the last few dynodes, some optimization of bleeder circuit may be needed to have sufficient linear response up to ~ 1 mA depending on the gain applied to PMTs.

The sensitivity of the PMT has a position dependence on its photocathode surface, reflecting the non-uniformity of the quantum efficiency and the amplification gain. Various factors such as the photocathode thickness, the irregularity of the focusing electric fields and the geometry of the first few dynodes affects the position dependence. Figure 4 shows a typical two dimensional response of the PMT gain. This response must be well understood because the accuracy of reconstruction of the event geometry and the longitudinal shower development relies on the measurement of the motion of the light spot emitted from the shower which sweeps the photocathode surface from edge to edge. In order to improve the accuracy, one can either measure the position dependence by scanning the pulsed LED source on the PMT window or place a wedge-shaped reflector at the PMT boundary such that the fluorescence light impinging on the insensitive band is reflected and guided to the central region of the photocathode minimizing the non-uniformity.

Front-end electronics to read out the air fluorescence signals need the capability of recording various profiles of pulses. The strength and duration of the air-fluorescence pulse depends largely on the geometry of the event, as illustrated in Fig. 5. A well defined short pulse with high amplitude will be produced by the nearby air shower traversing the camera's line of sight perpendicularly. A signal from a distant event with running away configuration, on the other hand, will be widely spread and is difficult to be separated from the night sky background. The recording of the signal time profile becomes increasingly important for events with a long duration, for which most of the information to determine

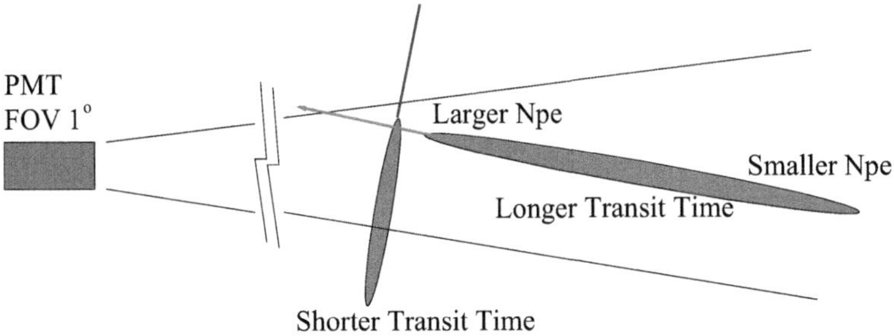

Fig. 5. Schematic illustration to show the relation between the event geometry and the resultant pulse profile. "Npe" is the number of photoelectrons

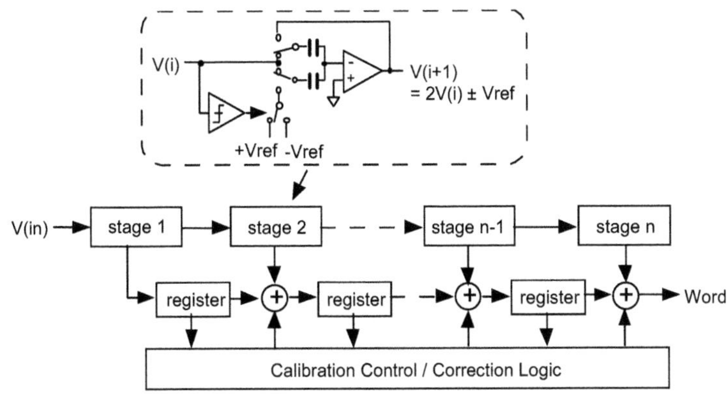

Fig. 6. Pipelined ADC

the geometry (particularly the direction) is included in the signal timing profile. Therefore the continuous digitization and recording of the signal wave form by FADC or pipelined ADC as shown in Fig. 6 is much preferred compared to the analog integration of the signal with a fixed gate width, or a fixed time constant. For the triggering, however, the traditional method of the analog integration with two or three various time constants works well enough to trigger the events without significant bias.

5 Systematic Errors

Let us summarize here the sources of systematic uncertainty in energy estimation.

- Uncertainties in the PMT quantum efficiency, the PMT/preamp gain, and the mirror reflectivity
  The absolute gain and sensitivity of the detector must be calibrated and monitored. In the ground array method, signals from local muons passing

through the particle counter give the reference for the absolute calibration. This would be more difficult in the case of the air fluorescence detectors, however, because there is no absolute light candle in nature. The standard procedure is to compare the signals produced by the stable light sources such as UV LEDs, xenon flashers, and YAP pulsars equiped in the optical detectors in field to those processed in the absolutely calibrated PMT/electronics. Holding the overall uncertainty below 10 % is the current reasonable goal for the calibration.

- Uncertainty in the fluorescence yield
  The measurement of the energy deposition in the EAS relies on the number of the fluorescence yield as directly expressed in Eq. 19. The yield has been well measured by a laboratory experiment [7]. However, the yield intensity for low energy electrons below 1 MeV is not quite understood, and the uncertainty is given as $\sim$ 10 %.
- Missing Energy
  As already described, the energy carried away by neutrinos and high energy muons in air shower cascades cannot be directly measured and must be estimated by Monte Carlo simulations. The correction factor depends on primary mass and there would be 5 % difference between proton and iron-induced showers [6] which leads to unresolved systematic uncertainty in the primary energy estimation. We note in this context that energy loss into invisible channels which could open up due to new particle physics at high center of mass energies (i.e., production of supersymmetric particles) would normally lead to *underestimating* the energy.
- Atmospheric extinction
  Measurement of the atmospheric extinction length determined the dominant conversion factor from the number of photoelectrons recorded by the detector to the number of fluorescence photons radiated from the shower. As already noted, this would be the largest correction in the energy determination. The current goal is 10 % of accuracy of determination of the extinction length.

Acknowledgments

The author wishes to thank the School organizers for their warm hospitality at this meeting. He also wishes to acknowledge very valuable discussions with Dr. Zhen Cao of the University of Utah. He is grateful to the Telescope Array collaboration for the continuous support on the R& D efforts to build the next generation fluorescence detectors. This works was partly supported by the Inamori Foundation at Kyoto.

References

1. P. Sokolsky: *Workshop on Observing Giant Cosmic Ray Air Showers from* $\geq 10^{20}$ *eV Particles from Space*, USA, (November 13-15, 1997), eds. J. F. Kizmanic, J. F. Ormes, R. E. Stretmatter (AIP, Woodbury NY 1997), pp. 65–75

2. S. Yoshida, H. Dai, C. C. H. Jui, P. Sommers: Astrophys. J. **479**, 547 (1997)
3. R. Cray , et al.: Proc. *26th International Cosmic Ray Conference*, eds. D. Kieda, M. Salamon, B. Dingus (University of Utah, Salt Lake City 1999) **5** pp. 353–356
4. M. Chikawa , et al.: Proc. *26th International Cosmic Ray Conference*, eds. D. Kieda, M. Salamon, B. Dingus (University of Utah, Salt Lake City 1999) **5** pp. 404–407
5. L. Wiencke, et al.: Nucl. Instr. Meth. A **428**, 593 (1999)
   L. Wiencke, et al.: *Ultraviolet Atmospheric and Space Remote Sensing: Methods and Instrumentation*, eds. R. Huffman, K. Dymond, (SPIE 1999) **3818** pp. 59
6. C. Song, Z. Cao, B.R. Dawson, B.E. Fick, P. Skolsky, X. Zhang: Astropart. Phys. **14**, 7 (2000)
7. F. Kakimoto, E.C. Loh, M. Nagano, H. Okuno, M. Teshima, S. Ueno: Nucl. Instr. Meth. A **372**, 527 (1996)
8. D. Heck, J. Knapp, J.N. Capdevielle, G. Schatz, T. Thouw: *CORSIKA: A Monte-Carlo Code to Simulate Extensive Air Showers.* (Report FZKA 6019, Forschungszentrum Karlsruhe 1998)

Fermi Acceleration of Astroparticles

Guy Pelletier

Institut Universitaire de France, Laboratoire d'Astrophysique de l'Observatoire de Grenoble, France

Abstract. The whole cosmic ray spectrum measured from GeV to 100 EeV energies could result from the same Fermi acceleration process at work in the supernova remnants of our Galaxy for energies from GeV to PeV, in extragalactic objects such as Active Galactic Nuclei, RadioGalaxies and Gamma Ray Bursts beyond PeV energies. Moreover the galactic or extragalactic sources of synchrotron radiation contain a population of ultrarelativistic electrons that have a powerlaw distribution similar to the cosmic ray population and that are plausibly explained also by the Fermi process. These lectures are designed to explain the 1st and 2nd order Fermi processes in the nonrelativistic regime, to give the theoretical tools necessary to complete the task on the remaining important open questions. Then the goal of the astrophysical explanation of the UHE Cosmic Rays seems to require the extension of Fermi process in the relativistic regime together with cosmic ray scattering off strong magnetic turbulence. This new development of the theory, which is not yet stabilised, will be introduced as well. The performance of the various cosmic accelerators are also presented.

1 Introduction

During many years, the astrophysics of high energy phenomena has been divided into two domains of investigations, namely measurement of the cosmic ray spectrum, and observation of nonthermal radiations of synchrotron sources, such as radiogalaxies, radioquasars, supernovae remnants, and pulsars. A very plausible theoretical link between these separated topics was the particle acceleration process, the celebrated

Fermi process. New instruments bringing new crucial observations in the high energy domain have incited particle physicists and astrophysicists working on high energy phenomena to get together and investigate the same objects and media. These "high energy gatherings" happened almost everywhere in the world and has been qualified as a new discipline, the "astroparticle physics". The progress of gamma ray astronomy has been crucial to inaugurate the new development, with the beautiful campaign of the Compton Gamma Ray Observatory that brought major discoveries, namely, on the one hand, the intense and highly variable Blazar phenomenon, and on the other hand, the cosmological distribution of Gamma Ray Bursts. In the mean time, ground Čerenkov arrays have been built (Whipple, HEGRA, CAT/CELESTE) and have discovered the TeV emission of some Blazars (BL-Lac). The fast variability of the emission is the signature of the environment of a compact object (very likely a massive black hole).

Martin Lemoine, Günter Sigl (Eds.): LNP 576, pp. 58–89, 2001.

The gamma emission of compact object environments is often interpreted as resulting from the inverse Compton process with ultrarelativistic electrons colliding the photons emitted by the accretion disk or the synchrotron photons (Synchrotron Self Compton, SSC) and the fast variability is more easily accounted for with electron acceleration. However the gamma emission may also have an hadronic origin. In supernova remnants collisions of energetic protons isexpected to generate π^0 whose decay produce gamma rays. Ultra High Energy Cosmic Rays (UHE Cosmic Rays, hereafter) generate pions through collisions with background photons above some threshold; the pion decay (π^0 and π^+) produces as many high energy neutrinos and gamma photons. Collisions of UHE-protons with the Cosmological Microwave Background at 2.7K produce the GZK-effect (Greisen, Zatsepin, Kusmin 1966) above the threshold of 3×10^{19}eV. Because of the energy decay of the protons, the events recorded above this threshold cannot come from sources beyond 100 Mpc [1], see Fig. 2. Such events have already been recorded by the AGASA [2] and the Fly's Eye [3] experiments and the Pierre Auger Observatory will produce a spectrum and a map in this energy range.

The process of photo-production of pions should work also in Active Galactic Nuclei above a threshold of 10^{16}eV. A new window of high energy astronomy could be opened with neutrino observatories (AMANDA, ANTARES), that could detect the neutrino emission resulting from pγ-collisions. The new word "astroparticles" refers to a new astronomy based on high energy particles transmitters rather than photons.

The cosmic ray spectrum is often divided into three ranges, see Fig. 1, and see also the general introduction by P. Biermann and G. Sigl on cosmic rays in this volume. In the first range extending from GeV to PeV the spectrum is an isotropic powerlaw in $\epsilon^{-\gamma}$ with an index $\gamma \simeq 2.7$ (in fact, instead of the energy variable ϵ, the so-called "rigidity" variable pc/Z is actually used) and is generally explained as resulting from supernovae of our Galaxy. The second range extending from PeV to 10EeV is also an isotropic power law spectrum with an index $\gamma \simeq 3.1$; the only possible scattering agent is the magnetic field, collisional interactions being inefficient for suprathermal particles, but the Galactic magnetic field of 3μG is unable to produce the MHD-disturbances required to scatter particles of energies larger than 10^{15}eV. This second range is probably due to AGNs and RadioGalaxies. Beyond 10^{19}eV (UHE Cosmic Rays), there clearly exists another extragalactic component [2,3], see Fig. 4, which is the target of the Pierre Auger Observatory. As previously indicated, beyond the GZK-threshold, the UHE-protons come from sources within 100 Mpc; very few RadioGalaxies are located inside this sphere and $\gamma-$ray bursts (GRBs) could be better candidates, unless more exotic phenomena explain these events [4] in relation with cosmological phenomena.

Between the first and the second ranges around 10^{16}eV, there is a smooth connection with an excess of cosmic rays called the "knee". The "knee" range is still unexplained. Is it from Galactic or extragalactic origin? Superbubbles or

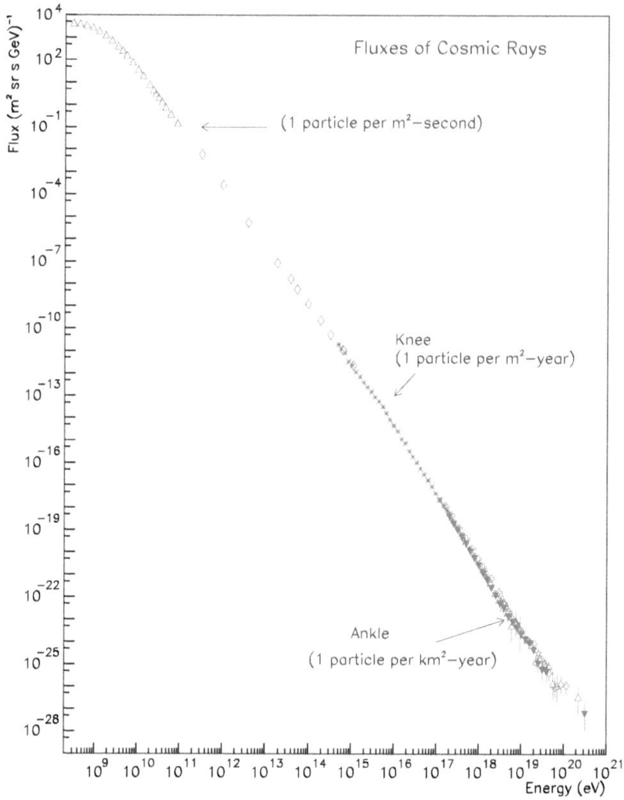

Fig. 1. From the "knee" to larger energies, the cosmic rays are probably of extragalactic origin.

Galactic Halo? Why is there so smooth a connection between two ranges: one Galactic, the other extragalactic?

1.1 Which Field for Acceleration?

Charged particles are obviously accelerated by an electric field. But which component is acting, electrostatic or electromotive? Cosmic plasmas do not allow to maintain high voltage drops. Regarding 10^{20}eV cosmic rays, one cannot find cosmic sites where such huge voltage would stand. The highest voltage drops considered in astrophysics take place in pulsar polar caps, where a potential drop on the order of 10^{13}V is presumed. However efficient particle acceleration in neutron star winds is nevertheless expected (see the contribution by B. Rudak in this volume). In turn, the magnetic field is obiquitous in cosmic objects and media; and their variations in space and/or time generate electromotive forces that can lead to extreme energies. Consider a field of typical intensity B over a scale R, considered also as the variation scale of B, with motions measured in fraction of the velocity of light; the work done by the electromotive force allows

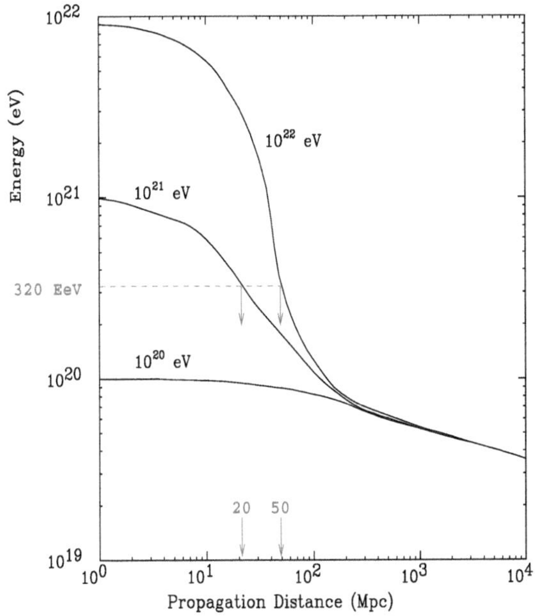

Fig. 2. Limitation of the travelling distance of the UHE-protons suffering the GZK-effect.

a relativistic particle of charge Ze to reach a significant fraction of the energy ϵ_{cl} such that

$$\epsilon_{\mathrm{cl}} = ZecBR \simeq \left(\frac{B}{1\mathrm{G}}\right)\left(\frac{R}{1\mathrm{pc}}\right) 10^{20}\mathrm{eV} \ . \tag{1}$$

This maximum energy corresponds to the energy of a particle whose Larmor radius is as large as the size of the accelerator; a particle of higher energy would not be confined, therefore the definition of this maximum energy as the confinement limit.

Other limitations of the particle energy can occur below this maximum, as will be seen later on, for instance radiation losses, particularly important for electrons, escape or finite duration of the acceleration, particularly for protons and other nuclei. However the criterion provided by ϵ_{cl} is interesting to select the possible sources of UHE Cosmic Rays by looking at the largest values of the product BR. The celebrated and very useful Hillas plot, Fig. 3, is based on this simple criterion.

Active Galactic Nuclei and their extensions should be important sites of cosmic ray acceleration because the accretion flow towards the central black hole concentrates magnetic field over a large volume. A magnetic field of few kilogauss is expected in the vicinity of a black hole of 10^8 solar masses on scales of few astronomical units. Some AGNs have large jets of several hundred kpc length, revealed by their synchrotron radiation, and the most powerful of them have terminal shocks producing synchrotron hot spots and extended lobes. The values of the product BR is favorable for the three regions, but this deserves to be more carefully analysed later on.

The formation of large scale structures of the Universe could have been an opportunity to produce UHE Cosmic Rays because they have given rise to converging flows of 10Mpc size. To achieve that, the magnetic field must reach submicrogauss size [5], which is not proven yet.

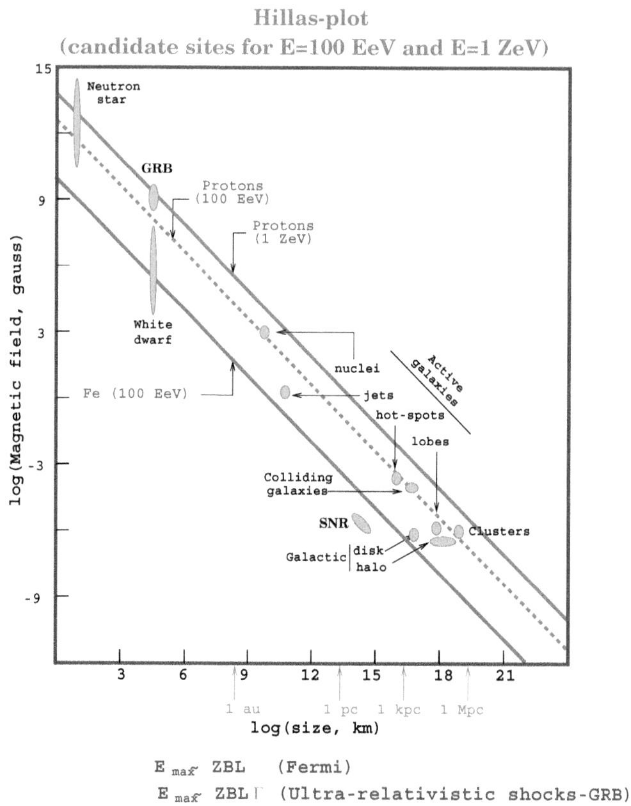

Fig. 3. The possible UHE cosmic ray sources are selected by their high product BR corresponding to high electromotive forces that can develop in them.

GRBs are good candidates for the production of UHE-protons thanks to the alleged ultrarelativistic flow with bulk Lorentz factor $\Gamma = 10^2 - 10^3$ (see the contribution by E. Waxman in this volume). Therefore a magnetic field of $10^2 - 10^3$G within a shell of 10^{-5}pc width would be enough, for the observed energy is the comoving energy multiplied by the large Lorentz factor Γ.

These possibilities together with the other cosmic sites of particle acceleration will be examined with more details in Sect. 5.3.

1.2 Are there Universal Nonthermal Spectra?

Synchrotron spectra of radio galaxies, radio quasars, supernovae remnants reveal nonthermal powerlaw energy spectra of relativistic electrons in these sources. The indices of the energy distribution are comprised between 2.5 and 3, in the same range as the cosmic ray spectrum measured at Earth. The spectrum of the cosmic rays is a power law from 10^9eV to 10^{20}eV with a slight break in the "knee" region around 10^{15}eV. Blazars exhibit nonthermal powerlaw distribution of relativistic electrons also, gamma ray bursts as well. Why and how nonthermal distributions are maintained for ever? Are those distributions produced by the same acceleration process? Are the powerlaws more or less universal?

The main purpose of these lectures is to present the principle of particle acceleration by the electromotive forces developed by moving magnetic disturbances in cosmic environments, which has been opened for the first time by Enrico Fermi in 1949, then to examine its modern versions in nonrelativistic as well as relativistic regimes and to estimate its performance in the various cosmic sites of acceleration.

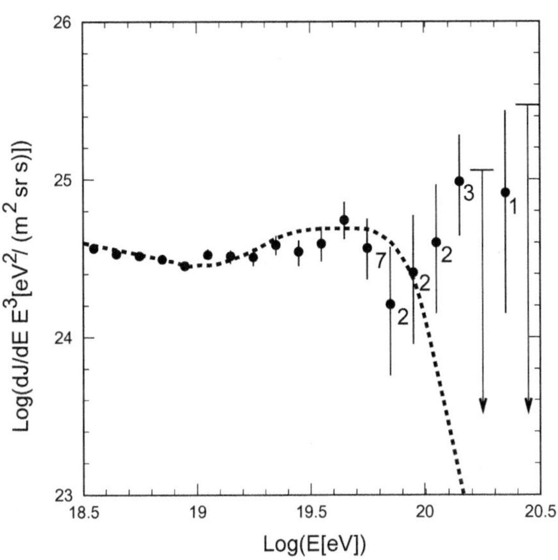

Fig. 4. Events beyond the GZK-threshold recorded by AGASA.

2 Scattering off Magnetic Disturbances and Fermi Acceleration

Magnetic field disturbances are the main cause of momentum scattering of high energy particles. Moreover the magnetic field is frozen in most cosmic media (large magnetic Reynolds number). Therefore the cosmic ray population is coupled with the ambient medium essentially through the magnetic field and this coupling is simply described by a scattering frequency ν_s that will be defined later on. In the frame of a magnetic disturbance, the electric field vanishes and energetic particles undergo momentum scattering and an ensemble of almost static disturbances in some frame tends to isotropize their distribution with respect to the comoving frame. A flow carrying MHD turbulence entrains the cosmic ray population that is isotropized in the flow frame. However in the restframe of a plasma, most of the MHD perturbations are moving in the form of Alfvén waves that propagates along the mean field B_0 at a velocity (when non relativistic)

$$V_{\mathrm{A}} = \frac{B_0}{\sqrt{\mu_0 \rho}} , \tag{2}$$

where ρ is the mass density of the medium. If all the waves were moving in the same direction, the suprathermal particle population would be roughly isotropized with respect of the wave-frame. Of course this is not generally the case and therefore one cannot find a frame were the electric field vanishes. The electric field modifies the magnetic force by a correction of order V_{A}/c. Therefore, for energetic particles, nonrelativistic magnetic disturbances in a plasma play the role of massive scattering centers, and are considered as static at first approximation or slowly moving. In turn the cosmic rays can absorb energy of magnetic disturbances, but usually at a low rate, at least when they interact with large scale perturbations.

2.1 Scattering and Diffusion of Cosmic Rays

Scattering properties depend on the characteristics of the disorganised magnetic field. Let us make it precise and first of all introduce some useful definitions. When it makes sense to discriminate a mean field $\boldsymbol{B}_0$ and a disorganized field $\delta\boldsymbol{B}$ relatively to the scale of the Larmor radii of the considered particles, it is convenient to write $\boldsymbol{B} = \boldsymbol{B}_0 + \delta\boldsymbol{B}$ and to define the turbulence ratio $\eta \equiv \langle\delta\boldsymbol{B}^2\rangle/\langle\boldsymbol{B}^2\rangle$. In order to keep definitions that still hold when there is no mean field (i.e. $\eta = 1$), Larmor radius (with no angular dependence) and Larmor time will be defined with $\bar{B} \equiv \langle\boldsymbol{B}^2\rangle^{1/2}$:

$$r_{\mathrm{L}} \equiv \frac{pc}{Ze\bar{B}} \simeq 3.33 \times 10^6 Z^{-1} \left(\frac{\epsilon}{1\mathrm{GeV}}\right) \left(\frac{B}{1\mathrm{G}}\right)^{-1} \mathrm{cm} \tag{3}$$

$$t_{\mathrm{L}} \equiv \bar{\omega}_{\mathrm{L}}^{-1} \equiv \frac{\epsilon}{Ze\bar{B}c} = 10^{-4} Z^{-1} \left(\frac{\epsilon}{1\mathrm{GeV}}\right) \left(\frac{B}{1\mathrm{G}}\right)^{-1} \mathrm{sec} \tag{4}$$

When $\eta \ll 1$, these defined quantities tend towards the usual ones (except that the usual Larmor radius is this one multiplied by the pitch angle sinus), whereas in strong turbulence (i.e. $\eta \to 1$)they characterize scalings for a particle of energy ϵ. The scattering time t_s (of order ν_s^{-1}) is long compared to the Larmor time for weak turbulence and can be as short as the Larmor time in strong turbulence, depending on the Larmor radius. In fact, for a given turbulent spectrum $S(k) \propto k^{-\beta}$ of coherence length $L_{\max}$ (the largest scale of magnetic disturbances for $\beta > 1$), the scattering efficiency depends on both the turbulence level η and the "rigidity" $\rho \equiv 2\pi r_{\rm L}/L_{\max} \propto p/Z$.

In weak turbulence theory with a mean field ($\eta \ll 1$), it is convenient to introduce the pitch angle α of the particle with respect to the mean field $\boldsymbol{B}_0$. The scattering frequency is defined as the rate of change of the pitch angle variance. Consider a variation $\Delta\alpha$ of the pitch angle during a time Δt longer than the correlation time of force experienced by the particle and shorter than the scattering time. The scattering frequency is then defined as

$$\nu_s \equiv \frac{\langle \Delta\alpha^2 \rangle}{\Delta t} . \tag{5}$$

The main aspect to bear in mind is that the interaction between the particles and the magnetic field is selective. Indeed a particle interacts with the field when its gyro-motion resonates with a Fourier component of wavelength equal to its Larmor radius. It is insensitive to shorter wavelength modes that make fast oscillations such that $\langle \Delta\alpha^2 \rangle = 0$ and adiabatically follows modes of larger wavelength without any jump of the pitch angle. To be more precise, the resonance occurs when the helical path of the particle is tuned with the helical variation of a specific circularly polarized wave of the spectrum (all the Fourier components can be casted in circularly polarized planed waves, even if these waves are not necessarily eigenmodes of the plasma) (see appendix B). It turns out that no resonance is possible within a pitch angle interval around 90^0. This scattering frequency is a function of the particle rigidity shaped by the turbulence spectrum through a scattering function $g(\rho,\eta) \equiv \nu_s/\bar{\omega}_{\rm L}$, see Fig. 5 and Ref. [6].

For an increasing turbulence level, the resonances broaden and mirroring of particle trajectories becomes easier, specially for particles of pitch angle close to 90^0 which do not resonate. It turns out that, for low enough rigidity, the power law dependence as a function of ν_s predicted by quasi-linear theory still applies. At larger rigidity ($0.1 < \rho < 1$), the scattering frequency saturates at a value close to Larmor frequency. For details see Refs. [7,8].

The scattering process also produces spatial diffusion. Indeed consider a particle of velocity $v_\parallel = v\mu$ (one notes $\mu \equiv \cos\alpha$) along the mean magnetic field. Random jumps of μ produce random variations of the position of the particle along the mean field line. Assuming that $\mu(t)$ is a stationary random process, then one easily derives the parallel diffusion coefficient in term of the scattering time t_s:

$$D_\parallel \equiv \frac{\langle \Delta x_\parallel^2 \rangle}{2\Delta t} = \frac{1}{3} v^2 t_s , \tag{6}$$

where the scattering time is rigorously defined as the correlation time of the pitch angle cosine as follows:

$$t_s \equiv \frac{1}{\langle \mu(t)^2 \rangle} \int_0^\infty \langle \mu(t)\mu(t-\tau) \rangle \mathrm{d}\tau \tag{7}$$

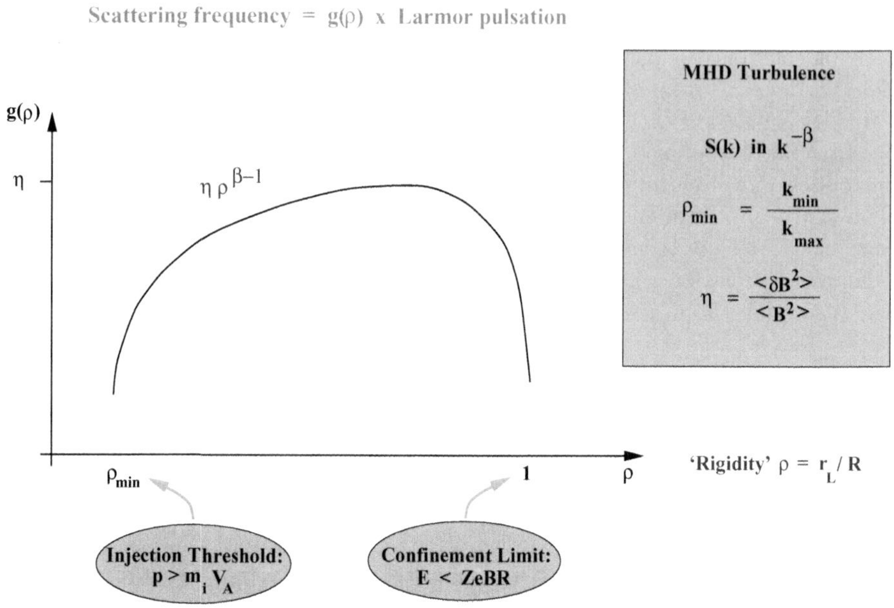

Fig. 5. The scattering frequency depends on the rigidity of the particle through the characteristics of the turbulence spectrum, its amplitude η and spectral index β (which equals 5/3 in Kolmogorov theory, 3/2 in Kraichnan theory with a mean field).

In the statistical theory, the average is supposed to be done over the ensemble of realisations of the random force, i.e. the initial phases of the field, which is assumed ergodic and equivalent to a space average. In modern approaches, the random phase assumption is not necessary; orbit instability ("chaos") is sufficient and the average must be done over the chaotic subset of phase space. The formula Eq. (6) for the parallel diffusion coefficient holds for an integration time Δt much larger than the scattering time. The scattering time $t_s \sim \nu_s^{-1}$ and it can even be shown that the equality holds in weak turbulence theory.

A transverse diffusion coefficient can be defined as well, involving the self-correlation function of the pitch angle sinus. When the mean field is strong enough,

$$D_\perp = D_\parallel \frac{1}{1 + (\omega_{\rm L} t_s)^2} \tag{8}$$

as long as the self-correlation of the pitch angle cosine or sinus exponentially decays over the scattering time; in Eq. (8) $\omega_{\rm L}$ is the usual Larmor pulsation defined with the mean field. When no mean field exists $D_\perp = D_\parallel$, whereas for significant mean field, $D_\perp \ll D_\parallel$. See Ref. [8] for a detailed discussion about the transverse diffusion coefficient and the relevance of Bohm estimate, i.e. $D \sim v r_{\rm L}$, for some interval of $\rho \sim 1$ and $\eta \sim 1$.

2.2 Fermi's Viewpoint 1949

Based on the scattering properties of magnetic disturbances on cosmic rays, in 1949 Fermi [9] devised a celebrated acceleration mechanism for cosmic rays. A magnetic perturbation is considered as a racket moving at a subrelativistic velocity $\boldsymbol{u}_0$ reflecting cosmic rays. Thus in the racket frame $\mathcal{R}'$, the particle energy ϵ' is conserved, only the normal component of its momentum changes its sign $\boldsymbol{p}'_n \mapsto -\boldsymbol{p}'_n$. In the observation frame $\mathcal{R}$, $\epsilon = \epsilon' + \boldsymbol{u}_0 \cdot \boldsymbol{p}$ and the particle energy changed by an amount $\Delta\epsilon = -2\boldsymbol{u}_0 \cdot \boldsymbol{p}_1$, $\boldsymbol{p}_1$ being the momentum before scattering.

Fermi realized that, if the MHD disturbances have converging motions, then the cosmic rays have systematic energy gains proportional to the flow velocity difference. This has been called the *first order Fermi process.* However in the forties, it was difficult to figure out whether this kind of event could be so frequent in the interstellar medium to account for the generation of the cosmic ray spectrum. Then Fermi devised the *second order process* which consists in considering a set of random MHD disturbances; he was thinking about interstellar clouds. The mean free path of cosmic rays between two collisions with magnetic disturbances being $\bar{l}$, during Δt, the cosmic ray energy has diffused according to:

$$\langle \Delta\epsilon^2 \rangle = \frac{4}{3} \langle \boldsymbol{u}_0^2 \rangle p^2 \frac{c}{\bar{l}} \Delta t \ . \tag{9}$$

Unfortunately the irregular motion of the clouds is too slow; taking $u_0 = 10\text{km/s}$, $\bar{l} = 30\text{pc}$, the typical time for the 2nd order Fermi acceleration,

$$t_2 = \frac{3\bar{l}c}{4u_0^2} \ , \tag{10}$$

is too long, for $t_2 \simeq 10^{11}\text{yrs}$ is longer than the age of the Universe. Nevertheless there are more efficient versions of the 2nd order Fermi process in sites where intense Alfvén waves are excited, as will be seen later on.

The 1st order Fermi process has been remodeled at the end of the seventies by an adaptation to strong shocks in a supersonic flow.

3 Acceleration at a Non-Relativistic Shock

During the same year, several authors produced the theory of 1st order Fermi acceleration at a shock: Krimsky (1977) [10], Bell (1978) [11], Axford, Leer, and Skadron (1977) [12], and Blandford and Ostriker (1978) [13]. The tenets of the theory are the following:

- A thermal plasma in supersonic motion experiences an adiabatic shock.
- The plasma carries a frozen in magnetic field; the field has a regular component $\boldsymbol{B}_0$ and a disorganized one $\delta\boldsymbol{B}$.
- A population of suprathermal particles is also transported by the plasma flow; these particles undergo frequent elastic scatterings off magnetic irregularities.

3.1 Shock

The slightest perturbation in a compressible fluid in supersonic motion amplifies as a shock. A shock is a transition layer where the velocity field of the fluid suddenly decreases over a width determined by the dissipative processes, and most of the kinetic energy flux is converted into thermal energy flux. It is convenient to describe the shock with respect to the front frame; in the simplest description, the shock transition layer is supposed to be plane and the flow stationary; however the extension to spherical shocks is quite easy. In this frame, the velocity field decays from its upstream supersonic value $u_1 > C_{s1}$, to a subsonic one downstream, $u_2 < C_{s2}$, where the indices 1 et 2 refer to upstream (unshocked flow) and downstream (shocked and thus heated flow) quantities respectively, and C_s is sound speed. The strength of a shock is measured by the Mach number: $\mathcal{M} \equiv u_1/C_{s1}$.

Since the flow is assumed stationary, the three conservation laws (mass, momentum, energy) lead to the so-called Rankine-Hugoniot jump relations for density, velocity and pressure. Mass conservation across the front reads, in term of the mass density ρ:

$$\rho_1 u_1 = \rho_2 u_2 \ . \tag{11}$$

The compression ratio r is defined as $r \equiv \rho_2/\rho_1$; the flow is thus slowed down so that $u_2 = u_1/r$. Momentum conservation law implies a variation of kinetic pressure P across the shock front such that:

$$\rho_1 u_1^2 + P_1 = \rho_2 u_2^2 + P_2 \ . \tag{12}$$

The pressure increase depends on the magnitude of the compression ratio:

$$P_2 - P_1 = \frac{r-1}{r} \rho_1 u_1^2 \ . \tag{13}$$

In an adiabatic shock where the radiation energy loss is unimportant, a strong heat flux proportional to the temperature gradient develops in the shock layer, but vanishes on both sides where the fluid is homogeneous. The third Rankine-Hugoniot relation corresponding to energy conservation involves the variation of the specific enthalpy h:

$$\rho_1 u_1 \left(\frac{1}{2} u_1^2 + h_1\right) = \rho_2 u_2 \left(\frac{1}{2} u_2^2 + h_2\right) \ . \tag{14}$$

For a perfect gas, $h(\rho) = \frac{\gamma_a}{\gamma_a - 1} P/\rho$. The three jump relations allow to derive the relation of the compression ratio with the Mach number:

$$r = \frac{\gamma_a + 1}{\gamma_a - 1 + 2/\mathcal{M}^2} \ ; \tag{15}$$

In a strong shock ($\mathcal{M} \gg 1$), if the shocked plasma pressure is non-relativistic, the adiabatic index $\gamma_a = 5/3$ and the compression ratio tends to 4. A plasma with relativistic pressure has an adiabatic index $\gamma_a = 4/3$ and the compression ratio tends to 7, see Fig. 6.

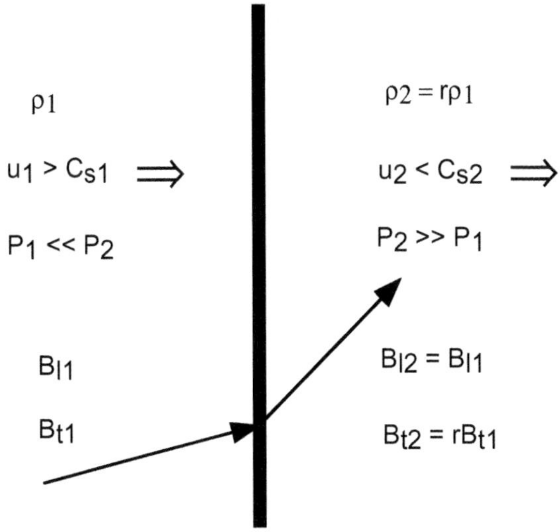

Fig. 6. In an adiabatic shock, the Rankine-Hugoniot relations allow to calculate the jumps of the three main hydrodynamic quantities in term of the compression ratio r.

In this simple analysis, the variation of the magnetic field was not considered. When the field is parallel to the flow it suffers no jump (flux conservation). When oblique, it can be shown that the transverse component is compressed by a factor r; in this case the pressure jump is modified, see Eq. (13).

3.2 Diffusion and Residence Time

Now consider a suprathermal particle that scatters off magnetic irregularities on both sides of the shock front. Assume the scattering efficient enough; because much faster than the shock $v \gg u_1$, the particle crosses the front many times before escaping downstream. The time between the first front crossing and the escape can be estimated as follows. The presence probability of the particle downstream has a gaussian density centered at the distance $u_2 t$ from the front and a standard deviation $\sqrt{2Dt}$, where D is the diffusion coefficient. The residence time in the downstream flow, t_{r2}, can be defined as the average time for a particle, that crossed the front from upstream to downstream, to return to the front after diffusion in the downstream flow. Similarly a residence time upstream, t_{r1}, can be defined, but clearly it is much shorter than the downstream one. Therefore one will retain the downstream value which exactly is :

$$t_r = \frac{2D}{u_2^2} \ . \tag{16}$$

The standard deviation around this average return time equals $\sqrt{2}t_r$. The diffusion coefficient D is $D_{\|}$ if the magnetic field is parallel to the flow, otherwise $D = D_{\|}\cos^2\theta + D_{\perp}\sin^2\theta$ in general case, θ being the angle of the field lines with respect to the shock normal.

3.3 Energy Gain and Escape Probability

A fast particle of velocity v coming from upstream, scattered downstream and coming back upstream, has gained an energy amount (averaged over pitch angle for quasi isotropic distribution):

$$\delta p = \frac{2}{3}\frac{u_1 - u_2}{v} p \; . \tag{17}$$

A similar calculation can be done for a particle crossing the front from downstream, scattered upstream and coming back downstream, and the result is the same. Because $v \gg u_1$, the relative gain is small and an averaged gain per complete Fermi cycle can be defined as:

$$\Delta p_{\text{cycle}} = \frac{4}{3}\frac{u_1 - u_2}{v} p \; . \tag{18}$$

The frequency of the Fermi cycles is known if the escape probability η in the downstream flow is known. This probability is obtained as the ratio of the particle flux flowing away downstream over the flux of particles crossing the front at the velocity v. Again assuming the distribution quasi isotropic, the result is (see Appendix):

$$\eta = \frac{4u_2}{v} \; . \tag{19}$$

Knowing the probability for a particle to undergo n cycles and only n, namely $(1-\eta)^n\eta$ (simplified version where $v = c$ is assumed and thus η is constant), the averaged number of cycles n_c is then:

$$n_c = \sum_{n=1}^{\infty} n\eta(1-\eta)^n = \frac{1-\eta}{\eta} \; . \tag{20}$$

The larger the ratio v/u the larger the number of cycles and it is worth noticing that the standard deviation about this averaged number is large also, since $\sqrt{\langle(n-n_c)^2\rangle} = \sqrt{1-\eta}/\eta$. Clearly all these reasonings are no longer correct for relativistic shocks. The averaged frequency of Fermi cycle is therefore:

$$\nu_{\text{cycle}} = n_c/t_r \simeq \frac{vu_2}{8D} \simeq \frac{3u_2}{8v}\nu_s \; . \tag{21}$$

The last estimate is obtained for the parallel diffusion coefficient estimation Eq. (6), it shows that the rate of cycles is lower than the scattering rate. It is now possible to provide an estimate of the first order Fermi acceleration at a shock in the form of an effective force:

$$\frac{\langle \Delta p \rangle}{\Delta t} = \nu_{\text{cycle}} \Delta p_{\text{cycle}} = \frac{r-1}{3t_r} p \; ; \tag{22}$$

The last expression, particularly simple, results from inserting the escape probability Eq. (19) and the gain Eq. (18) in the acceleration rate; it shows that the acceleration characteristic time is proportional to the escape time (i.e. the residence time). That remarkable proportionality between acceleration and escape is the origin of the formation of universal power law spectra as will be seen in the next subsection.

3.4 Spectrum

The spectrum can be derived from the probability for a particle of initial energy p_0 to reach an energy larger than p. This event implies at least n cycles such that $p_n \geq p$, and its probability is $Pr_n = \prod_{k=1}^{k=n}(1-\eta_k)$. Since the energy after n cycle

$$p_n = p_0 \prod_{k=1}^{k=n}(1+\frac{4}{3}\frac{u_1-u_2}{v_k}) = p_0 \prod_{k=1}^{k=n}(1+\frac{r-1}{3}\eta_k) \ , \tag{23}$$

the following ratio is obtained:

$$\frac{\ln Pr_n}{\ln p_n/p_0} = \frac{\sum_{k=1}^{k=n}\ln(1-\eta_k)}{\sum_{k=1}^{k=n}\ln(1+\frac{r-1}{3}\eta_k)} \simeq -\frac{3}{r-1} \ . \tag{24}$$

Therefore the probability for the energy to reach at least the value p is given by the following power law:

$$Pr[\geq p] \propto \left(\frac{p}{p_0}\right)^{-3/(r-1)} \tag{25}$$

and the energy spectrum is derived from the probability density:

$$S(p) \propto p^{-\frac{r+2}{r-1}} \ . \tag{26}$$

This result [11] holds also when the particles are not relativistic, p strictly being the momentum. The energy $\epsilon = pc$ for ultra relativistic particles. Therefore, downstream a strong non relativistic shock dominated by non relativistic pressure, so that the compression ratio $r = 4$, a power law spectrum of relativistic particles is set up with a universal index; the energy distribution function is proportional to ϵ^{-2}.

This spectrum is generally not observed. This spectrum is injected in the surrounding medium and then particles suffer various losses which inflex the spectrum. Cosmic rays of our Galaxy are likely generated by the shock of supernova remnants up to 10^{15}eV. The observed flux between 10GeV and 3×10^6GeV is such that $I_{cr}(>\epsilon) \simeq 1.0(\epsilon/1\mathrm{GeV})^{-1.7}$part./cm^{2}sr. At 1GeV, the energy density of cosmic rays is 1eV/cm^3; 10^{-2} less for cosmic ray electrons. The energy of a supernovae is $E_{SN} \simeq 3\times10^{50}$erg, the galactic rate is 3×10^{-2}per yr, so that the power density input is $Q_{\mathrm{inj}} \simeq 10^{-26}$erg/cm^{3}s. Thus a conversion of 10 percent of the super novae energy into cosmic ray energy is enough to account for the observed flux. Then the cosmic rays escape from the Galaxy to the halo mostly through diffusion across the magnetic field. At low energy the escape is due to convection, whereas at high energy the escape time is estimate as $t_{\mathrm{esc}} = h^2/(2D)$, where h is the width of the galactic plane. The escape probability is measured through the flux of secondary nuclei produced by spallation, such as ^{10}Be, whose lifetime, $\tau = 2.2\times10^6 E/mc^2$ yrs, is comparable to the escape time. These measurements at GeV energies leads to $t_{\mathrm{esc}} \propto \epsilon^{-0.6}$. Therefore the "leaky box" model [14] explains the measured spectrum in $\epsilon^{-2.7}$ as

$$S_{\mathrm{obs}}(\epsilon) = Q_{\mathrm{inj}}(\epsilon)\tau_{\mathrm{esc}}(\epsilon) \propto \epsilon^{-2.6} \ . \tag{27}$$

Regarding the electron spectra of synchrotron and inverse Compton sources, they are known through the simple relation between the radiated spectrum, $S(\nu) \propto \nu^{-\alpha}$, and the energy distribution of the relativistic electrons, $\rho(\epsilon) \propto \epsilon^{-\gamma}$, namely $\gamma = 2\alpha + 1$. The radiative losses are responsible for the high energy cut off and also for a spectral break due to incomplete radiative cooling: $S_e(\epsilon) = Q_{\rm inj}(\epsilon)\tau_r(\epsilon)$, where the radiative cooling time $\tau_r(\epsilon) \propto \epsilon^{-1}$; thus $\alpha \mapsto \alpha + 1/2$; which is compatible with observations.

However the theory works too good, because the pressure of accelerated particles tends to diverge since the distribution in ϵ^{-2} leads to a relativistic pressure:

$$P_* = \frac{1}{3}\int pcS(p)dp \propto \ln\left(\frac{p_{\rm max}}{p_0}\right) . \tag{28}$$

A high energy cut off is necessarily set up because of escape or radiative losses. Nevertheless, in spite of the cut off, the pressure of the accelerated particles can easily exceed the thermal pressure. Thus this test particle theory must be reconsidered by taking into account the modification of the shock caused by the cosmic rays.

3.5 Obliquity Effect

The presented theory seems insensitive to the magnetic field obliquity. This must be made more precise, because when the field is not aligned with the flow, there exists a compensating electric field that insures the electric equilibrium against the electromotive field developed by the frozen in magnetic field:

$$\boldsymbol{E} + \boldsymbol{u} \times \boldsymbol{B} = \boldsymbol{0} . \tag{29}$$

Let first take the opportunity to mention an important result derived from this relation Eq. (29) that the transverse component of the magnetic field is amplified by a compression factor r. Thus the inclinaison of field lines with respect to shock normal is amplified such that $\tan\theta_2 = r\tan\theta_1$.

The Fermi process works when there exists a frame where this electric field vanishes. Otherwise, the electric field would be able to inhibate the Fermi acceleration. The change of frame (De Hofmann-Teller transformation) is such that $\boldsymbol{u} = \boldsymbol{u'} + \boldsymbol{V}$ with $\boldsymbol{E} + \boldsymbol{V} \times \boldsymbol{B} = \boldsymbol{0}$, see Fig. 6. Thus the new velocity field $\boldsymbol{u'}$ is aligned with the magnetic field, because $\boldsymbol{u'} \times \boldsymbol{B} = \boldsymbol{0}$. The following relations are otained: $u_1 B_1^t = u_2 B_2^t = -E = B^l V$. Therefore the transformation involves an entrainment velocity $\boldsymbol{V}$ of norm $V = u_1 \tan\theta_1$. Clearly the transformation is possible only if $u_1 \tan\theta_1 < c$. This requires that the magnetic field is not too much oblique. Acceleration of particles at almost perpendicular shocks (called "superluminal"), where the electric field cannot be removed, is not simply governed by a Fermi process whose relevance in this case is not clear; another effect called "drift acceleration" must be taken into account.

In non-relativistic quasi perpendicular shocks, energetic particle cross the shock in undergoing many Larmor gyrations. The dynamics of the particles are almost insensitive to the gyro-phase at the initial crossing point. Thus the adiabatic invariant is almost preserved and $p_\perp^2/B$ is constant. When crossing the shock the particles increase their momentum and thus their energy such that $p_{\perp 2}^2 = rp_{\perp 1}^2$; this effect was shown by Evry Schatzmann [15].

When the de Hofmann-Teller transformation is possible, the previous results are not changed. However the effective diffusion coefficient is composed of both parallel and perpendicular coefficients: $D = D_\| \cos^2\theta_2 + D_\perp \sin^2\theta_2$.

The lower the diffusion coefficient, the faster the Fermi acceleration. Since the perpendicular diffusion coefficient is usually much smaller than the parallel one, except for strong turbulence where $\eta \simeq 1$, it is tempting to consider quasi-perpendicular shocks; which is correct as long as the shock remains subluminal. However this requires $D_\|/D_\perp < \tan^2\theta_2 < c^2/u_2^2$; which is almost impossible in non relativistic shocks. In that case, Bohm's estimate for $D_\perp$ is often assumed, which has in fact a restrictive range of validity [8]. Because of the assumption of Bohm's scaling, many high energy cut off are overestimated in the litterature. It has also been proposed that subdiffusion could occur downstream a perpendicular shock, which would change the efficiency and the spectrum index [16].

4 Transport of Cosmic Rays

The simple theory that has been presented allows to understand most of the Fermi acceleration process at shocks. A transport equation is necessary not only to master the evolution of the distribution out of the sources, but also to perform more elaborate calculations in the source itself. The solution that was presented with little means can also be more accurately obtained by solving the transport equation.

The transport equation is a Fokker-Planck type equation, generalising the diffusion equation, which governs the evolution of the distribution function in phase space, that also includes the description of fluid motion, radiative losses and phase space diffusion (2nd order Fermi process).

4.1 Spatial Diffusion and Energy Diffusion

The transport equation can be introduced by simply extending the usual spatial diffusion equation. Let consider a spatial random coordinate $x(t)$ of a particle diffusing in a fluid of bulk velocity u. During Δt, very short compared to the diffusion time, the particle position varied of an amount $\Delta x = u\Delta t + \delta x$; the first contribution is due to the bulk motion of the scattering medium and the second one δx is due to purely random diffusion of vanishing average and of variance proportional to Δt:

$$\langle \delta x^2 \rangle = 2D\Delta t \; ; \tag{30}$$

this behaviour is typical of a random motion with short correlation time like brownian motions. The probability density g describing the location of the particle in the flow is governed by the following equation:

$$\frac{\partial}{\partial t}g = -\frac{\partial}{\partial x}ug + \frac{\partial}{\partial x}D\frac{\partial}{\partial x}g \ . \tag{31}$$

Consider now the energy variable p. It suffers a radiative decrease (ordered variation), a systematic gain by first order Fermi process (ordered variation in average) and also random variations δp at the scattering time scale caused by the second order Fermi process on Alfvén waves. Thus the variation is of the form:

$$\Delta p = A\Delta t + \delta p \ , \tag{32}$$

with $\delta p = \pm\beta_* p\delta\mu$ where $\beta_* = V_\mathrm{A}/c$. For each mirroring effect $\delta\mu = -2\mu_0$ and for scatterings $\langle\delta\mu^2\rangle \propto \nu_s\Delta t$. The energy diffuses according to the following law:

$$\langle\delta p^2\rangle = 2\beta_*^2 p^2\langle\delta\mu^2\rangle \propto \beta_*^2 p^2\nu_s\Delta t \ . \tag{33}$$

The second order Fermi process is thus described by an energy diffusion coefficient Γ such that

$$\Gamma \equiv \frac{\langle\delta p^2\rangle}{2\Delta t} \sim \beta_*^2\nu_s p^2 \ . \tag{34}$$

The transport equation is an evolution equation for the isotropic part $\bar{f}(p,x)$ of the complete distribution, assuming weak anisotropy; the function is normalized such that the number density of cosmic rays $n_* = \int f 4\pi p^2 \mathrm{d}p$. The transport equation reads [17]:

$$\frac{\partial}{\partial t}\bar{f} + \frac{\partial}{\partial x}u\bar{f} = -\frac{1}{p^2}\frac{\partial}{\partial p}p^2 A\bar{f} + \frac{1}{p^2}\frac{\partial}{\partial p}p^2\Gamma\frac{\partial}{\partial p}\bar{f} + \frac{\partial}{\partial x}D\frac{\partial}{\partial x}\bar{f} \ . \tag{35}$$

The "friction" term A describes not only all the various kinds of energy loss, but also the energy gain by first order process. The radiation loss of a relativistic particle of Lorentz factor γ comes from forward photon emission in a narrow cone of half-angle γ^{-1} with respect to its momentum, which leads to a friction force in opposite direction to its momentum. Synchrotron and inverse Compton radiative losses contribute to A as follows:

$$A_\mathrm{rad} \equiv \left.\frac{\langle\Delta p\rangle}{\Delta t}\right|_\mathrm{rad} = -\frac{4}{3}\sigma_T\left(\frac{m_e}{m}\right)^2 (W_\mathrm{m} + W_\mathrm{ph})\gamma^2 \ , \tag{36}$$

where W_m is the magnetic energy density (synchrotron) and W_ph the energy density of the low energy photons (Compton effect in Thomson regime). Usually the radiative losses are considered for the electrons only because of the very small ratio $(m_e/m)^2$ for other particles; however for UHE-protons, these radiative losses can be important as will be examined further on. The contribution of the first order Fermi process is obtained by inserting the average power delivered to the particle through the convergence of the scattering medium. Indeed the first order Fermi process can be described as a non inertial entrainment due to the deceleration of the scattering medium. In this physical situation, the inertial force is $F_j = -p_i(\partial u_j/\partial x_i)$, and its accelerating power

$$P_\mathrm{acc} = -\langle v_j p_i\rangle\frac{\partial u_j}{\partial x_i} = -\frac{pv}{3}\nabla\cdot\boldsymbol{u} \ . \tag{37}$$

Only a compressed flow ($\nabla \cdot \boldsymbol{u} < 0$) produces a first order acceleration. Its contribution to the simplified Eq. (35) is thus

$$A_{\rm acc} = -\frac{p}{3}\frac{\partial u}{\partial x} \ . \tag{38}$$

4.2 Acceleration at a Shock again

The Fermi acceleration at shocks has been derived by several authors [18], in particular by Blandford and Ostriker [13], from the cosmic ray transport equation. This straightforward derivation deserves to be presented, because the necessary assumptions clearly appear as well as the necessary extensions. Neglecting the second order Fermi process and the radiative losses in the transport equation, and assuming a sudden shock transition (i.e. shock width much smaller than the diffusion length of the cosmic rays) such that $u(x) = u_1 + (u_2 - u_1)\theta(x)$ and $\partial u/\partial x = (u_2 - u_1)\delta(x)$, the downstream distribution function can be calculated in term of the upstream distribution function. Indeed the stationary solution is such that

$$u\frac{\partial}{\partial x}\bar{f} - \frac{1}{3}\frac{\partial u}{\partial x}p\frac{\partial}{\partial p}\bar{f} = \frac{\partial}{\partial x}D\frac{\partial}{\partial x}\bar{f} \ . \tag{39}$$

This equation is easily solved on both sides of the shock front and the continuity of the distribution function is stipulated. Thus the distribution function is necessarly uniform downstream as long as losses are neglected. Upstream the distribution function exponentially increases up to the front over a diffusion length ($\bar{l}_1 = D_1/u_1$). In particular, the relativistic pressure exponentially increases up to the front (precursor generation). Integrating over x from $-\infty$, where $f(x,p) \to f_1(p)$, to $+\infty$, where $f(x,p) \to f_2(p)$, both sides of Eq. (39), a simple differential equation for the downstream function $\bar{f}_2$ is obtained :

$$p\frac{\partial}{\partial p}\bar{f}_2 + q\bar{f}_2 = q\bar{f}_1 \ . \tag{40}$$

Its integration provides the relation between the downstream function and the given upstream function [13]:

$$\bar{f}_2(p) = qp^{-q}\int_{p_m}^{p} p'^q \bar{f}_1(p')\frac{\mathrm{d}p'}{p'} \ . \tag{41}$$

A powerlaw is found again at high energy with the index $q \equiv 3r/(r-1)$. Since the energy distribution is proportional to $p^2\bar{f}(p)$, the expected index is found again for it varies like $\epsilon^{-(r+2)/(r-1)}$.

This theory was recognized as successful for two reasons: on the one hand, it predicts powerlaw spectra with an universal index close to the observed values of the cosmic ray spectrum and of the synchrotron sources spectra, on the other hand, it can account for the high energy cutoff against losses in most cases, the exception could be the case of the UHE Cosmic Rays...

5 Maximum Energy Achieved and Loss Limitations

In a cosmic accelerator of effective size R, the Fermi process cannot produce particles of energy larger than $\epsilon_{\mathrm{cl}} = ZeBR$, since beyond this energy the Larmor radius of the particle becomes larger than R and the particle escapes. This upper bound of achievable energy can be higher for an observer if the source is moving at relativistic speed towards the observer with a bulk Lorentz factor Γ; the confinement limit energy is therefore $\epsilon_{\mathrm{cl}} = ZeBR\Gamma$. But other limitations often impose a smaller energy cut off, as for instance the radiation losses for relativistic electrons. Generally the proton energy is limited by the size or by the age of the accelerator. However UHE cosmic rays can suffer radiation losses that can produce a cut off before the confinement limit. The cut off energy is determined by equating the Fermi time scale and the loss time scale. It is convenient to measure the Fermi time in term of the Larmor time: $t_a = \eta_a^{-1} t_{\mathrm{L}}$. As previously seen, η_a depends on the particle rigidity ρ and the magnetic turbulence intensity η, namely

$$\eta_a \sim \beta_a^2 g(\rho, \eta) \ll 1 , \tag{42}$$

where the scattering function $g(\rho, \eta) \sim \eta \rho^{\beta-1}$ for $\rho < 0.1$ and a "Bohm plateau" is reached for $\rho = 0.1 - 1$ with $g \simeq 0.3\eta$ but for η sufficiently close to unity [8]. The velocity β_a is the shock velocity in the case of the 1st order Fermi process or the Alfvén velocity in the case of the 2nd order Fermi process. For protons that can reach large rigidities smaller but of order 1, η_a takes its lowest value which is of order 0.1. Electrons have much lower rigidities at their cut off energy and η_a can be much smaller, 10^{-4} say.

5.1 Electrons Cut Off: Radiative Losses

Usually the main energy limitation of electrons is due to synchrotron loss; when they suffer inverse Compton radiation loss, like in AGNs for instance, the soft photon energy density is not much larger than magnetic energy density, thus the synchrotron time is sufficient to estimate the radiative cut off. The limit on the Lorentz factor of a particle of mass m caused by the synchrotron loss is:

$$\gamma_{\mathrm{syn}} = 10^8 \times \sqrt{\eta_a} \frac{m}{m_e} \left(\frac{B}{1\mathrm{G}} \right)^{-1/2} \tag{43}$$

Table 1 gives a rough estimate of this cut off for electrons in various sources.

5.2 Proton Cut Off

The proton maximum energy in a supernova remnant depends on its age, for its radius determines the maximum Larmor radius achievable by the acceleration process [19]. Optimistic estimates put the cut off at 10^{15}eV ($B \sim 10^{-6}$G and $R \sim 10$pc). The precise maximum energy for a shock velocity of 3000km/s is

$$\epsilon_{\max} = 10^{14} \mathrm{eV} \eta_a \left(\frac{t_{\mathrm{age}}}{300\mathrm{yrs}} \right) \left(\frac{B}{1\mu\mathrm{G}} \right) . \tag{44}$$

Table 1. Performance table of the cosmic accelerators. "Large" means much greater than the confinement limit. The numbers are just rough indications, of course.

					protons	electrons	
					$(\eta_a = 0.1)$	$(\eta_a = 10^{-4})$	
	B	R	Γ	$\epsilon_{\rm cl}$	ϵ_{syn}	ϵ_{syn}	Remark
SNRs	10^{-6}G	10pc	1	10^{15}eV	large	10^{14}eV	Radio synchr. Gamma SSC
AGNs							
-Nucleus	10^{3}G	10^{-4}pc	1	10^{19}eV	10^{18}eV	10^{10}eV	pγ-process $\tau_{\gamma\gamma} > 1$
-Relat. Jets	1G	10^{-3}pc	10	10^{18}eV	10^{20}eV	10^{13}eV	pγ-process Gamma SSC
-Hot Spots	10^{-4}G	10^{3}pc	1	10^{19}eV	large	10^{14}eV	Radio synchr. X-rays SSC
-Radio Lobes	10^{-6}G	10^{6}pc	1	10^{20}eV	large	10^{15}eV	Radio synchr.
GRBs	10^{3}G	10^{-5}pc	10^{3}	10^{21}eV	10^{22}eV	10^{13}eV	pγ-process Gamma SSC
Large cosm. Structures	10^{-7}G	10^{7}pc	1	10^{20}eV	large	10^{16}eV	t_a quite long
Pulsar Winds	10^{-5}G	10^{-3}pc	10^{6}	10^{17}eV	large	large	Synchr. Radio-X
Plerions							SSC

Regarding UHE cosmic rays, if the magnetic field is too low, the acceleration process is too long, and if the magnetic field is too strong, synchrotron loss will kill them. To get $\epsilon > 10^{20}$eV, the magnetic field must be smaller than $0.25\eta_a^{-1}$G, if there is no relativistic motion of the source; and the acceleration time is necessarily longer than one year. These constraints make difficult to get UHE cosmic rays in AGNs, even in blazar jets with $\Gamma \simeq 10$ where intraday variability is observed. Jet hot spots and extended lobes are possible accelerators of UHE cosmic rays, see Table 1. Gamma Ray Bursts with bulk Lorentz factors that could be larger than 100 are better candidates, since the constraints are $B < 0.25\eta_a^{-1}\Gamma$ and $t_a > \Gamma^{-1}$yr.

5.3 Performances of Cosmic Accelerators

The following table gives rough estimates of the cut off energy in all the astrophysical objects that are known to produce high energy particles. Clearly it turns out that the Fermi acceleration of electrons can account for all the observations of non thermal radiations in the form of both synchrotron (radio) and inverse Compton (hard X and gamma).

- SNRs. Regarding the cosmic ray spectrum, the supernova contribution can cover the range up to energies of 10^{14}eV, and does not reach the knee easily. The absence of detection of π^0 decay by Čerenkov arrays at TeV energies in several SNRs except in Cassiopeia A [20] puts a limit on the efficiency of the Fermi acceleration [21]. Compton emission by electrons could also dominate the gamma range and thus hide the π^0 contribution.

- AGN. AGN and radiogalaxies can contribute up to 10^{19}eV [22,23], and a bump is expected at the GZK-threshold of 3×10^{19}eV. Beyond the GZK-threshold, the origin of the UHE Cosmic Rays must be found within 100Mpc, where very few Radio Galaxies are located. However synchrotron loss in the nuclei where $B > 100$G kills the UHE-protons. The threshold of photo-production of pions is at 10^{16}eV, however the gamma rays cannot escape because the nuclei are optically thick to pair creation, but the high energy neutrinos can escape. Jets, hot spots and extended lobes could be sites of production of UHE-protons if the Fermi process achieves its maximum efficiency ($\eta_a \sim 0.1$ or maybe more in the relativistic regime [24,25]), see Sect. 7.

- GRBs. The contribution of Gamma Ray Bursts is in favour for the energy budget. Indeed with an energy $E_{\mathrm{GRB}} \sim 10^{51} - 10^{53}$ergs at a rate $10^{-8}/\mathrm{Mpc}^3/\mathrm{yr}$, a conversion of 10 percent of the energy in the cosmic ray component would be enough to account for the UHE cosmic ray flux [26,27], see also the contribution by E. Waxman in this volume. The most successful model, that convincingly account for the "afterglow" stage, namely the "fireball" model [28,29], explains the phenomenology of GRBs with an explosion involving a very high enthalpy that generates an ultrarelativistic wind undergoing a free expansion with a bulk Lorentz factor $\Gamma = 10^2 - 10^3$, followed by a deceleration stage governed by a strong relativistic shock [30]. The range of 100EeV for accelerated protons is reachable with conservative numbers, namely a comoving magnetic field of $10^2 - 10^3$G (close to equipartition) in a shell of width $R/\Gamma \sim 10^{-5}$pc just before deceleration, the proton energy in the comoving frame reaches at least 10^{17}eV and the high bulk Lorentz factor Γ allows to go beyond the GZK-threshold in the observer frame. But what is seen till now is the radiation of the accelerated electrons through synchrotron and inverse Compton emission during about 0.1 seconds for some events, 10 seconds for the others; they display a power law energy distribution with an index between 2 and 3. The observability of UHE-Cosmic Rays from GRBs is, of course, a crucial issue [31,32].
- Large Scale Structures. The magnetic field in cosmological large scale structures is unknown. But if it could reach submicrogauss intensities, they could contribute to the cosmic ray spectrum up to 10^{19}eV [5].
- Pulsar Winds. The pulsar wind, that contains $e^+ - e^-$ pairs and maybe a tiny ion component, is considered to have very large bulk Lorentz factor, possibly 10^6 [33], see also the contribution by B. Rudak in this volume. Its violent terminal shock takes place at $0.01 - 0.1$pc and the magnetic field can reach 10μG. Large particle energy can be reached, but not larger than 10^{17}eV.

6 Open Questions in the Non-Relativistic Regime

In spite of its success, the theory of Fermi acceleration at non relativistic shocks and the modelisation of supernovae remnants have still shadow points, that have remained unsolved for 20 years.

6.1 Injection

The injection problem can be simply adressed as follows: Fermi acceleration works for particles that already have enough energy. There are several reasons, but the main one is that the particles must be scattered by magnetic disturbances, the acceleration time scale being controled by the time for momentum turn over. If a particle has a Larmor radius much smaller than the scale of variation of the magnetic field, its adiabatic invariant is preserved and the particle can undergo mirror reflections but not pitch angle scattering. Efficient pitch angle scattering is required and it works for Larmor radii larger than the smallest scale of MHD turbulence, namely $r_0 \equiv V_{\rm A}/\omega_{cp}$ (ω_{cp} being the gyro-pulsation of protons), which puts an interaction threshold on the momentum: $p > m_p V_{\rm A}$. Injection of protons above this threshold does not seem difficult, bearing in mind the counter streaming instability generated by the reflection of incoming protons on the potential barrier at collisionless shock front. The threshold is severe for electrons that need to be ultrarelativistic already, especially in extragalactic sources where the Alfvén velocity is quite high. The solution of this problem is again necessarily in the frame of kinetic theory of plasma [34,35].

6.2 Radiative Shocks

The effect of losses is easily treated in the theory if, on the one hand, it merely introduces a high energy cut off and can be neglected at energies lower than the cut off and, on the other hand, if the power involved in the loss does not ruin the adiabatic shock approximation (i.e. the luminosity much smaller than the kinetic energy flux). When the radiative power becomes comparable to the kinetic flux, things complicate and universal laws seem unachievable.

6.3 Comparison 1st Order - 2nd Order

In these new versions of Fermi acceleration, the "scattering centers" are essentially the Alfvén waves and therefore it is useful to reconsider the relative importance of first and second order Fermi acceleration. The acceleration time of the first order Fermi process is given by, see Eq. (22):

$$t_1^{-1} \sim \frac{(u_1 - u_2)u_2}{v^2} \nu_s \ . \tag{45}$$

The acceleration time of the second order Fermi process is given by the energy diffusion time, see Eq. (33):

$$t_2^{-1} \sim \frac{V_{\rm A}^2}{c^2} \nu_s \ . \tag{46}$$

When the downstream magnetic pressure is not very low as compared to the kinetic pressure, which is often the case, the Alfvén velocity, the sound velocity and the downstream velocity u_2 are of the same order of magnitude. Therefore the efficiency of both Fermi processes is comparable [24,36,37,38]. This comes from the fact that the frequency of Fermi cycles at a shock is much lower than the scattering frequency by a factor u_2/c. The second order Fermi acceleration is often unduly neglected because of a negative prejudice, whereas it cannot be ruled out from the viewpoint of efficiency. Its main defect is its lack of producing a universal index like the first order process does [39]... when it works alone....

A noteworthy remark is that both processes are more efficient when the flow speed and Alfvén speed are close to the velocity of light. But in the relativistic regime, the expansion in first order and second order Fermi processes does not make sense and the theory must be reconsidered.

6.4 Nonlinear Effects

We have seen that the pressure of cosmic rays tends to diverge downstream of a strong shock and so modifies the shock structure. It generates a preshock that spreads over a typical diffusion length of cosmic rays. The effective compression ratio of the shock increases and thus eventually the divergence gets worse, since the energy distribution index becomes smaller than 2! Detailed calculations have been done for both quasi parallel and quasi perpendicular magnetic field [40,41]. An analytical solution has been derived in the case where the shock is completely dominated by the relativistic particles, with a smooth transition over a diffusion length without any subshock, and with an energy independent diffusion coefficient [42]. For this very particular case, one finds again an energy distribution in ϵ^{-2}.

Another aspect of the nonlinear theory is the generation of turbulence upstream caused by the cosmic rays themselves [43,44,19,45]. Because the turbulence contributes to heat the thermal plasma, its excitation reduces the efficiency of cosmic ray acceleration [40,46], especially at low energy. Due to these nonlinear effects, the modified spectrum would be softened at low energies and hardened at high energies.

In spite of the development of big Monte Carlo simulations [47,48], the detailed description of the shock broadening through particle acceleration has not been yet completed.

7 Fermi Acceleration in Relativistic Regime

Relativistic fronts and/or large amplitude magnetic disturbances can lead to an acceleration time as short as the Lorentz time. This is required in Blazars and GRBs to reach the energy range beyond the GZK-threshold. Therefore this is an important field of investigation. The relativistic regime of Fermi acceleration occurs in relativistic flows producing a shock or in a plasma where the Alfvén velocity is relativistic. Therefore relativistic shocks and wavefronts should be relevant in AGNs relativistic jets where $\Gamma \sim 10$, in microquasars [49] where similar jets are observed, in pulsar winds where $\Gamma \sim 10^6$ as been argued [33], in GRBs expanding shell ($\Gamma \sim 10^2 - 10^3$). The theory of relativistic Fermi acceleration is not yet stabilised and will not be detailed. The Fermi acceleration scheme can be extended to the case where the disturbances propagate at relativistic speed and the salient points will be sketched. The specific aspects of the theory that will be commented are:

- At each scattering the particle energy jump can be large.
- The acceleration time can be shorter than the scattering time for strong enough magnetic disturbances.
- The distribution functions are strongly anisotropic.

The first two points are obviously very interesting for getting an efficient acceleration mechanism. The third point is an unavoidable theoretical complication, inherent to relativity, see Fig. (7).

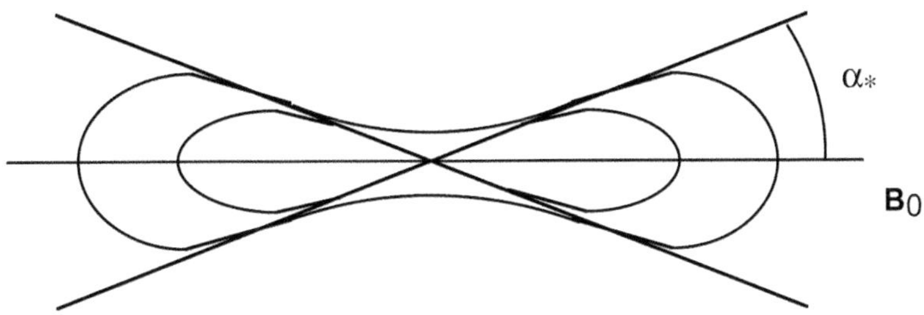

Fig. 7. The distribution function of the particles accelerated by relativistic disturbances is concentrated in a cone of half-angle $\alpha_* \simeq 1/\gamma_*$

The Fermi process can be formulated as follows in the spirit of relativity. Consider an electromagnetic perturbation that propagates at the velocity V_* (one notes $\beta_* \equiv V_*/c$ and $\gamma_* \equiv (1-\beta_*^2)^{-1/2}$). During an elastic scattering off the magnetic perturbation, the particle energy and pitch angle cosine (defined, here, as the angle of the momentum with respect of the propagation direction) are modified according to:

$$(p_1, \mu_1) \xrightarrow{L_{\beta_*}} (p_1', \mu_1') \xrightarrow{S} (p_2', \mu_2') \xrightarrow{L_{\beta_*}^{-1}} (p_2, \mu_2) \tag{47}$$

The Lorentz transform L_{β_*} is such that:

$$p_1'^0 = \gamma_*(p_1^0 - \beta_* p_1^{\|}) \tag{48}$$

$$p_1'^{\|} = \gamma_*(p_1^{\|} - \beta_* p_1^0) \tag{49}$$

and taking account of the ultrarelativistic approximation, $p^0 = p$ and $p^{\|} = p\mu$, the following simple relations are obtained:

$$p'_1 = \gamma_*(1 - \beta_*\mu_1)p_1 \tag{50}$$

$$\mu'_1 = \frac{\mu_1 - \beta_*}{1 - \beta_*\mu_1} \tag{51}$$

The scattering S does not change the energy, it randomly changes the pitch angle: $p'_2 = p'_1$ and $\mu'_1 \mapsto \mu'_2$. Then after inverse Lorentz transform $L_{-\beta_*}$,

$$p_2 = \gamma_*(1 + \beta_*\mu'_2)p'_1 \tag{52}$$

$$\mu_2 = \frac{\mu'_2 + \beta_*}{1 + \beta_*\mu'_2} \tag{53}$$

and the particle energy has changed according to :

$$p_2 = \gamma_*^2(1 - \beta_*\mu_1)(1 + \beta_*\mu'_2)p_1 \sim \gamma_*^2 p_1 \ . \tag{54}$$

The energy gain is by a factor γ_*^2 when the pitch angle is not smaller than $1/\gamma_*$. Thus the energy jump can be large at each scattering; it can be written as:

$$\Delta p^+ = \beta_* \frac{\mu_2 - \mu_1}{1 - \beta_*\mu_2} p_1 \ , \tag{55}$$

which shows the big difference between the non relativistic regime $\beta_* \ll 1$, which allows a Fokker-Planck description, and the relativistic regime $\beta_* \simeq 1$, which allows a "Markovian" description (short memory random process) but not a Fokker-Planck one, that is valid for small jumps only. Expansion in first and second order processes does not make sense in relativistic regime.

Two kinds of relativistic Fermi acceleration can be considered: i) acceleration through several Fermi cycles at a relativistic shock, the magnetic disturbances of both side of the shock front being almost static. ii) acceleration through crossing forward and backward Alfvén fronts propagating at relativistic speed.

7.1 Acceleration at Relativistic Shocks

When a relativistic shock propagates in the upstream medium with a large Lorentz factor Γ_s like in GRBs, one could guess that two or three Fermi cycles are sufficient to bring the proton energy above the GZK-threshold, because of the gain factor of order Γ_s^2 at each cycle. However as recently shown by Gallant and Achterberg [50], only the first half cycle "udu" can do it, because a particle that comes back upstream necessarily has a pitch angle that differs from π by less than $1/\Gamma_s$ in order to cross the front. To gain another factor Γ_s^2, its momentum should very rapidly come out of the cone of half-angle $1/\Gamma_s$. But for any reasonable deflection time, the particle is catched up by the shock front before having significantly changed its pitch angle. Therefore after the first half-cycle "udu" that provides a gain of $2\Gamma_c^2$, the next half-cycles provide a gain by a factor 2 only. This is nevertheless better than the small gain provided by a non relativistic shock.

Now the problem is to know if a cosmic ray can do many cycles, because the escape probability is much larger than in the non relativistic case (between

0.3 and 0.5). This sensitively depends on the incident pitch angle and thus of the angular distribution function that tends to be set up [51]. If enough cycles are experienced by the cosmic rays, then a power law distribution is obtained with an index 2.2. This was recently derived analytically by Kirk et al. [51] and numerically by Bednarz and Ostrowski [52].

If cosmic rays undergo a single crossing, no universal power law is obtained but an efficient acceleration is realized anyway [53]. An interesting calculation has been done by Begelman and Kirk [54] for quasi perpendicular relativistic shocks. Contrary to the case of a non relativistic shock where a particle undergoes many Larmor gyrations across a perpendicular shock front (as previously seen), a cosmic ray has crossed a relativistic front before having done a few gyrations. The adiabatic invariant is thus broken and a larger amplification of the particle energy is possible as compared to the adiabatic case. An upstream distribution function $\bar{f}_1$ of index α_0 gives rise to a downstream distribution such that

$$\bar{f}_2 \simeq \frac{(2\Gamma_s)^{\alpha_0}}{\alpha_0 + 1} \bar{f}_1 \ . \tag{56}$$

7.2 Acceleration with Relativistic Wavefronts

In a relativistic plasma of pressure P, that is magnetically confined such that the magnetic pressure dominates the kinetic pressure $P_m > P$, hydromagnetic waves propagate at a relativistic velocity and can accelerate cosmic ray efficiently [55,25]. The Alfvén wave propagates at a speed depending of the energy mass density e ($e = 3P$ for a perfect relativistic gas) and given by

$$V_* = \frac{c}{\sqrt{1 + \frac{e+P}{2P_m}}} = \frac{c}{\sqrt{1 + 2\frac{P}{P_m}}} \ . \tag{57}$$

Fast magnetosonic waves propagates faster with a Lorentz factor $\gamma_F = \sqrt{3/2}\gamma_*$. Long waves and intense localised fronts can efficiently scatter cosmic rays of Larmor radii smaller than their width (or wavelength) during a few Larmor time. Relativistic localised fronts, and a fortiori shocks, have a larger width than non relativistic front; the width is of order $r_0\langle\gamma^2\rangle/\langle\gamma\rangle$ instead of r_0 [55]. Fermi acceleration occurs only if there are wavefronts propagating in opposite directions. The particle experiences an energy gain by a factor $2\gamma_*^2$ at each scattering with an incoming front. If there are enough front crossings, the acceleration is efficient although γ_* is not expected to be very large like the GRB bulk Lorentz factor. The distribution function elongates inside a cone of half-angle $1/\gamma_*$ in momentum space. The acceleration time is shorter than the scattering time by a factor γ_*^{-2}.

These events can occur in all the relativistic flows mentioned previously provided they are magnetically confined. Backward waves or fronts are produced by interaction with the ambient medium. Forward waves are produced either by flares in the jet source or by backscattering of the backward waves. This latter case should happen during the "free" relativistic expansion of a shell or a cloud that is pervaded by the ambient medium along the flow axis which is also the magnetic axis; which generates an intense wave by converting the energy-momentum influx in the shell. The large amplitude backward wave triggers a parametric instability by coupling with the relativistic sound that backscatters a sizable fraction of the mother wave [56]. An efficient acceleration process can be expected with relativistic waves and fronts, but it does not lead to a universal power law distribution. However it is worth mentioning that a spatial superposition of local exponential distributions with increasing width can give rise to an integrated power law spectrum... Fronts of various sizes are actually observed in all the previously mentioned relativistic flows.

When a relativistic shock is set up (which is obtained as the continuous sharpening of a front until wave breaking), the next wavefronts in the downstream flow still continue to scatter and accelerate cosmic rays that have already been accelerated by a first half-cycle (the $2\Gamma_s^2$ amplification). Thus the wave fronts can achieve the expected acceleration alone or by helping a strong relativistic shock as well.

8 Status of this Research and Prospective

Globally the Fermi acceleration processes seem capable to account for all the nonthermal phenomena in the observed Universe, such as the Cosmic Ray spectrum from GeV up to 100 EeV, and the high energy electron distributions in compact object environments that radiate synchrotron and inverse Compton emission. The Fermi process has been successfully investigated in the interplanetary shocks, but of course at much lower energies than implied by astrophysical shocks. It has been investigated in numerical simulations as well, especially Monte Carlo simulations [47,48]. But of course, the large dynamical scales implied by astrophysical shocks cannot be set in those simulations. However many questions are still opened either at the theoretical level or at the level of performance estimation for each astrophysical source. Even the low energy part of the cosmic ray spectrum, which was thought to be well explained, still has shadow points. The opened questions in the theory of Fermi acceleration in the non-relativistic regime has been stated. There are still two important measurement problems for the low part of the CR-spectrum: one comes from the non-detection of the gamma ray emission expected from π^0 decay [21] except from Cassiopeia A [20], and the other from the measured escape probability as a function of energy, used in the leaky box model, that does not correspond to the expected theoretical law infered from a Kolmogorov turbulence spectrum. The low energy cosmic ray spectrum seems satisfactorily explained by supernova remnants [45]. However the high energy cutoff is probably overestimated, which would explain the non detection of TeV gamma emission. Radiation by electrons seems to dominate in the gamma range. Moreover, as briefly mentioned, there is a nonlinear effect in shocks that tends to reduce the energy of accelerated particles in favour of thermal energy.

Regarding the UHE Cosmic Rays, the theoretical opened questions are mostly related to the relativistic regime of Fermi acceleration. They are the following.

- The discrimination between thermal and nonthermal plasmas in relativistic shocks is not clear at all. A single crossing of a high Lorentz factor shock creates a population of highly relativistic particles already.
- The ultrarelativistic particles make the shock transition quite large and its width could be related to the Larmor radius of the most energetic particles [55]. Electrons, that have small Larmor radii, are probably not accelerated by the Fermi process at the shock and the observed powerlaw distribution of GRBs cannot clearly be interpreted as resulting from the Fermi process at shock. Moreover these electrons, whose energy is not larger than GeV, experience a large electrostatic potential variation that occurs in the shock transition [55].
- Even if the Fermi process works at relativistic shocks, it is not yet clear how many cycles can be undergone by the particles because of the large escape probability (between 0.2 and 0.5), and how many particles are involved in several cycles. If many, then a nice universal powerlaw is obtained with an index close to 2.2 [51,52].
- Strong anisotropy is inherent to relativistic kinematics and a detailed description of the scattering and diffusion processes in the appropriate magnetic turbulence must be fully investigated.
- The theory can be completely checked only by heavy Monte Carlo simulations, as those developed by Bednarz and Ostrowski [57,52].
- Besides the acceleration at shocks, the investigation of the nonlinear dynamics in a relativistic plasma [55,25,56], which is still poorly done, should be pursued.

The Fermi acceleration at relativistic shocks or wavefronts is potentially efficient enough in GRBs to explain the origin of the UHE-Cosmic Rays. Although they can produce very high energy cosmic rays, the other astrophysical sources, such as AGNs, RadioGalaxies, Large Scale Structure shocks, Pulsars, seem less promising to go beyond the GZK threshold, unless unexpected physical effects. The alternative would be the particle production caused by interconnections and decays of topological defects [4,58].

The Pierre Auger Observatory will surely give the answer to the question of the galactic or extragalactic origin of the UHE cosmic rays with a good statistics on the super-GZK events and will provide an interesting map.

The detection of high energy neutrinos emitted by the sources of UHE-protons by the AMANDA and ANTARES neutrino observatories would be a great event [59]. In particular, coincidence of gamma and neutrino detections from GRBs would be helpful.

So, together with the possibility of gravitational waves detection, these are new windows and new fascinating frontiers that are opened to astrophysics and to particle physics (the neutralino could also be detected by some of the new instruments). For the first time the transmitters of the cosmic information will be particles instead of the photons. This cross fertilisation of the two disciplines deserves the use of the neologism "Astroparticle Physics".

Acknowledgement

I sincerely thank the organizers of the UHECR 2000 meeting in Meudon for their invitation to give these lectures. I also feel especially grateful to Martin Lemoine for his careful reading of the manuscript and helpful comments.

9 Appendix

9.1 Average Gain and Escape Probability

The average energy gain per half-cycle can be stated as follows. Consider the half-cycle "udu" which increases the energy according to $\delta\epsilon = -2(\boldsymbol{u}_2 - \boldsymbol{u}_1).\boldsymbol{p}$, so that, with $\mu > 0$, the relative gain

$$\frac{\delta p}{p} = -2\frac{u_2 - u_1}{v}\mu > 0 \ . \tag{58}$$

The gain for the half-cycle "dud" is similar with $\mu < 0$,

$$\frac{\delta p}{p} = -2\frac{u_1 - u_2}{v}\mu > 0 \ . \tag{59}$$

The pitch angle average for the half-cycle "udu" is obtained by integrating over the distribution within a swept volume $Sv_{\|}\delta t$, S being an arbitrary cross section of the shock front. The distribution function $f = \bar{f} + \delta f$, and the anisotropic part δf is assumed much smaller than the anisotropic one $\bar{f}$. Therefore one gets:

$$\frac{\delta p}{p} = 2\frac{u_1 - u_2}{v}\frac{\int_0^1 \mu f 2\pi p^2 \mathrm{d}p S v \mu \delta t \mathrm{d}\mu}{2\int_0^1 f 2\pi p^2 \mathrm{d}p S v \mu \delta t \mathrm{d}\mu} = \frac{2}{3}\frac{u_1 - u_2}{v} \ . \tag{60}$$

The quasi-isotropy assumption is very important and holds as long as the scattering process is the fastest diffusion process.

The calculation of the escape probability is similar. It stems from the ratio of the particle flux carried by the downstream flow at the velocity u_2 over the flux of particles that cross the front at the speed v:

$$\eta = \frac{\bar{f}_2 4\pi p^2 \mathrm{d}p S u_2 \delta t}{\int_0^1 f_0 2\pi p^2 \mathrm{d}p S v \mu \delta t \mathrm{d}\mu} \ . \tag{61}$$

In this expression $\bar{f}_2$ denotes the isotropic part of the distribution function far downstream, whereas f_0 denotes the distribution function just behind the shock. Again, $f_0 \simeq \bar{f}_0$ is assumed, the shock width is assumed much shorter than the diffusion length of the particles and the distribution is rapidly homogeneized behind the shock so that $\bar{f}_0 \simeq \bar{f}_2$. The announced result is thus derived:

$$\eta = \frac{4u_2}{v} \ . \tag{62}$$

9.2 Gyro-Resonances and Pitch Angle Scattering

In weak turbulence theory [6], the particle momentum undergoes pitch angle scattering and energy diffusion through gyro-resonances with MHD-waves. It is instructive to explore these resonances in the simplest case of Alfvén waves propagating along the mean field; for oblique propagation, the calculations are tedious but lead to similar results. The method consists in perturbatively calculating the motions of a particle in the wave electromagnetic force experienced along the unperturbed orbit. The unperturbed orbit is described as follows:

$$z(t) = z(0) + v_{\|}t \tag{63}$$

$$\boldsymbol{v}_{\perp}(t) = v_{\perp}\left[\boldsymbol{e}_1 \cos(\omega_{\rm L}t + \psi) + \varepsilon_c \boldsymbol{e}_2 \sin(\omega_{\rm L}t + \psi)\right] \tag{64}$$

where $\varepsilon_c = +1$ for negative charge and -1 for a positive charge. For a mode of the form

$$\boldsymbol{A}(z,t) = A_0\left[\boldsymbol{e}_1 \cos(\omega t - kz + \phi) + \varepsilon \boldsymbol{e}_2 \sin(\omega t - kz + \phi)\right] \tag{65}$$

with $\omega = \pm k V_{\rm A}$, the perturbing force has a parallel component such that

$$\delta F_{\|} = q\boldsymbol{v}_{\perp} \cdot \frac{\partial \boldsymbol{A}}{\partial z} = q v_{\perp} k A_0 \cos(\Omega t + \theta_0) \tag{66}$$

where $\Omega = \omega - kv_{\|} - \varepsilon\varepsilon_c\omega_{\rm L}$. Therefore, the time integration of that force component to get the momentum variation in the parallel direction displays a resonant divergence for $\omega - kv_{\|} \pm \omega_{\rm L} = 0$. This defines a resonant wave number $k_0 = (r_{\rm L}|\mu|)^{-1}$ (for $V_{\rm A} \ll c$); and note that because the wave numbers of the MHD spectrum are smaller than $\omega_{cp}/V_{\rm A} \equiv r_0^{-1}$, resonances occur only in the cone $|\mu| > r_0/r_{\rm L}$. The particle energy varies much more slowly if $V_{\rm A} \ll c$ as can be seen from

$$\delta F_0 = -q\frac{\boldsymbol{v}_{\perp}}{c} \cdot \frac{\partial \boldsymbol{A}}{\partial t} = \pm\frac{V_{\rm A}}{c}\delta F_{\|} \ . \tag{67}$$

Thus the fastest process is the pitch angle scattering and $\dot{p}_{\|} \simeq p\dot{\mu}$. The scattering frequency defined by Eq. (5) must be derived from

$$\frac{\langle(\Delta\mu)^2\rangle}{\Delta t} = (1-\mu^2)\nu_s = \frac{2}{p^2}\int_0^{\Delta t} C_{\|}(\tau)\mathrm{d}\tau \ , \tag{68}$$

where $C_{\|}(\tau) \equiv \langle \delta F_{\|}(t)\delta F_{\|}(t-\tau)\rangle$. For a single mode, the self-correlation function of the parallel force displays permanent oscillations:

$$C_{\|}(\tau) = \frac{1}{2}q^2 v_{\perp}^2 A_0^2 k^2 \cos(\Omega\tau) \ ; \tag{69}$$

whereas for a continuum of independent modes, it decays on a time scale (the correlation time) governed by the spectrum band width:

$$C_{\|}(\tau) = q^2 v_{\perp}^2 \bar{B}^2 \int \frac{\mathrm{d}k}{2\pi} S(k) \cos(\Omega\tau) \ , \tag{70}$$

where $S(k)$ is the spectrum of the turbulent magnetic field normalised to the turbulence level η such that

$$\int S(k)\frac{\mathrm{d}k}{2\pi} = \eta \ . \tag{71}$$

The correlation time $\tau_\mathrm{c} \sim (\omega_\mathrm{L}\frac{\Delta k}{k_0})^{-1}$, where $k_0 \equiv \omega_\mathrm{L}/v_\parallel$ is the resonant wave number. This gives the expression of the pitch angle frequency in the case of weak turbulence [6]:

$$\nu_s = \frac{\omega_\mathrm{L}}{4}k_0 S(k_0) = \omega_\mathrm{L}\frac{\pi}{4}(\beta - 1)\eta(\rho|\mu|)^{\beta-1} \ . \tag{72}$$

In stronger turbulence, the gyro-resonances broaden, but the scaling law with rigidity ρ and turbulence level η (properly defined!) can be extrapolated [7,8]. But the assumption of a small jump $\mu \mapsto \mu + \Delta\mu$ during a correlation time no longer applies. The correlation time and the scattering time become comparable and the memory of an initial value of μ is ruled out; thus the variations of μ are the main cause of the resonance broadening.

References

1. F.A . Aharonian, J. W. Cronin: Phys. Rev. D **50**, 1892 (1994)
2. N. Hayashida, K. Honda, N. Inoue, et al.: Proc. *19th Texas Symposium on Relativistic Astrophysics and Cosmology*, eds. E. Aubourg, T. Montmerle, J. Paul, P. Peter, Nuc. Phys. B (Proc. Supp.) **80B** (2000)
3. P. Sokolsky: Proc. *19th Texas Symposium on Relativistic Astrophysics and Cosmology*, eds. E. Aubourg, T. Montmerle, J. Paul, P. Peter, Nuc. Phys. B (Proc. Supp.) **80B** (2000)
4. G. Sigl, D. N. Schramm, P. Battacharjee: Astropart. Phys. **2**, 401 (1994)
5. C. A. Norman, D. B. Melrose, A. Achterberg: Astrophys. J. **454**, 60 (1995)
6. J. R. Jokipii: Astrophys. J. **146**, 480 (1996); ibid. **313**, 842 (1987)
7. J. Giacalone, J. R. Jokipii: Astrophys. J. **520**, 204 (1999)
8. F. Casse, M. Lemoine, G. Pelletier: Phys. Rev. D to appear, astro-ph/0109223; to appear in Phys. Rev. D
9. E. Fermi: Phys. Rev. **75**, 1169 (1949)
10. G. F. Krimsky: Sov. Phys. Dokl. **23**, 327 (1977)
11. A. R. Bell: Mon. Not. R. Astron. Soc. **182**, 147 (1978)
12. W. I. Axford, E. Leer, G. Skhadron: Proc. *15th International Cosmic Ray Conference* (1977)
13. R. D. Blandford, J. P. Ostriker: Astrophys. J. **221**, L29 (1978)
14. V. S. Berezinskii, S. V. Bulanov, V. A. Dogiel, V. L. Ginzburg, V. S. Ptuskin: *Astrophysics of Cosmic Rays* (North Holland, Amsterdam, 1990)
15. E. Schatzman: Annales d'Astrophysique **26**, 234 (1963)
16. J. G. Kirk, P. Duffy, Y. A. Gallant: Astron. Astrophys **314**, 1010 (1996)
17. J. Skilling: Mon. Not. R. Astron. Soc. **172**, 557 (1975)
18. L. O' . C. Drury: Sp. Sc. Rev. **36**, 57 (1983)
19. P. O. Lagage, C. Cesarsky: Astron. Astrophys. **125**, 249 (1983)

20. F. A. Aharonian, et al.: Proc. *International Symposium on High Energy Gamma Ray Astronomy - Heidelberg2000*, p.749 (2000), astro-ph/0102391; to appear in Astron. Astrophys.
21. E. G. Berezhko, H. J. Völk: Astropart. Phys. **14**, 201 (2000)
22. P. L. Biermann: Astron. Astrophys. **271**, 649 (1993)
23. J. P. Rachen, P. L. Biermann: Astron. Astrophys. **272**, 161 (1993)
24. G. Henri, G. Pelletier, P. O. Petrucci, N. Renaud: Astropart. Phys. **11**, 347 (1999)
25. G. Pelletier: Astron. Astrophys. **350**, 705 (1999)
26. M. Vietri: Astrophys. J. **453**, 883 (1995)
27. E. Waxman: Astrophys. J. **452**, L1 (1995)
28. M. J. Rees, P. Mészáros: Mon. Not. R. Astron. Soc. **258**, 41 (1992)
29. P. Meszaros, M. J. Rees: Astrophys. J. **482**, L29 (1997)
30. R. D. Blandford, C. F. McKee: Phys. Fluids **19**, 1130 (1976)
31. E. Waxman: Phys. Rev. Lett. **75**, 386 (1995)
32. E. Waxman, J. N. Bahcall: Phys. Rev. D **59**, 023002 (1999)
33. F. A. Aharonian, A. M. Atoyan, T. Kifune: Mon. Not. R. Astron. Soc. **291**, 162 (1997)
34. B. R. Ragot, R. Schlickeiser: Astropart. Phys. **9**, 79 (1998)
35. M. E. Dieckmann, S. C. Chapman, K. G. McClements, R. O. Dendy, L. O' . C. Drury: Astron. Astrophys. **356**, 377 (2000)
36. R. Schlickeiser, A. Campeanu, L. Lerche: Astron. Astrophys. **276**, 614 (1993)
37. R. Schlickeiser: Astrophys. J. Suppl. **90**, 929 (1994)
38. R. Vainio, R. Schlickeiser: Astron. Astrophys. **343**, 303 (1999)
39. F. Jones: Astrophys. J. Suppl. **90**, 561 (1994)
40. G. Pelletier, J. Roland: Astron. Astrophys. **163**, 9 (1986)
41. G. Pelletier, J. Roland: Astron. Astrophys. **196**, 71 (1988)
42. L. O' . C. Drury, W. I. Axford, D. Summers: Mon. Not. R. Astron. Soc. **198**, 833 (1982)
43. H. J. Völk, L. O' . C. Drury, J. F. McKenzie: Astron. Astrophys. **130**, 19 (1984)
44. J. F. MacKenzie, H. J. Völk: Astron. Astrophys. **116**, 191 (1982)
45. L. O' . C. Drury, H. J. Völk, E. G. Berezhko: Astron. Astrophys. **299**, 222 (1995)
46. A. Decourchelle, D. C. Ellison, J. Ballet: Astrophys. J. **543**, 57 (2000)
47. M. G. Baring, D. C. Ellison, F. C. Jones: Astrophys. J. **409**, 327 (1993)
48. D. C. Ellison, M. G. Baring, F. C. Jones: Astrophys. J. **473**, 1029 (1996)
49. F. Mirabel, L. F. Rodriguez: Nature **371**, 46 (1994)
50. Y. Gallant, A. Achterberg: Mon. Not. R. Astron. Soc. **305L**, 6 (1999)
51. J. Kirk, A. W. Guthmann, Y. Gallant, B. Achterberg: Astrophys. J. **542**, 235 (2000)
52. J. Bednarz, M. Ostrowski: Mon. Not. R. Astron. Soc. **310**, L11 (1999)
53. J. G. Kirk, P. Schneider: Astrophys. J. **315**, 425 (1987)
54. M. Begelman, J. Kirk: Astrophys. J. **353**, 66 (1990)
55. G. Pelletier, A. Marcowith: Astrophys. J. **502**, 598 (1998)
56. G. Pelletier, E. Kersalé: Astron. Astrophys. **361**, 788 (2000)
57. J. Bednarz, M. Ostrowski: Phys. Rev. Lett. **80**, 3911 (1998)
58. S. Bonazzola, P. Peter: Astropart. Phys. **7**, 161 (1997)
59. R. J. Protheroe, T. Stanev T.: Proc. *High Energy Neutrino Astronomy*, eds. V.J. Stenger, et al. (World Scientific, Singapore 1992), p. 40

Rotation Powered Pulsars as Sources of High-Energy Particles

Bronisław Rudak[1,2]

[1] Nicolaus Copernicus Astronomical Center, Rabiańska 8, 87 100 Toruń, Poland
[2] Torun Centre for Astronomy, NCU

Abstract. Highly magnetised rapidly spinning neutron stars are widely considered to be natural sites for acceleration of charged particles. A powerful acceleration mechanism due to unipolar induction is thought to operate in the magnetospheres of isolated neutron stars, bringing the particles to ultrarelativistic energies at the expense of the neutron star rotational energy, with inevitable emission of high energy photons. The aim of this review is to present the basic ingredients of modern models of magnetospheric activity of rotation powered pulsars in the context of high-energy radiation from these objects. Several aspects of pulsar activity are addressed and related to spectacular results of pulsar observations with two major satellite missions of the past – CGRO and ROSAT. It is then argued that future high sensitivity experiments - GLAST, VERITAS, and MAGIC - will be vital for a progress in our understanding of pulsar magnetospheric processes. In a conservative approach rotation powered pulsars are not expected to be the sources of ultra high energy cosmic rays. However, several scenarios have been proposed recently to explain the ultra high energy cosmic ray events above the Greisen-Zatsepin-Kuzmin limit with the help of acceleration processes in the immediate surrounding of newly born pulsars. Major features of these scenarios are reviewed along with references to contemporary models of magnetospheric activity.

1 Introduction

High energy radiation from various classes of galactic and extragalactic objects has been observed for nearly 30 years. A large fraction of galactic sources is associated with neutron stars: rotation powered pulsars (RPP), accretion powered pulsars (APP), cooling neutron stars, and soft $\gamma-$ray repeaters. (SGR)Rotation powered pulsars like Crab, Vela and Geminga have a long history of successful observations with balloon-born and satellite $\gamma-$ray and X-ray experiments. The performance of old experiments has been, however, surpassed in terms of sensitivity, energy range, number of positive detections, or photon statistics per object by the COMPTON Gamma-Ray Observatory (CGRO) and the Röntgen Satellit (ROSAT). The spectacular results of observational campaigns of RPP with ROSAT and CGRO induced a new wave of interest in theoretical aspects of pulsar magnetospheric activity.

The pair creation paradigm is a pivotal element in any model of magnetospheric activity of RPP. Electron-positron pairs ($e^{\pm}$-pairs) are necessary since they are thought to be responsible for radio emission observed in radiopulsars

Martin Lemoine, Günter Sigl (Eds.): LNP 576, pp. 90–121, 2001.

which is interpreted as the coherent curvature radiation of $e^{\pm}$ plasma. Pairs can be produced in magnetospheric environments either via photon absorption in a dense field of soft photons (photon-photon collision) or via photon absorption in a strong magnetic field. In either case a supply of high-energy (HE) photons is required in order to fulfill stringent threshold conditions for the pair creation. It it quite reasonable then to assume that not all of those HE-photons would be subject to absorption. On the contrary, many HE photons will escape the magnetosphere without any attenuation. This argument leads us to expect that RPP (and all radiopulsars in particular) should be the sources of HE radiation. To make the production of HE photons possible, highly relativistic charged particles are to be injected into the magnetosphere. One may speculate that some of these particles will either retain their energy or regain it (under circumstances to be specified) upon escaping from the source. It is up to theoretical models of the RPP activity to show whether a rate of HE radiation and/or particles is interestingly high with respect to the sensitivity of recent and future HE detectors and telescopes.

This review focuses on RPP as sources of HE photons, presenting the most important observational results as well as their interpretation in terms of basic processes expected in the magnetospheres of RPP. The interpretation is offered by referring to a particular class of models of magnetospheric activity, known as polar gap models (or polar cap models). The name reflects the association of the accelerator (the gap) with a polar cap on the neutron star (NS) surface. Contrary to polar gap models, outer gap models [1] postulate the existence of accelerators located in regions where local corotation charge density reaches zero, close to the light cylinder. The $e^{\pm}$–pair creation occurs there either via one photon magnetic absorption (Crab-type outer gaps) or via photon-photon collisions (Vela type outer gaps). These models are relevant for both classical and millisecond pulsars with sufficiently high spin-down luminosity $L_{\rm sd}$. The outer-gap accelerators cease to produce $e^{\pm}$ pairs once the pulsar crosses the death-line $\log \dot{P} = 3.8 \log P - 11.2$ [2]. A modern version of the outer-gap accelerator, the so called "thick gap solution" [3], is however able to accommodate pulsars of longer spin periods, like Geminga and B1055-52, removing at the same time serious problems with the original model of Ref. [1]. A detailed review of outer gap models in the context of HE radiation is available [3] and therefore we will concentrate on polar gap models.

The review is organised in the following way: Sect. 2 defines basic quantities and introduces the assumptions used in pulsar physics. The status of X-ray and γ–ray observations of RPP and essential features of the radiation detected from several sources are presented in Sect. 3. Sect. 4 offers simple estimates of how effective a unipolar inductor (i.e. accelerator) can be when acting in the framework of a neutron star. Sect. 5 discusses the vertical structure of some polar gap accelerators. Sect. 6 presents the properties of the most important radiative processes induced by such inductors inside the RPP magnetosphere. With energetic arguments formulated in Sect. 4, Sect. 7 addresses a question of whether newly born and fastly spinning RPP might lead to generation of ultrarelativistic charged particles responsible then for the ultra high energy cosmic ray (UHECR) events observed above the Greisen-Zatsepin-Kuzmin (GZK) limit. Sect. 8 emphasizes the anticipated role of high sensitivity HE missions of the near future in contributing to the physics of RPP.

2 Basic Parameters

The aim of this section is to define basic quantities used throughout the review and to introduce their mutual relations. Several excellent monographs covering this subject in a detailed and sophisticated way are available e.g. Ref. [4] with a critical discussion.

The starting point are two fundamental quantities measured for pulsars – period P, interpreted as a period of rotation of a neutron star, and $\dot{P}$, its time derivative. Suppose, that a neutron star of radius $R_{\rm s}$ and moment of inertia I rotates with angular velocity $\Omega = 2\pi/P$ which decreases in time (for whatever reason) at a rate $\dot{\Omega} = -2\pi\, P^{-2}\dot{P} < 0$. The rotational energy and its time derivative then read

$$E_{\rm rot} = \frac{1}{2} I\,\Omega^2 \simeq 2\times 10^{46}\, I_{45} P^{-2}\ {\rm erg} \tag{1}$$

$$\dot{E}_{\rm rot} = I\,\Omega\,\dot{\Omega} \simeq -4\times 10^{31}\, I_{45}\dot{P}_{-15}P^{-3}{\rm erg\ s^{-1}} \tag{2}$$

where P is in seconds, $\dot{P}_{-15} \equiv \dot{P}/10^{-15}$ and $I_{45} \equiv I/10^{45}$ g cm^2. Instead of $\dot{E}_{\rm rot}$ one uses the so called spin-down luminosity $L_{\rm sd}$ defined as

$$L_{\rm sd} \equiv -\dot{E}_{\rm rot}. \tag{3}$$

The name *luminosity* is misleading since the carriers of the major part of $\dot{E}_{\rm rot}$ are not luminous for us: no one has ever managed to "see" them in a direct way by any type of detector (but see Sect. 7) and their nature remains unknown, at least for the time being. Therefore we need a model for the spin-down of a neutron star. Let us assume that a magnetic dipole is attached to the center of a neutron star, with its moment $\boldsymbol{\mu}_B$ inclined at angle Θ to the spin axis $\boldsymbol{\Omega}$, and let the mean strength of the field at the stellar surface be $B_{\rm s}$. The magnetic dipole, rotating in a vacuum will emit energy at the rate

$$L_{\rm magn} = \frac{2}{3c^3}\, B_{\rm s}^2 \sin^2\Theta\, R_{\rm s}^6\, \Omega^4 \tag{4}$$

suggesting thus the following model of the neutron star spin-down:

$$L_{\rm sd} = L_{\rm magn}. \tag{5}$$

The quantity $B_{\rm s}\sin\Theta$ can be inferred from P and $\dot{P}$ for a neutron star with known values of I and $R_{\rm s}$. For a large number of randomly oriented rotators the factor $\sin^2\Theta$ can be replaced with its averaged value of $2/3$.

Another model, where the dipolar radiation is replaced with a magnetospheric wind of particles [5], gives a similar result as (4) for an orthogonal rotator:

$$L_{\rm sd} = L_{\rm wind} \simeq \frac{1}{c^3} B_{\rm s}^2 R_{\rm s}^6 \Omega^4 \qquad (6)$$

and therefore is independent of the angle Θ. Since there exists no observational support for $\dot{P}$ depending on $\sin\Theta$ the standard approach is to apply the latter model to derive the strength of the dipolar component of the magnetic field

$$B_{12}^2 = 10^{15} I_{45} R_6^{-6} P \dot{P} \qquad (7)$$

where $B_{12} \equiv B_{\rm s}/10^{12}$G, and $R_6 \equiv R_{\rm s}/10^6$cm. Assuming that $B_{\rm s}$ does not change with time one can integrate (7) to obtain the characteristic spin-down time scale τ – a period of time elapsed since the pulsar was born with initial period $P_{\rm i}$

$$\tau = \frac{P}{2\dot{P}} \left[1 - \frac{P_{\rm i}^2}{P^2}\right] \ {\rm s.} \qquad (8)$$

As long as $P_{\rm i} \ll P$, which is thought to be satisfied for all classical pulsars and most of millisecond pulsars, the last factor in (8) plays no role and thus $\tau \simeq P/2\dot{P}$.

It is likely that neutron star magnetic fields contain high-order multipoles which may dominate the dipolar component at the surface level. Their relative amplitudes as well as distribution remain, however, unknown. It will be assumed throughout the paper that the dipolar magnetic field is not distorted by rotational effects or the presence of strong outflowing winds of particles (the latter effect has been recently invoked to decrease very high values of $B_{\rm s}$ inferred from P and $\dot{P}$ for two SGRs [6]; in consequence, their classification as magnetars became questionable).

The field is therefore approximated by an axisymmetric static dipole with field lines satisfying $r\sin^{-2}\theta = R_{\rm dc}$ in polar coordinates r and θ, with the dipole constant $R_{\rm dc}$. A dipole constant for which a rigid rotation with the angular velocity Ω reaches the speed-of-light limit (it occurs for $R_{\rm dc} = c/\Omega$ and this particular value is denoted as $R_{\rm lc}$) determines the so called light cylinder of radius $R_{\rm lc}$. All field lines which cross the light cylinder are then considered as open lines, and their footpoints on the stellar surface define two polar caps of radius $R_{\rm pc} \simeq R_{\rm s} \cdot (R_{\rm s}/R_{\rm lc})^{1/2}$, where the latter factor is the sine function of the polar coordinate θ for the outer rim of the polar cap: $\sin\theta_{\rm pc} = (R_{\rm s}/R_{\rm lc})^{1/2}$ (see Fig. 1).

3 Observational Overview of High-Energy Domain

A posteriori evidence that high-energy activity of pulsars must somehow draw from their rotational energy $E_{\rm rot} = I\Omega^2/2$ comes from a simple finding that, essentially, the success of detection of a particular pulsar in X-rays and/or γ–rays was strongly correlated with its position in the lists of targets ranked by spin-down flux values $L_{\rm sd}/D^2$.

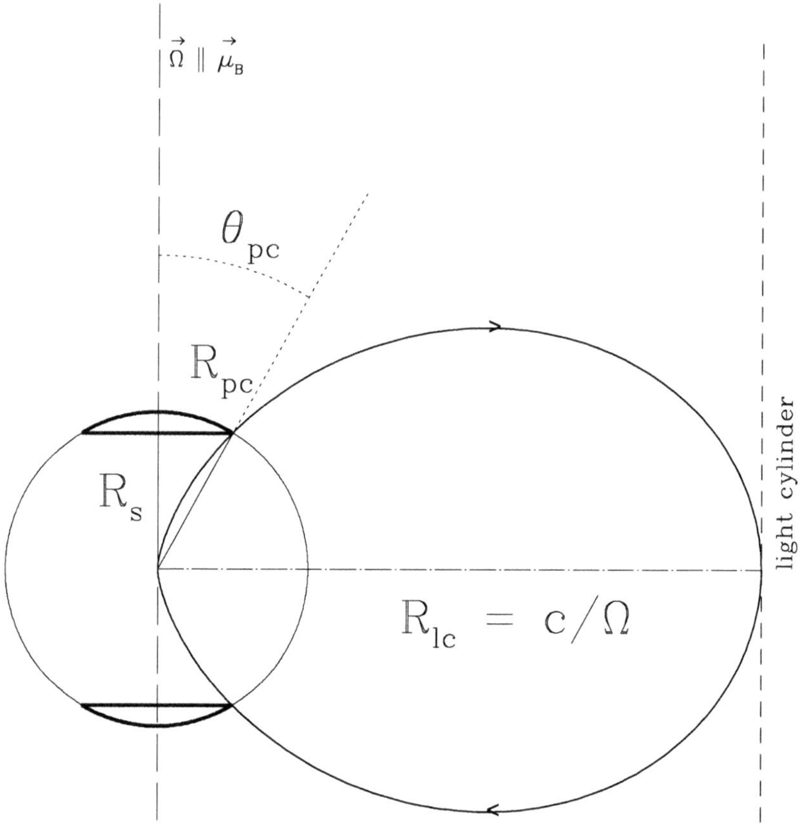

Fig. 1. Two identical polar caps of radius R_{pc} are shown on the surface of a neutron star of radius R_s for the aligned case, i.e. with the spin axis parallel to the magnetic moment axis. Their outer rims are determined by the set of dipolar lines for which the dipole constant R_{dc} equals the light cylinder radius $R_{lc} = c/\Omega$. The opening angle of each cap θ_{pc} satisfies then $\sin\theta_{pc} = (R_s/R_{lc})^{1/2}$. Therefore, whenever $R_s \ll R_{lc}$, the polar cap radius is approximated by $R_{pc} \simeq R_s \cdot (R_s/R_{lc})^{1/2}$.

The aim of this section is to review the recent observational status of RPP in the HE domain, with X-rays included. The HE domain is hereafter arbitrarily defined as extending from a fraction of keV up to about 30 GeV. Nevertheless, more emphasis is put on $\gamma-$ray results. $\gamma-$ray detections are particularly precious since their interpretation is thought to be less ambiguous in comparison to X-ray detections. In the latter case (especially for very young objects) contributions from initial cooling, internal friction, or other factors of *a priori* unknown magnitude may dominate the X-ray emission [1].

[1] indeed, four pulsars – Vela, Geminga, B1055-52 and B0656+14 – are classified as initial cooling candidates [11] since their X-ray emission is dominated by a component which may be modelled by a blackbody emission from a NS surface

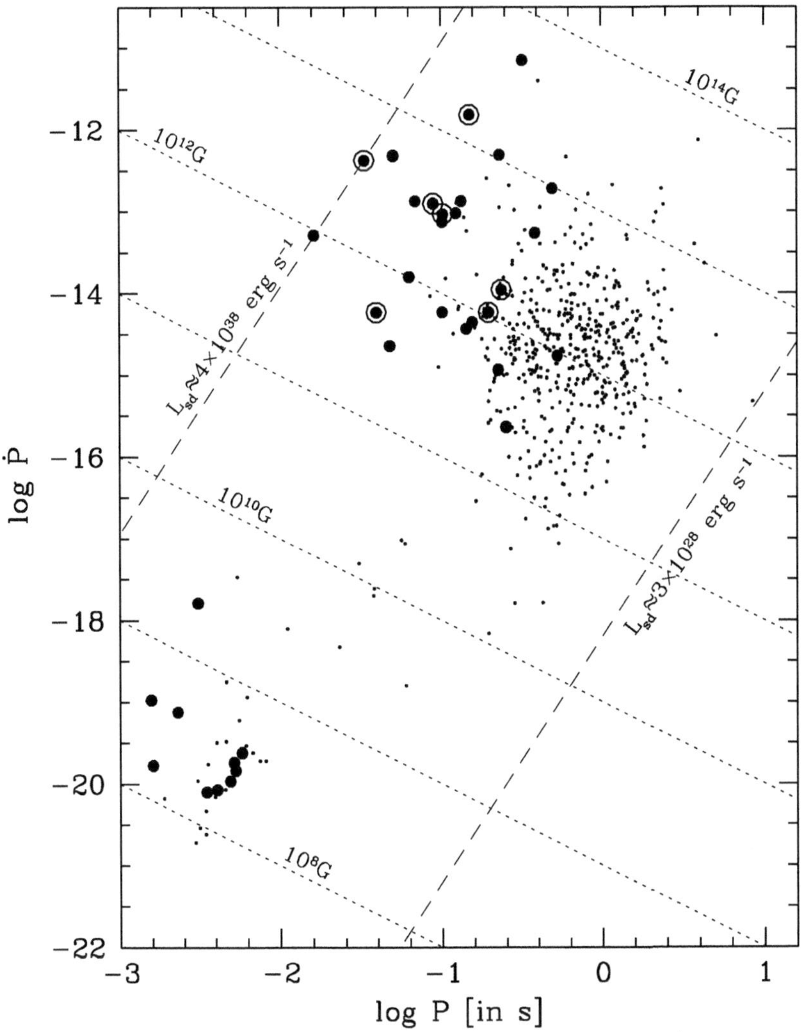

Fig. 2. $P-\dot{P}$ diagram for Rotation Powered Pulsars. The pulsars detected exclusively in radio are indicated with dots; they are taken mostly from the data base of Ref. [7]. Thirty five pulsars emitting X-rays are indicated with bullets. These include two objects recently discovered with RXTE: J0537-6910 in SNR N157B in LMC [8] is the fastest young pulsar known, spinning twice as fast as the Crab pulsar but with similar value of spin down luminosity; J1846-0258 in SNR Kes-75 [9], with $P = 0.32$s, the highest $\dot{P}$ among all RPP and no radio counterpart so far. Seven bullets in circles indicate seven γ−ray pulsars. Dashed lines correspond to constant values of the spin down luminosity $L_{\rm sd}$ given by (1) and (3) with $I_{45} = 1$. The upper line ($L_{\rm sd} \simeq 4 \times 10^{38}$erg s^{-1}) includes the Crab pulsar and J1846-0258, the lower one ($L_{\rm sd} \simeq 3 \times 10^{28}$erg s^{-1}) includes J2144-3933 – the slowest ($P = 8.5$s) radio pulsar detected so far [10]. Dotted lines correspond to constant values of the dipolar component of the surface magnetic field as inferred from P and $\dot{P}$ through (7) with $R_6 = 1$.

For more than 1500 pulsars known to date only about 35 positive detections in X-rays and no more than 10 detections in $\gamma-$rays have been achieved. There are firm detections by CGRO of 7 pulsars (dubbed *Seven Samurai*) and another 3 cases classified as "likely" detections. The $\gamma-$ray sources were identified by virtue of flux pulsations with previously known P and $\dot{P}$. Crab and Vela are the only pulsars seen by all three instruments of CGRO. No trace of pulsed signal in the very high energy (VHE) range (300 GeV – 30 TeV) has been found so far for the $\gamma-$ray pulsars [12,13,14]. However, strong steady VHE emission is associated with 3 out of 10 $\gamma-$ray pulsars. Two plerionic sources of the steady VHE radiation – The Crab Nebula and the plerion around B1706-44 – may serve as standard candles, with "grade A" according to [14]. A third plerion – around the Vela pulsar – was given "grade B" in the same ranking. All 10 $\gamma-$ray pulsars are strong X-ray emitters.

The positions of these HE pulsars are shown in the $P-\dot{P}$ diagram of Fig. 2 along with positions of about 700 radio pulsars for which $\dot{P}$ values were available. A remarkable fact is that the location of X-ray sources does not correlate with the inferred strength of magnetic field $B_{\rm s}$; at least not in a naively anticipated way that high-B objects would emit HE radiation, whereas low-B objects would not. In particular, 10 millisecond pulsars – about thirty percent of all millisecond pulsars (the objects with $P \lesssim 0.01$ s and $\dot{P} \lesssim 10^{-17}$, i.e. with low B values: $B_{\rm s} \lesssim 10^9$G) known to date – have been detected as X-ray sources. So far, millisecond pulsars eluded the detection in gamma rays and just upper limits have been available for a handful of them from EGRET observations [15]. In the case of J0437-4715 the upper limit is interestingly tight – in disagreement with the empirical relation $L_\gamma \propto L_{\rm sd}^{1/2}$ (see Fig. 4). Very recently, however, the likely detection of pulsed $\gamma-$ray emission from J0218+4232 has been reported [16].

Spectral analysis for pulsars detected with ROSAT PSPC (0.1 keV to 2.4 keV) shows that in most cases a power-law spectral model provides acceptable fits to the data [11]. Moreover, an intriguing empirical relation between inferred X-ray luminosity and spin down luminosity was found, $L_X \simeq 0.001\, L_{\rm sd}$, confirming the rotational origin of most of the X-ray activity. An interesting point is that the relation was obtained for all the sources regardless their temporal characteristics (about 50 % of all pulsars detected with ROSAT are unpulsed sources). Figure 3 presents these results in a somewhat different way and for slightly different values for L_X (compiled by the author). A complementary empirical relation was found for pulsed emission from 19 pulsars observed with ASCA (0.6 keV to 10 keV). Assuming the opening angle of X-rays to be one steradian, the inferred pulsed X-ray luminosity correlates with spin-down luminosity as

$$L_X = 10^{34} \left(\frac{L_{\rm sd}}{10^{38}\,{\rm erg\ s}^{-1}} \right)^{3/2} {\rm erg\ s}^{-1}, \qquad (9)$$

according to Ref. [22].

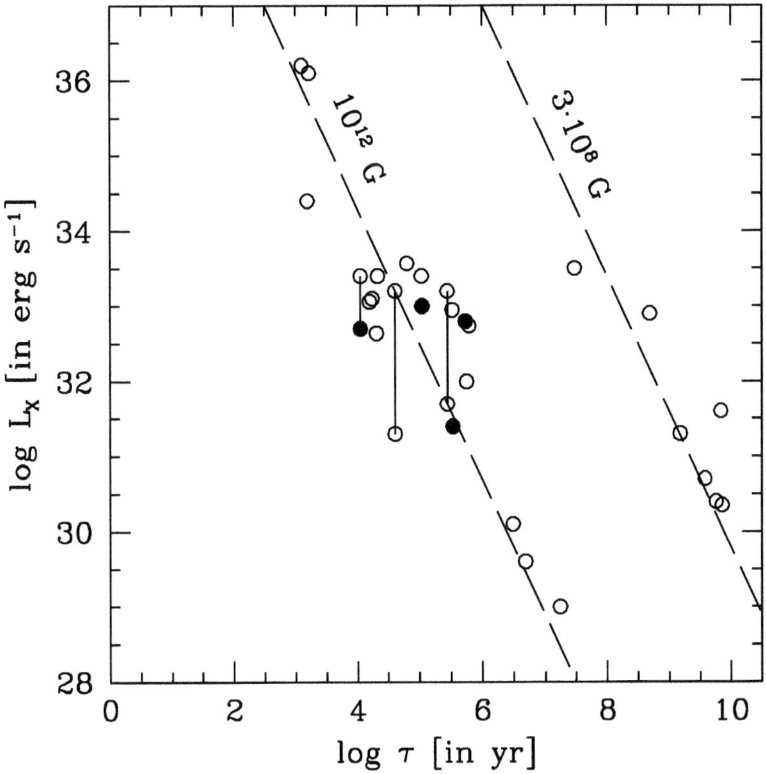

Fig. 3. X-ray luminosity versus spin-down age τ for 29 out of 35 pulsars detected with ROSAT, ASCA and RXTE. Isotropic emission into 4π steradians was assumed for pulsed and unpulsed sources to infer L_X. The empirical relation $L_X \simeq 0.001\, L_{sd}$ found by Ref. [11] can be rewritten as $L_X \propto B_s^{-2}\tau^{-2}$, see (6) and (7). If all classical and millisecond pulsars were to have the surface magnetic field B_s of a fixed value 10^{12}G and 3×10^8G, respectively, they would follow the two dashed lines labelled with B_s. Four filled circles are the initial cooling candidates; in increasing τ these are: B0833-45 (Vela) [17], B0656+14 [18], J0633+17 (Geminga) [19], and B1055-52 [20]. In three cases the fitting of the data with either blackbody or power law spectral model was equally justified, but inferred X-ray luminosities are strongly model-dependent (circles connected with vertical bars); in increasing τ these objects are: B0833-45 [17], B2334+61, and B0114+58 [21].

A similar power-law empirical relation holds for γ–rays (cf. Fig. 4), but with a different power-law index (e.g. Ref. [23]):

$$L_\gamma \simeq 10^{35} \left(\frac{L_{sd}}{10^{38}\,\mathrm{erg\ s^{-1}}} \right)^{1/2} \mathrm{erg\ s^{-1}}. \qquad (10)$$

An important conclusion from Figs. 3 and 4 is that neither L_X nor L_γ becomes a sizable fraction of L_{sd}. The most efficient conversion of spin-down lumi-

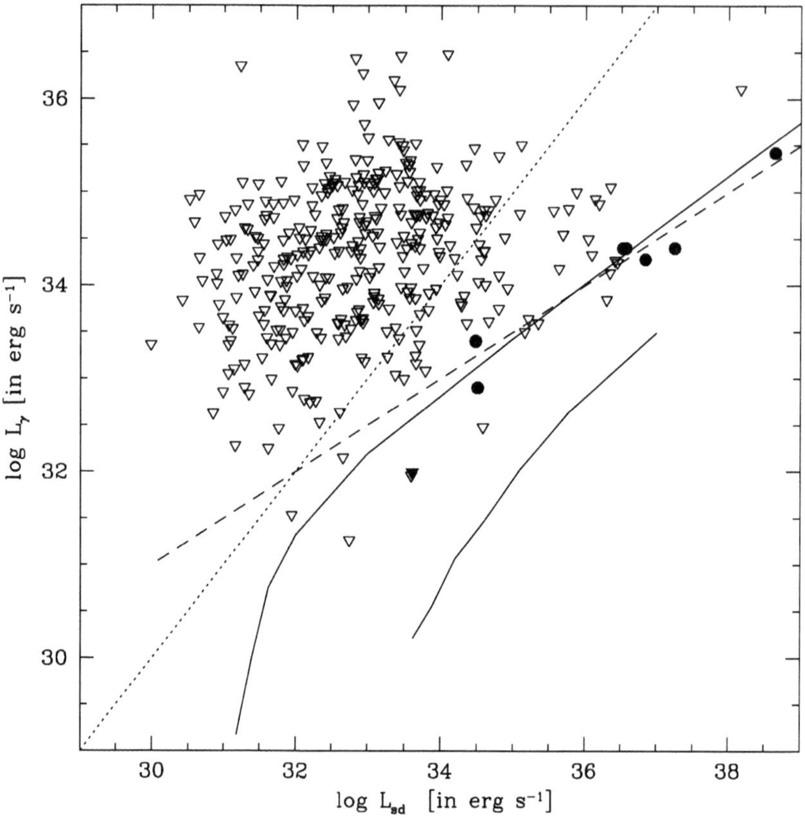

Fig. 4. γ–ray luminosity versus spin-down luminosity for seven pulsars (filled dots) detected with the CGRO instruments. An opening angle of one steradian was assumed for the γ–ray emission. Open triangles are the EGRET upper limits after Ref. [15] for 350 objects, including seven millisecond pulsars. The filled triangle indicates the upper limit for J0437-4715 [24]. Note, that most of the upper limits are well above the maximum possible value for L_γ set by $L_\gamma = L_{\rm sd}$ (dotted line). The dashed line marks the empirical relation derived for the CGRO pulsars: $L_\gamma \propto L_{\rm sd}^{1/2}$. Solid lines show evolutionary tracks for a classical pulsar with $B_{\rm s} = 10^{12}$G (upper line) and a millisecond pulsar with $B_{\rm s} = 10^9$G (lower line) according to the phenomenological model of Ref. [25].

nosity into high-energy radiation is taking place for B1055-52 – the oldest pulsar among the *Seven Samurai* – with $L_\gamma \simeq 0.1\, L_{\rm sd}$.

Broadband energy spectra per logarithmic energy bandwidth extending from radio, optical and UV, to X-rays and γ–rays, constructed for pulsed phased-averaged components of the *Seven Samurai* are particularly impressive [23] and instructive. Figure 5 reveals substantial spectral differences among the objects, which became a subject of theoretical debates and speculations.

Short spectral characteristics of all 9 γ–ray pulsars (including 2 "likely" sources) are given below after Ref. [23] for EGRET, Ref. [26] for COMPTEL,

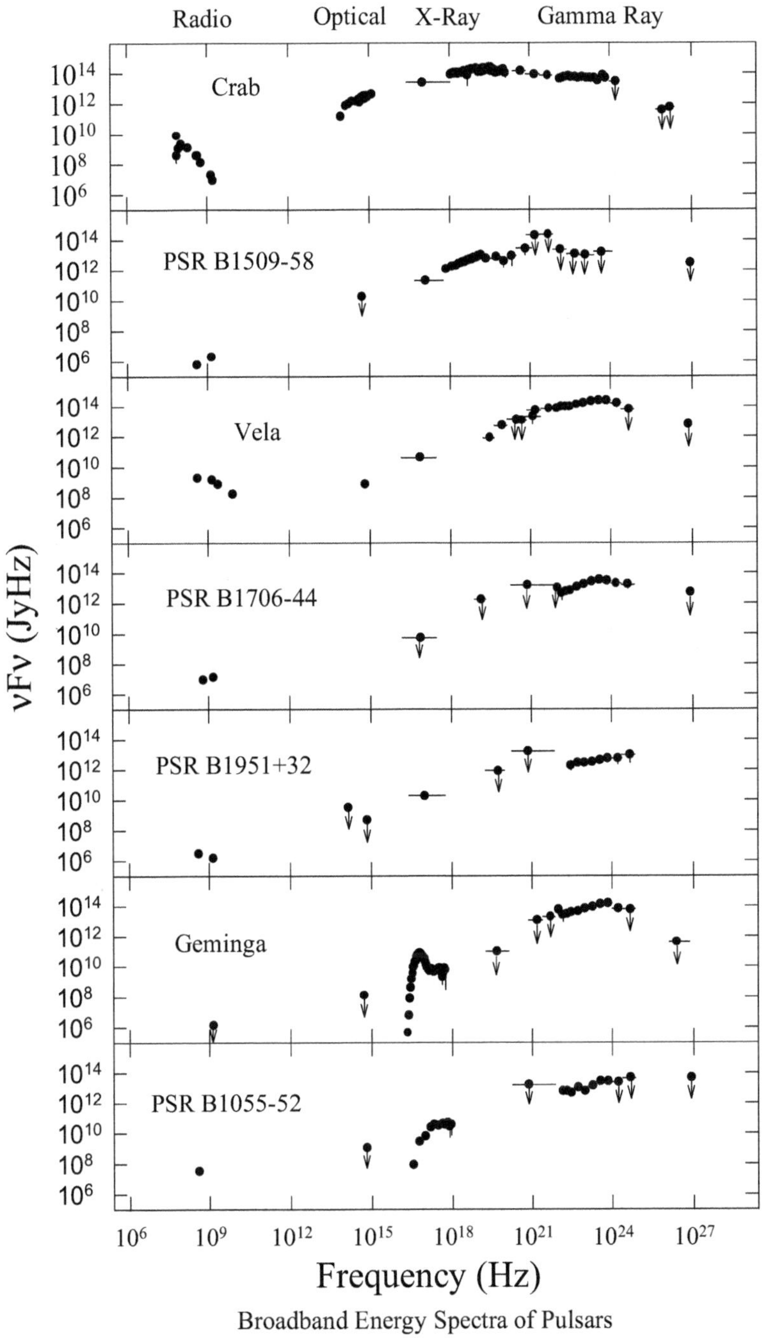

Fig. 5. Phase-averaged spectra for seven γ–ray pulsars (Courtesy D.J. Thompson).

and Ref. [27] for OSSE (see also references therein). These instruments operating on board CGRO [28] covered the following parts of the HE domain: the Oriented Scintillation Spectrometer Experiment (OSSE) was operating in the energy range 50 keV – 10 MeV, the Imaging Compton Telescope (COMPTEL) – in the energy range 0.75 MeV – 30 MeV, and the Energetic Gamma Ray Experiment Telescope (EGRET) in the energy range 50 MeV – 30 GeV.

1) B0531+21 (Crab) – Detected by EGRET, COMPTEL, and OSSE. Its $\gamma-$ray flux consists of pulsed and unpulsed components, the latter one coming from the Crab Nebula. The overall phase-averaged photon spectrum in the range between 50 keV and 10 GeV is described satisfactorily by a broken power-law shape $\mathrm{d}N_\gamma/\mathrm{d}\varepsilon \propto \varepsilon^{-s}$ with a break at $\varepsilon_{\mathrm{br}} \simeq 120$ keV, and photon power-law index $s = 1.71$ for $\varepsilon \leq \varepsilon_{\mathrm{br}}$, and $s = 2.21$ for $\varepsilon > \varepsilon_{\mathrm{br}}$. The energy flux is $f_\gamma \simeq 7.3 \times 10^{-9}\mathrm{erg\ s^{-1}cm^{-2}}$.

2) B1509-58 – Detected by COMPTEL and OSSE. The initial COMPTEL detection was of marginal significance ($\sim 3\sigma$-detection) in a narrow (0.75 – 1 MeV) energy band. However, recent analysis shows that the spectrum extends to higher energies with a cutoff around 10 MeV [29] (Note: this new finding is not marked in Fig. 5). The energy flux at 1 MeV may be as high as 1.4×10^{-9} erg $\mathrm{s^{-1}cm^{-2}}$, but the COMPTEL point stands above the corresponding OSSE point by a factor of 4. The OSSE spectral fit between 50 keV and $\sim$ 5 MeV with $\mathrm{d}N_\gamma/\mathrm{d}\varepsilon \propto \varepsilon^{-1.68}$ yields $f_\gamma \simeq 5.6 \times 10^{-10}$ erg $\mathrm{s^{-1}cm^{-2}}$. EGRET put strong upper limits for photon flux above 100 MeV and 1 GeV , which clearly fall below simple power-law extrapolation of the OSSE spectral fit. That indicates a presence of spectral roll-over at several MeV, in agreement with the cutoff claimed by Ref. [29]. The pulsar has the highest inferred magnetic field ($B_{\mathrm{s}} \simeq 1.5 \times 10^{13}$G) among seven $\gamma-$ray pulsars, an essential point for explaining the cutoff at 10 MeV as due to photon-splitting effect [30].

3) B0833-45 (Vela) – Detected by EGRET, COMPTEL, and OSSE. Its phase-averaged photon spectrum between 30 MeV and 2 GeV can be reproduced as a power law with $s = 1.7$, and a strong spectral break above $\sim$ 4 GeV. The spectrum flattens out in the OSSE range with $s = 1.3$. Estimated energy flux reaches $\sim 9 \times 10^{-9}$ erg $\mathrm{s^{-1}cm^{-2}}$ (the brightest object in the $\gamma-$ray sky).

4) B1706-44 – Young Vela-like pulsar, detected by EGRET. The spectrum extends from 50 MeV beyond 10 GeV and may be approximated with a broken power law, with photon index s changing from 1.27 to 2.25 at 1 GeV.

5) B1951+32 – Detected by EGRET and COMPTEL. The spectrum extends from 0.75 MeV up to 30 GeV and may be approximated with a single power law with photon index $s = 1.89$. It has an extremely sharp cut-off, with no apparent decline in the flux level. But an extrapolation towards TeV falls 2 order of magnitude above an upper limit (not shown in Fig. 5) set by the Whipple group.

6) J0633+17 (Geminga) – Confirmed detection by EGRET only. The photon spectrum may be approximated by a single power law, with $s = 1.50$, extending from 30 MeV to a roll-off at 2 GeV.

7) B1055-52 – Detected by EGRET above 70 MeV. Its spectrum can be represented by a single power law with photon index $s = 1.73$, and a possible break around 1 GeV.

8) B0656+14 – A 3σ-detection by EGRET had been reported. The pulsar with its parameters (P, $\dot{P}$, and particularly $L_{\rm sd}$) resembles Geminga and B1055-52. Its photon spectrum estimated for a low number of events may be represented between 10 MeVand 10 GeV as a very steep power-law with the index $s = 2.8$.

9) J0218+4232 – Marginal detection by EGRET (at $3.5\,\sigma$ level) of pulsed emission has been reported recently in the energy range 100 MeV – 300 MeV for this distant ($D \gtrsim 5.85\,$kpc) millisecond ($P = 0.0023$ s) pulsar in a binary system with a low-mass white dwarf [16]. The inferred luminosity of the pulsed emission for 1 steradian opening angle reaches $L_\gamma \simeq 1.64 \times 10^{34}$erg s$^{-1} \simeq 0.07\,L_{\rm sd}$.

10) B1046-58 – Likely candidate for a $\gamma-$ray pulsar, reported recently [31], based on analysis of the EGRET source 2EG J1049-5847 spatially coincident with this radiopulsar. The estimated L_γ above 400 MeV and 1 steradian opening angle reaches $0.011 L_{\rm sd}$ for a distance $D = 3$kpc.

$\gamma-$ray light curves differ significantly from those in X-rays, optical, and radio. Their most striking feature are relatively long duty cycles as well as phase shifts in comparison to the radio pulses. Only for the Crab pulsar the peaks in $\gamma-$rays as well as in radio wavelengths occur at the same rotational phases. The light-curve shapes fall into two categories. The Crab pulsar, Vela and Geminga show two sharp pulses separated in phase by $0.4 - 0.5$ and connected by an interpulse bridge of considerable level. B1706-44 shows two peaks separated by 0.2 in phase, with some hints of a third component in between. Other pulsars exhibit broad single pulses. Unknown opening angles for $\gamma-$ray emission introduce a factor of uncertainty when inferring the $\gamma-$ray luminosities. Broad peaks in $\gamma-$ray pulses do not necessarily mean large opening angles for $\gamma-$ray emission. Polar cap models, which rely on purely dipolar magnetic fields postulate nearly aligned rotators, where the inclination of magnetic axis to spin axis is comparable to the angular extent of the polar cap [32].

With a dicovery of sharply peaked pulsed X-ray emission in the fastest millisecond pulsar B1937+21 an apparently separate group of millisecond X-ray pulsars emerges, with its members – B1821-24 [22], J0218+4232 [16], and B1937+21 [33] – being scaled-down versions of the Crab pulsar as far as sharp pulse profiles and hard power-law X-ray spectra are concerned. An astonishing common feature within the group is the same strength of the magnetic field estimated at the light cylinder and the fact that it matches the strength of the Crab pulsar magnetic field at the light cylinder.

4 Unipolar Induction – A Toy Model

4.1 Vacuum Rotator

Let us begin with a frequently invoked order-of-magnitude estimate advertising rotating neutron stars as potentially powerful accelerators and thus good candidates to explain UHECR (the problem addressed in Sect.7). Consider a neutron star of radius R_s and surface magnetic field B_s as a perfect conductor. For the rotating star with its dipolar magnetic field immersed in a vacuum an external quadrupole electric field $\boldsymbol{\mathcal{E}}$ develops, with non-zero component along magnetic field lines at the surface. The corresponding electrostatic potential Φ in polar coordinates r and θ reads [4]

$$\Phi(r,\theta) = -\frac{Q}{r^3}\left(3\cos^2\theta - 1\right), \tag{11}$$

where $Q = \pi/3\, B_s/(cP)\, R_s^5$ is the quadrupole moment. The maximal electromotive force will then be induced between one of the two poles of the star and its equator:

$$\Delta\Phi_{\text{equator}} = \frac{1}{2c}\, B\, \Omega\, R_s^2. \tag{12}$$

The corresponding voltage drop

$$\Delta V_{\text{equator}} \simeq 3 \times 10^{16} B_{12} R_6^2 P^{-1} \mathrm{V} \tag{13}$$

reaches huge values. If such a unipolar inductor was to operate in the Crab pulsar, B1509 or J1846-0258 (see Fig. 2), it would bring a fully ionized atom of iron ($Z = 26$) close to the energy of 10^{20}eV. However, the assumption about a vacuum surrounding the entire star is not correct. The space containing field lines closed within the light cylinder is expected to fill in quickly with trapped charged particles supporting the electric field which forces them to corotate with the star (cf. the subsection below). Therefore, the only regions on the stellar surface appropriate for the unipolar induction to act are those containing open field lines only, i.e. the polar caps. From (11) and (12) the potential difference between the pole and the outer rim of the polar cap follows as

$$\Delta\Phi_{\text{pc}} = \Delta\Phi_{\text{equator}} \left(\frac{R_{\text{pc}}}{R_s}\right)^2, \tag{14}$$

with voltage drop of

$$\Delta V_{\text{pc}} \simeq 7 \times 10^{12}\, B_{12}\, P^{-2}\, \text{Volt}\,. \tag{15}$$

4.2 Rotating Magnetosphere

We will concentrate hereafter on a class of models where the supply of charged particles from a neutron star surface along open field lines is not limited by binding or cohesive energy of the particles and therefore can reach the so-called Goldreich–Julian rate at the surface. Such a supply of charges was dubbed "Space Charge Limited Flow" (SCLF) [34,35] or "free emission". A concise but to-the-point account of essential properties of the SCLF models has been presented recently by [36].

Three boundary conditions essential for the electrodynamics above the polar cap are [37]:
1) $\boldsymbol{\mathcal{E}} \cdot \boldsymbol{B} = 0$ for the magnetosphere within the closed field lines,
2) $\Phi = 0$ at the surface and at the interface between the closed magnetosphere and the open field lines,
3) $\mathcal{E}_{\|} = 0$ at the surface level,
where $\mathcal{E}_{\|}$ is the electric field component parallel to the magnetic field.
Last but not least, it is assumed that the outflow is stationary and the magnetosphere remains axisymmetric.

The electric field $\boldsymbol{\mathcal{E}}$ required to bring a charged particle into corotation satisfies the following equation

$$\boldsymbol{\mathcal{E}} + \frac{1}{c}\left[(\boldsymbol{\Omega} - \boldsymbol{\omega}_{\mathrm{LT}}) \times \boldsymbol{r}\right] \times \boldsymbol{B} = 0 \tag{16}$$

where the inertial-frame dragging effect is included [37] with $\boldsymbol{\omega}_{\mathrm{LT}} = \kappa_{\mathrm{g}}(R_{\mathrm{s}}/r)^3\,\boldsymbol{\Omega}$, where $\kappa_{\mathrm{g}} = I/(M_{\mathrm{s}}R_{\mathrm{s}}^2) \cdot R_{\mathrm{g}}/R_{\mathrm{s}}$, and $R_{\mathrm{g}} = 2GM_{\mathrm{s}}/c^2$.

The charge density necessary to support this local $\boldsymbol{\mathcal{E}}$ is

$$\varrho_{\mathrm{corot}} = \frac{1}{4\pi}\boldsymbol{\nabla} \cdot \boldsymbol{\mathcal{E}} \simeq -\frac{\boldsymbol{\Omega} \cdot \boldsymbol{B}}{2\pi c}\left[1 - \kappa_{\mathrm{g}}\left(\frac{R_{\mathrm{s}}}{r}\right)^3\right] . \tag{17}$$

The charge density ϱ_{corot} due to SCLF at $r = R_{\mathrm{s}}$ is called the Goldreich–Julian charge density and is labelled "GJ":

$$\varrho_{\mathrm{GJ}} \simeq -\frac{\boldsymbol{\Omega} \cdot \boldsymbol{B}}{2\pi c}\left[1 - \kappa_{\mathrm{g}}\right] . \tag{18}$$

As charged particles flow out along the open field lines a deviation of the local charge density ϱ_{local} from the local corotation density ϱ_{corot} develops. By using now two relations satisfied in the dipolar structure, $B(r) \propto r^{-3}$ and $\varrho(r) \propto r^{-3}$, one obtains a simple formula for the local deviation from the corotation charge density:

$$\varrho_{\mathrm{local}} - \varrho_{\mathrm{corot}} \simeq \frac{\boldsymbol{\Omega} \cdot \boldsymbol{B}}{2\pi c}\,\kappa_{\mathrm{g}}\left[1 - \left(\frac{R_{\mathrm{s}}}{r}\right)^3\right] . \tag{19}$$

Accordingly, the accelerating potential drop in SCLF reads

$$\Delta\Phi_{\|} \simeq \Delta\Phi_{\mathrm{pc}}\,\kappa_{\mathrm{g}}\left[1 - \left(\frac{R_{\mathrm{s}}}{r}\right)^3\right] , \tag{20}$$

where $\Delta\Phi_{\mathrm{pc}}$ is given by (14) and $\kappa_{\mathrm{g}} = 0.15\,I_{45}$. This is a remarkable result obtained by Ref. [37]: due to the inertial-frame dragging effect the particles drop through the potential which is significantly larger than in the slot gap model [35] (which offers $\Delta\Phi_{\mathrm{slot}} \simeq 0.01P^{-1/2}\Delta\Phi_{\mathrm{pc}}$) provided $P \gg 0.004$ s. For example, for $P \sim 1$ s $\Delta\Phi_{\|}$ becomes about 10 times larger than $\Delta\Phi_{\mathrm{slot}}$. Moreover, the electric field $\mathcal{E}_{\|}$ develops now along all open lines regardless of their orientation with respect to the spin axis (the effect is not shown in this simplified presentation). Therefore, all open field lines are "favourable", in contrast to Ref. [35].

5 Electric Field Structure in SCLF Gaps

The model with frame dragging effects [37], presented in a simplified form in subsection 4.2, does not take into account possible feed-back effect due to $e^{\pm}$-pairs formed via photon absorption within open magnetic field lines. Copious pair formation occurs in a relatively thin layer called for this reason a pair formation front (PFF). The creation of pairs leads to screening of the accelerating field $\mathcal{E}_{||}$ within the layer of PFF. A detailed picture of this effect would require to follow the dynamics of electrons and positrons in a self-consistent way. Instead, it is reasonable to assume, that the field is shorted out at the height were the first $e^{\pm}$-pair is created (hereafter denoted as h_c): $\mathcal{E}_{||} = 0$ for $h \geq h_c$.

The problem of electric field structure in the context of SCLF with boundary condition $\mathcal{E}_{||} = 0$ set at $h_0 = 0$ (stellar surface) and at $h \geq h_c$ (PFF) was formulated and solved by Ref. [38]. The solution is rather lengthy and includes special functions. It is however possible to obtain simple but quite accurate analytical approximations. As long as the length h_c of the accelerator is of the order of the polar cap radius $R_{\rm pc}$, the accelerating electric field may be approximated according to Ref. [39] as

$$\mathcal{E}_{||} \simeq -1.46 \, \frac{B_{12}}{P^{3/2}} \, h \left(1 - \frac{h}{h_c}\right) f_1(\xi) \cos\chi \text{ Gauss}, \tag{21}$$

where $B_{12} = B_{\rm s}/10^{12}$G, P is the spin period in seconds, χ is the angle between the spin axis and the magnetic moment of the rotating star, h is expressed in cm, and $M = 1.4\,M_{\odot}$, $R_{\rm s} = 10^6$cm. The magnetic colatitude $\xi \equiv \theta/\theta(\eta)$ is scaled with the half-opening angle of the polar magnetic flux tube $\theta(\eta)$, where $\eta \equiv 1 + h/R_{\rm s}$. The magnetic colatitude function $f_1(\xi)$ is a monotonically decreasing function, with $f_1(0) \simeq 1$ and $f_1(1) = 0$.

The vertical structure of the electric field depends (via the location of PFF) on radiative processes which induce the pair creation: curvature radiation (CRV) and inverse Compton scattering (ICS) on soft X-ray photons from the stellar surface (brief characteristics of these processes are presented in the next section). An interesting effect was noticed in this context [38]: Suppose that a small fraction of positrons is stopped by a residual (non-zero) electric field at the site of their creation and then forced to flow towards the stellar surface (this effect was noticed already in Ref. [34]). The backflowing positrons are expected to induce the formation of an additional PFF, which would short out the electric field at h_0. These positrons cool upon the action of ICS and CRV. With reasonable surface temperatures ($T_{\rm s} \sim 5\times10^5$K) it is ICS which dominates the cooling of upward moving electrons and downward moving positrons. Therefore, the $e^{\pm}$-pair creation will be induced by upscattered photons from the stellar surface, rather than by curvature photons. The situation is not symmetrical, however, for electrons and positrons since the field of soft photons is not symmetrical with respect to both types of particles. In consequence, the positrons cool more efficiently, the ICS-induced cascades are easier to achieve for positrons than electrons and thus a lower PFF (where $\mathcal{E}_{||} = 0$) tends to be located above the stellar surface.

Such a situation is not stable, therefore. However, elevating the accelerator up to altitude $h_0 \sim 1\,R_s$ above the surface diminishes the role of the ICS; the CRV cooling dominates here and a stable accelerator is possible. No self-consistent calculations of such a "sandwich-like" accelerator exist at present, but the results obtained in Ref. [38] with an approximate treatment of the problem look promising indeed.

6 Radiative Processes in Pulsar Magnetospheres

Cooling of ultrarelativistic electrons via curvature radiation (CRV) and magnetic inverse Compton scattering (ICS) are the most natural ways of producing hard γ−rays capable of inducing cascades of $e^{\pm}$-pairs and secondary HE photons. These two processes dominate within two distinct ranges of Lorentz factors γ of the primary electrons.

When $\gamma \lesssim 10^6$, magnetic inverse Compton scattering plays a dominant role in braking the electrons and it is the main source of hard γ−ray photons [40]. Energy losses due to resonant ICS limit the Lorentz factors of the particles to a level which depends on electric field strength $\mathcal{E}_{\|}$, temperature T and size of the hot polar cap, and magnetic field strength B_s [41,42]. The Lorentz factors can then be limited even to $\sim 10^3$. This stopping effect becomes more efficient for stronger magnetic fields, and it was suggested as an explanation for the observed cutoff at ~ 10 MeV in the spectrum of B1509-58 [42].

However, in their modern versions the accelerators of particles are strong enough to outpower the ICS cooling. In consequence, very high Lorentz factors – $\gamma \gtrsim 10^6$ – are achieved by electrons, limited by CRV. The first detailed scenario of radiative processes in CRV-induced cascades was presented in Ref. [43] and despite many modifications and additions its basic features remain valid. The model assumes that primary electrons accelerated to ultrarelativistic energies emit curvature photons which in turn are absorbed by the magnetic field and $e^{\pm}$-pairs are created. These pairs cool off instantly via synchrotron radiation process. Whenever the SR photons are energetic enough they may lead to further creation of pairs, etc.. ICS can still be incorporated to the models with CRV-induced cascades as the process relevant for $e^{\pm}$-pairs, since typical Lorentz factors of theirs do not exceed $\sim 10^3$. According to the analytical model of Ref. [44] the empirical relations for X-ray and γ−ray luminosities of pulsars (presented in Sect.3) can be reproduced satisfactorily when the ICS involving $e^{\pm}$-pairs is included.

Processes relevant for production and transfer of HE radiation in pulsar magnetospheres are, therefore:
- Curvature Radiation,
- Magnetic Inverse Compton Scattering,
- Magnetic Pair Creation $\gamma + B \rightarrow e^{\pm}$
- Synchrotron Radiation,
- Photon splitting $\gamma \rightarrow \gamma + \gamma$,
- Photon-photon Pair Creation $\gamma + \gamma \rightarrow e^{\pm}$.

Basic properties of these processes are briefly reviewed below. The last process in the list has been omitted. The reason is that, whenever recalled in the context of pulsar magnetospheres, photon-photon pair creation is treated exactly as in free space. Such treatment is justified in models of "thick outer gaps" [3] but within the framework of polar cap models this is not the case, in general. However, no handy formula is available for the cross section of this process in the limit of high B and standard non-magnetic formulae are in use, e.g. Ref. [45].

In order to illustrate the significance of these processes in forming HE spectra extending over many decades in energy, the numerically calculated effects due to the first 4 processes in the list will be presented in Fig. 6 (after Ref. [46]) along with overlaid data points for the Vela pulsar. The dipolar field in the Vela pulsar does not exceed 10^{13}G and thus photon splitting is not competitive to magnetic pair creation; its effects are negligible and will not shown. The electric field structure of the accelerator used in these calculations is taken after Ref. [38] and appropriate rescaling due to $h_0 > 0$. However, the case calculated for Fig. 6 was chosen with $h_0 = 3R_\mathrm{s}$, i.e. much higher than in Ref. [38], in order to better reproduce the phase averaged spectrum of the Vela pulsar.

6.1 Curvature Radiation

A relativistic electron of energy γmc^2 (we take $\gamma \gg 1$) sliding along the magnetic field line of curvature ϱ_cr will emit photons with a continuum energy spectrum peaked at

$$\varepsilon_\mathrm{peak} \simeq 0.29\,\varepsilon_\mathrm{cr}\,, \tag{22}$$

where

$$\varepsilon_\mathrm{cr} = \frac{3}{2}\,c\,\hbar\,\frac{\gamma^3}{\varrho_\mathrm{cr}}, \tag{23}$$

is called the characteristic energy of CRV. The radius ϱ_cr for a purely dipolar line attached to the outer rim of the polar cap can be approximated not far away from the NS surface as $\varrho_\mathrm{cr} \simeq \sqrt{R_\mathrm{s} \cdot R_\mathrm{lc}} \simeq 10^8\sqrt{P}$cm. The cooling rate of that electron is

$$\dot{\gamma}_\mathrm{cr} = -\frac{2}{3}\,\frac{e^2}{mc}\,\frac{\gamma^4}{\varrho^2_\mathrm{cr}}. \tag{24}$$

For a monoenergetic injection function of electrons $Q(\gamma) \propto \delta(\gamma - \gamma_0)$ and their cooling due solely to CRV the electrons will assume a single power-law distribution in energy space $N_\gamma(\mathrm{el.}) \propto \gamma^{-4}$ for $\gamma < \gamma_0$ as long as they stay within the region of the cooling. Their escape introduces a natural low-energy cutoff γ_cutoff in $N_\gamma(\mathrm{el.})$. Therefore, the unabsorbed CRV energy spectrum $f_\varepsilon(\varepsilon)$ due to the injected electrons has a broken power-law shape, with a high-energy limit set by γ_0 and the break at some energy $\varepsilon_\mathrm{break}$. For $\varepsilon > \varepsilon_\mathrm{break}$ the energy spectrum is $f_\varepsilon(\varepsilon) \propto \varepsilon^{-2/3}$, and $f_\varepsilon(\varepsilon) \propto \varepsilon^{+1/3}$ for $\varepsilon < \varepsilon_\mathrm{break}$. Since nonthermal spectra cover ususally many decades in energy it is more convenient to use $\varepsilon f_\varepsilon(\varepsilon)$ for easy comparison of power in different parts of energy space (see Figs. 5 and 6). Accordingly, $\varepsilon f_\varepsilon(\varepsilon) \propto \varepsilon^{+1/3}$ above the break, and $\propto \varepsilon^{+4/3}$ below the break.

The cutoff limit $\gamma_{\rm cutoff}$ can be found by comparing the characteristic cooling time scale $t_{\rm cr} \equiv \gamma/|\dot{\gamma}_{\rm cr}|$ with the estimated time of escape $t_{\rm esc}$, which we take as $t_{\rm esc} \simeq \varrho_{\rm cr}/c$. Therefore, using Eqs. (22), (23), and (24),

$$\varepsilon_{\rm break} \simeq 0.29 \cdot \frac{9}{4}\,\hbar\,\frac{c}{r_0} \simeq 44 \text{ MeV}, \tag{25}$$

where r_0 is the classical electron radius. Note, that the photon energy $\varepsilon_{\rm break}$ at which the spectral break occurs does not depend on any pulsar parameters.

The spectrum of CRV calculated numerically to model the Vela pulsar [46] is shown in Fig. 6 as the dot-dashed line. A high-energy cutoff due to one-photon magnetic absorption occurs around 10 GeV. Note the importance of $\gamma-$ray detectors capable to operate above 10 GeV for (in)validating the model. The low-energy CRV spectral break $\varepsilon_{\rm break}$ is prominent at $\sim$ 40 MeV. Below $\varepsilon_{\rm break}$ the power of CRV decreases and eventually becomes unimportant at $\sim$ 1 MeV where the synchrotron component takes over.

6.2 Magnetic Pair Creation

Pair creation via magnetic photon absorption ($\gamma + \boldsymbol{B} \rightarrow e^{\pm} + \boldsymbol{B}$) is kinematically correct since the magnetic field can absorb momentum. To ensure high probabilities for the process to occur it is not enough for a photon propagating at a pitch angle ψ to the local $\boldsymbol{B}$ to satisfy the energy threshold condition, $\sin\psi \cdot \varepsilon \geq 2mc^2$, but high optical thickness $\tau_{\gamma B}$ within the magnetosphere is required. In fact the condition $\tau_{\gamma B} = 1$ has been used as a criterium for the so called death-line for radiopulsars in the $P - \dot{P}$ diagram. Maximal values of $\sin\psi$ for curvature photons in the dipolar field do not exceed $\sim 0.1 \sin\theta_{\rm pc} \simeq 0.0014\, P^{-1/2}$ [34], so both conditions are difficult to meet for long-period rotators.

The absorption coefficient for the process as described in Ref. [47] and used to calculate $\tau_{\gamma B}$ reads

$$\eta(\varepsilon) = \frac{1}{2}\frac{\alpha}{\lambda\!\!\!^{-}_{\rm c}}\frac{B_\perp}{B_{\rm crit}} T\,(\chi) \tag{26}$$

where α is the fine structure constant, $\lambda\!\!\!^{-}_{\rm c}$ is the Compton wavelength, $B_{\rm crit} = m^2c^3/e\hbar \simeq 4.4\times10^{13}$G, $B_\perp$ is the component of the magnetic field perpendicular to the photon momentum, and $\chi \equiv (B_\perp/B_{\rm crit})\varepsilon/(2m_ec^2)$ is the Erber parameter. The function $T(\chi)$ is then approximated as $T(\chi) \simeq 0.46 \exp\left(-4f/3\chi\right)$, which is valid for $\chi \lesssim 0.2$; for $\chi \gtrsim 0.2$ this approximation starts to overestimate η. The function f is the near-threshold correction introduced in Ref. [48] important in particular in the case of classical pulsars. Electron-positron pairs created through magnetic absorption radiate then via the synchrotron process (SR) described below.

6.3 Synchrotron Radiation

Consider a particle of energy γmc^2 gyrating around a local field line at a pitch angle ψ. Let $\gamma_\parallel$ denote the Lorentz factor of the reference frame comoving with the center of the gyration. As long as $\gamma_{\|} \gg 1$ it relates to the pitch angle ψ via $\sin\psi \simeq \gamma_{\|}^{-1}$. The energy available for synchrotron emission at the expense of the particle is $\gamma_{\perp} mc^2$, and $\gamma = \gamma_{\perp}\,\gamma_{\|}$.

The rate of SR cooling reads

$$\dot{\gamma}_{\rm sr} = -\frac{2}{3}\frac{r_0^2}{m_{\rm e}c}\,B^2\gamma_{\perp}^2 = -\frac{2}{3}\frac{r_0^2}{m_{\rm e}c}\,B^2\sin^2\psi\,\gamma^2. \tag{27}$$

In comparison to the CRV cooling it is enormous (due to much smaller curvature radius).

The critical photon energy [analogous to (24)] reads

$$\varepsilon_{\rm sr} = \frac{3}{2}\hbar\,\frac{eB}{m_{\rm e}c}\gamma^2\sin\psi. \tag{28}$$

For a monoenergetic injection function of particles ($e^{\pm}$-pairs in the context of this review) and their cooling due to SR the energy spectrum of SR spreads between a high-energy limit $\varepsilon_{\rm sr}(\gamma_0)$ set by γ_0 of the injected (created) particles, and a low-energy turnover $\varepsilon_{\rm ct}$ determined by the condition $\gamma_{\perp} \sim 1$:

$$\varepsilon_{\rm ct} \equiv \varepsilon_{\rm sr}(\gamma = \gamma_{\|}) = \frac{3}{2}\hbar\,\frac{eB}{m_{\rm e}c}\,\frac{1}{\sin\psi}. \tag{29}$$

The spectrum assumes a single power-law shape $f_\varepsilon(\varepsilon) \propto \varepsilon^{-1/2}$ (and accordingly – $\varepsilon f_\varepsilon(\varepsilon) \propto \varepsilon^{+1/2}$) above the turnover. Below $\varepsilon_{\rm ct}$, the spectrum f_ε changes it slope, asymptotically reaching $\propto \varepsilon^{+2}$. It is built up by contributions from low-energy tails emitted by particles with $\gamma_{\perp} \gg 1$, and each low-energy tail is assumed to cut-off at the local gyrofrequency, which in the reference frame comoving with the center of gyration is $\omega_B = eB/(m_{\rm e}c\,\gamma_{\perp})$.

The spectrum of SR calculated with Monte-Carlo method to model the Vela pulsar is shown in Fig. 6 as a dashed line. The low-energy part of the SR spectrum at $\varepsilon_{\rm ct}$ seems to be essential for reproducing an interpolation between the RXTE and the OSSE data.

6.4 Magnetic Inverse Compton Scattering

Consider an electron with a Lorentz factor γ moving along a magnetic field line $\boldsymbol{B}$ and a photon of energy $\varepsilon = \epsilon mc^2$ moving at angle $\arccos\mu$ to the field line. In the reference frame comoving with the electron (primed symbols) the counterpart of the free-space Compton formula, due to energy-momentum conservation appropriate for collisions with the electron at the ground Landau level both in the initial and final state, reads

$$\epsilon'_s = \left(1 - {\mu'_s}^2\right)^{-1}\Big\{1 + \epsilon'(1 - \mu'\mu'_s) + \\ - \left[1 + 2\epsilon'\mu'_s(\mu'_s - \mu') + {\epsilon'}^2(\mu'_s - \mu')^2\right]^{1/2}\Big\} \tag{30}$$

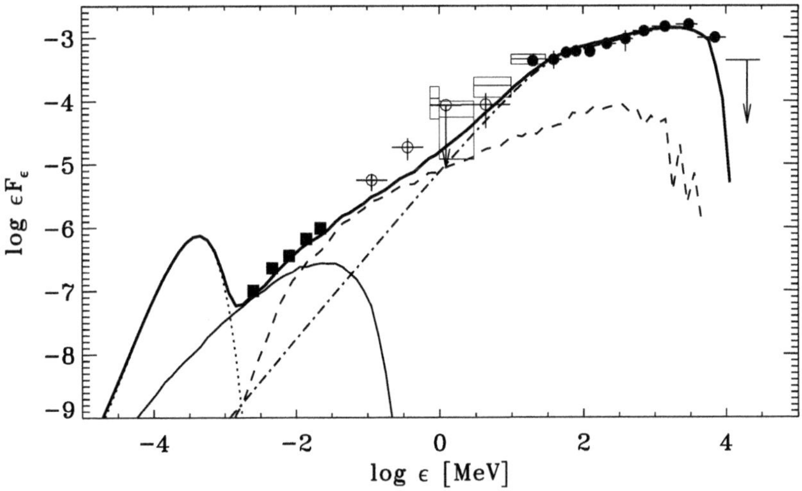

Fig. 6. The model energy spectrum calculated in Ref. [46] to reproduce the spectral features of the Vela pulsar ($P = 89$ ms, $B_s = 6 \times 10^{12}$G). The accelerator is located at $h_0 = 3R_s$ above the surface (see Sect.6). The broad–band spectrum consists of four components due to: curvature radiation of primary electrons (dot-dashed), synchrotron radiation of secondary $e^{\pm}$–pairs (dashed), inverse Compton scattering of surface X-ray photons on the $e^{\pm}$–pairs (thin solid) and the blackbody surface emission (dotted). A surface temperature $T_s = 1.26 \times 10^6$K was assumed. The total spectrum is given by the thick solid line. Phase-averaged data points for Vela from different satellite experiments are indicated. Filled squares – RXTE [49]; open circles – OSSE [50]; open squares – COMPTEL [26]; filled circles plus upper limit just above 10 GeV – EGRET [51]. Vertical axis is in log of MeV cm^{-2} s^{-1} units.

where $\epsilon' = \epsilon\gamma(1 - \beta\mu)$ [52], and symbols with no subscript and with the subscript s refer to the state before the scattering and after the scattering, respectively. A longitudinal momentum of the electron in the electron rest frame changes due to recoil from zero to $(\epsilon'\mu' - \epsilon'_s\mu'_s)mc$.

The polarization-averaged relativistic magnetic cross section in the Thomson regime may be approximated with a nonrelativistic formula [53]:

$$\sigma = \frac{\sigma_T}{2}\left(1 - \mu'^2 + (1 + \mu'^2)\left[g_1 + \frac{g_2 - g_1}{2}\right]\right) \tag{31}$$

where σ_T is the Thomson cross section, and g_1 and g_2 are given by

$$g_1(u) = \frac{u^2}{(u+1)^2}, \qquad g_2(u) = \frac{u^2}{(u-1)^2 + a^2} \tag{32}$$

where $u \equiv \epsilon'/\epsilon_B$, $a \equiv 2\alpha\epsilon_B/3$, and $\epsilon_B \equiv \hbar eB/m^2c^3$.

The resonance condition for the scattering is therefore the cyclotron resonance $\epsilon' = \epsilon_B$. The factor a represents a "natural" broadening of the resonance due to finite lifetime at the excited Landau level.

In the Klein-Nishina regime ($\epsilon' > 1$) the relativistic magnetic cross section for the $|\mu'| \simeq 1$ case becomes better approximated by the well known Klein-Nishina relativistic nonmagnetic total cross section σ_{KN} [54,53].

The rate $\mathcal{R}$ of scatterings subject to an electron moving across the field of soft photons, measured in the lab frame is

$$\mathcal{R} = c \int d\Omega \int d\varepsilon \; \sigma \left(\frac{dn_{\rm ph}}{d\varepsilon d\Omega} \right) (1 - \beta\mu) \tag{33}$$

where $\Omega = d\mu d\phi$ is the total solid angle subtended by the source of soft photons, $\mu = \cos\theta$, σ is a total cross section for the process, and $dn_{\rm ph}/d\varepsilon/d\Omega$ is the local density of the soft photons.

The properties of the field of soft photons are usually simplified by taking $dn_{\rm ph}/d\varepsilon/d\Omega$ as for the blackbody radiation. This simplification should be taken with care since magnetised atmospheres of neutron stars introduce strong anisotropy as well as spectral distortions to the outgoing radiation [55]. Effectively it means that the ICS effects obtained with this simplification are just upper limits to the actual effects.

To estimate the electron cooling rate $\dot{\gamma}_{\rm ICS}$ due to the ICS the differential form of (31) is necessary:

$$\begin{aligned} \frac{d\sigma}{d\Omega'_s} = \frac{3\sigma_T}{16\pi} \Big[&(1 - \mu'^2)(1 - {\mu'_s}^2) + \\ &+ \frac{1}{4} (1 + \mu'^2)(1 + {\mu'_s}^2)(g_1 + g_2) \Big] \end{aligned} \tag{34}$$

(e.g. Ref. [56]), where $d\Omega'_s = d\phi'_s d\mu'_s$ is an increment of solid angle into which outgoing photons with energy ϵ'_s in the electron rest frame are directed. The mean electron energy loss rate then reads

$$\begin{aligned} \dot{\gamma}_{\rm ICS} = -c \int d\epsilon \int d\Omega \left(\frac{dn_{\rm ph}}{d\epsilon d\Omega} \right) (1 - \beta\mu) \times \\ \times \int d\Omega'_s \left(\frac{d\sigma}{d\Omega'_s} \right) (\epsilon_s - \epsilon) \end{aligned} \tag{35}$$

where $\epsilon_s = \epsilon'_s \gamma (1 + \beta\mu'_s)$ is the scattered photon energy in the lab frame (e.g. Ref. [53]).

The spectrum of magnetic ICS calculated numerically to model the Vela pulsar is shown in Fig. 6 with thin solid line. The blackbody soft photons originating at the stellar surface (dotted line) are upscattered by secondary $e^{\pm}$-pairs at the expense of their "longitudinal" energy $\gamma_{||} mc^2$, assumed to remain unchanged during the burst of synchrotron emission. Without the ICS component due to the $e^{\pm}$-pairs the RXTE data for the Vela pulsar would be difficult to reproduce within the model. It is worth to note that the magnetic ICS component due to primary electrons (not shown in Fig. 6) is energetically insignificant comparing to the CRV component because the case of strong accelerating field was used in this particular model.

6.5 Photon Splitting

Photon splitting into two photons in the presence of magnetic field B is a third-order QED process with no energy threshold [57]. The attenuation coefficient, after averaging over the polarization states, reads [30]

$$T_{\rm split}(\epsilon) \simeq \frac{\alpha^3}{10\pi^2}\frac{1}{\lambda\!\!\!^{-}_{\rm c}}\left(\frac{19}{315}\right)^2\left(\frac{B\sin\theta_{\rm kB}}{B_{\rm crit}}\right)^6 \epsilon^5\ {\rm cm}^{-1}, \tag{36}$$

(provided B does not exceed $B_{\rm crit}$ substantially) where ϵ is the photon energy in units of mc^2, and $\theta_{\rm kB}$ is the angle between photon momentum vector and the local magnetic field. The process, therefore, strongly depends on magnetic field strength B.

Photon splitting has attracted substantial interest in recent years due to the discovery of neutron stars with supercritical magnetic fields (i.e. magnetars; see Fig. 6) [58]. It has been analysed in details in Ref. [30] and incorporated in a Monte Carlo code tracing the propagation of electromagnetic cascades in the magnetospheres of high-B pulsars. The effect was found to explain satisfactorily the unusual cut-off observed in the $\gamma-$ray spectrum of B1509-58 (see Sect.3). Generally, it becomes competitive to the magnetic pair creation for dipolar magnetospheres with $B_{\rm s} \gtrsim 0.3B_{\rm crit}$ [30]. The degradation of photon energy in the course of splitting inhibits also any development of electromagnetic cascades. In consequence, high-B RPP should not emit coherent radio emission. Indeed, there exists a high-B region in the $P - \dot{P}$ diagram (Figs. 1,6) void of radiopulsars. Even though two recently discovered (during The Parkes Multibeam Pulsar survey) high-B radiopulsars [59] are located above the limiting line derived in Ref. [60], the general argument for magnetars expected to be radio-quiet RPP remains valid [61].

7 RPP as Sources of UHECR Generated Beyond the Light Cylinder

The hypothesis that cosmic ray events above $\sim 5 \times 10^{19}$eV, i.e. above the GZK cutoff, are due to charged particles accelerated by strongly magnetised neutron stars is being kept under consideration [62,63,64] apparently for two reasons. First, within bottom-up scenarios which rely on conventional physics a list of classes of objects satisfying a necessary condition to generate such cosmic rays is rather short according to the appealing Hillas diagram [65], see the contribution by G. Pelletier in this volume. Second, no compelling breakthrough has been achieved in promoting other candidates in this context, like central engines of AGN (which suffer from substantial energy losses of the particles due to pion production in the intense radiation fields) or jets extending from Faranoff-Riley II radio galaxies. For a more detailed discussion of these acceleration scenarios see the contributions by G. Pelletier and P. Biermann and G. Sigl in this volume.

In the case of neutron stars the most promising and natural reservoir of required energy is their rotational energy, a point reiterated on numerous occasions, e.g. Ref. [66], and assumed in most models. There are three fundamental features discriminating the models:
1) the site of acceleration with respect to the neutron star,
2) the mechanism allowing to tap the rotational energy by charged particles,
3) the nature and origin of the charged particles subject to acceleration.

Simple but sound arguments supported by observed HE-radiation properties of RPP make the first point rather clear: the process of particle acceleration into the UHE domain should take place beyond the light cylinder. The potential advantage of this choice over acceleration process within the light cylinder is twofold. First, the full potential drop across open field lines $\Delta\Phi_{\rm pc}$ (14) is available, at least in principle, for particles outside the light cylinder, whereas the capability of an accelerator inside the magnetosphere is severely constrained by copious formation of electron-positron pairs which short out the electric field $\mathcal{E}_{\|}$ easily, i.e. $\Delta\Phi_{\|} < \Delta\Phi_{\rm pc}$, see Eq. (20). Second, unlike inside the magnetosphere, the acceleration of charged particles is not limited by any radiation losses – a point especially important in the context of UHECR. Therefore, a particle of charge Ze reaches the maximal possible energy

$$E_{\rm max} \simeq Ze\,\Delta\Phi_{\rm pc} = 6 \times 10^{19}\, Z\, B_{13}\, P_{\rm ms}^{-2}\, {\rm eV}, \tag{37}$$

(where $B_{13} \equiv B_{\rm s}/10^{13}{\rm G}$) which is substantially higher than the energy $E_{\rm p}$ which can be attained in the accelerating field of either the polar gap or outer gap: $E_{\rm p} \ll Ze\,\Delta\Phi_{\|} < E_{\rm max}$.

The idea of UHECR events due to an accelerator or a converter of the rotational energy into kinetic energy of particles, located beyond the light cylinder has been pursued recently in a couple of diametrically different models. Historically the first model considered the fate of charged particles injected into the plerionic nebula powered by a shocked relativistic wind from a central pulsar [67,68]. This model was motivated by the theoretical analysis of magnetohydrodynamics within the Crab Nebula [69,70,71]. The acceleration was proposed to occur in the electric field of the shocked wind. A simple structure of the electric field was derived as induced by the relativistic radial wind crossing a toroidal magnetic field (which originates inside the light cylinder) with assumed radial and angular dependence after Ref. [69] and Ref. [71]. Charged particles entering the nebula are subject to the $\boldsymbol{\mathcal{E}} \times \boldsymbol{B}$ drift as well as to the ∇B drift (due to strong inhomogeneity in $\boldsymbol{B}$) along various paths, and may gain energy at the expense of $\boldsymbol{\mathcal{E}}$ before exiting. Since the electric field is potential here, the net change in the particle energy does not depend on the path but solely on the points of entry and exit. In particular, the maximal gain ΔE of energy is due to the potential difference between the pole (fixed by the rotation axis of the pulsar) and the equator (cf. Eq. (3) in Ref. [68]), and actually it equals $E_{\rm max}$ given by Eq. (37). In order for the particle to penetrate the nebula from outside it should be already highly relativistic, with $E_{\rm init} \sim 10^{15}$eV, presumably pre-accelerated by diffusive acceleration at the outer shock where the supernova remnant meets the interstellar medium. Otherwise the Larmor radius of the particle is too small for the ∇B inward drift at the pole (the case relevant for one combination of signs of $\boldsymbol{B}$ and Ze) to counter the $\boldsymbol{\mathcal{E}} \times \boldsymbol{B}$ outward drift. Though very attractive, the model is unable to make any strong predictions about spectral properties of its UHECR without reliable assumptions about the spatial distribution and directional properties of the pre-accelerated particles.

All other models make use of charged particles, either protons or iron nuclei, going from inside of the light cylinder i.e. supplied by a neutron star itself. A recently proposed phenomenological model of UHECR within the Galaxy incorporates iron Fe^{26} nuclei ($Z = 26\, Z_{26}$) accelerated in a relativistic MHD wind flowing out of very young, rapidly rotating highly magnetised neutron stars [72]. Let the number of nuclei crossing the light cylinder per unit time is equal to the Goldreich–Julian rate at the polar cap

$$\dot{N}(\mathrm{Fe}^{26}) \simeq \dot{N}_{\mathrm{GJ}} \simeq A_{\mathrm{pc}} \frac{\varrho_{\mathrm{GJ}}}{Ze}\, c, \tag{38}$$

where $A_{\mathrm{pc}} \simeq \pi R_{\mathrm{s}}^3 R_{\mathrm{lc}}^{-1}$ and ϱ_{GJ} is defined in (18). Suppose now that the electron-positron pairs formed within the magnetosphere do not dominate the flow in terms of the rest mass. This is a reasonable assumption because otherwise the number of created $\mathrm{e}^{\pm}$ pairs per nucleus would have to exceed $m_{\mathrm{Fe}}/2m_{\mathrm{e}} \simeq 5 \times 10^4$ which is unlikely in very strong magnetic fields $B_{\mathrm{s}} > 10^{13}$G due to photon-splitting effects [30]. A key postulate of the model is that somewhere beyond the light cylinder (i.e. in the MHD-wind zone) a sizeable fraction ($\lesssim 1$) of the spindown flux is transferred into the kinetic energy flux of the ions, producing thus ultra-high energy ions of the UHECR. At the light cylinder, the spindown flux (i.e. the rotational energy loss rate per unit area) is dominated by the Poynting energy flux (due to the toroidal magnetic field component induced by the outflowing particles). In terms of the so called magnetization parameter σ_{m} which is defined as the ratio of the Poynting energy flux to the particle kinetic energy flux [71] this means that for some reason σ_{m} does not remain constant in the wind: $\sigma_{\mathrm{m}} > 1$ at the light cylinder converts further out to $\sigma_{\mathrm{m}} \ll 1$, i.e. the wind must depart from an ideal MHD flow case.

This picture is motivated again by the wind models for the Crab Nebula (e.g. Ref. [69,73,70]; see also Ref. [74] for calorimetric properties of the Crab Nebula, and Ref. [75] for an account of the problem of the coupling of RPP to plerionic nebulae) which give $\sigma_{\mathrm{m}} \simeq 0.003$ at $r \simeq 0.1$pc. No consensus about likely mechanisms responsible for the dissipation of the Poynting flux has been reached so far, though several models have been proposed (e.g. Ref. [71,76]). For a single iron nucleus the postulated conversion means acceleration up to

$$E_{\mathrm{UHECR}} \simeq \frac{L_{\mathrm{sd}}}{\dot{N}\,(\mathrm{Fe}^{26})} \simeq Ze\, \Delta\Phi_{\mathrm{pc}} \simeq 10^{21} Z_{26}\, B_{13}\, P_{\mathrm{ms}}^{-2}\, \mathrm{eV}, \tag{39}$$

where Eqs. (38), (6), (18), (12), and (14) were used, with R_{lc} and R_{pc} as given at the end of section 2. Similarly, therefore, as in the previous model, the energy gain is in fact directly related to the full potential drop $\Delta\Phi_{\mathrm{pc}}$ across the open

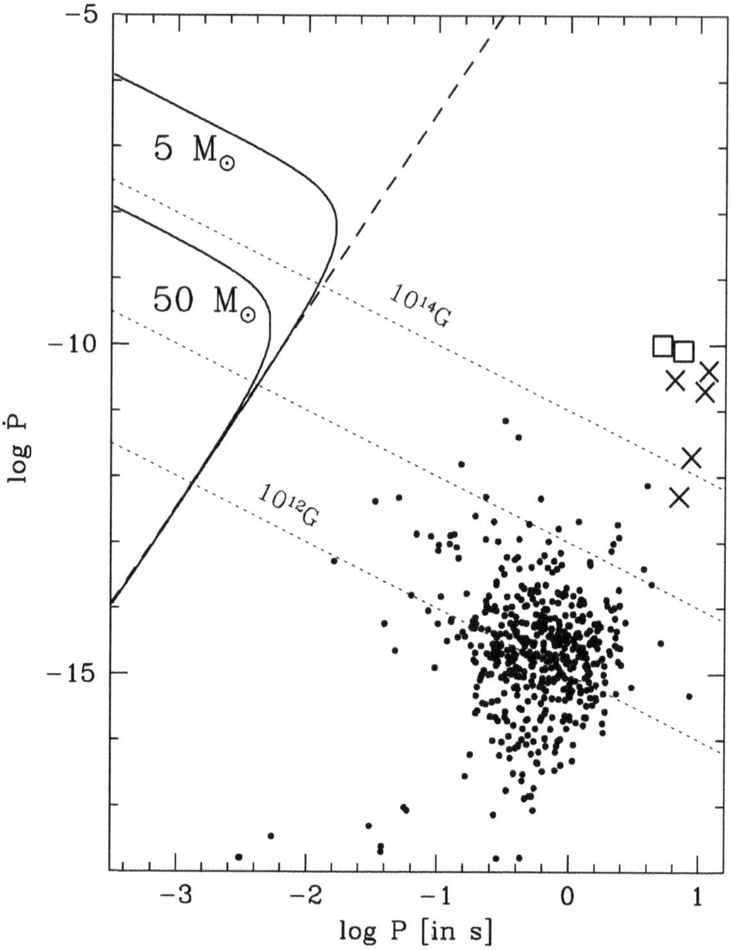

Fig. 7. The high–B_s part of the $P - \dot{P}$ diagram (see Fig. 2). Dashed line indicates the full potential drop $\Delta\Phi_{pc}$ across open field lines, Eq. (14), equal to $\frac{1}{26} \times 10^{20}$V. The allowed region in the model of Ref. [72] to accelerate a fully ionized atom of iron ($Z = 26$) to the energy of 10^{20}eV and let it traverse the pre-supernova envelope without spallation effects lies to the left of two solid lines which are drawn for two values of the mass of the envelope: $5M_\odot$ and $50M_\odot$. A group of magnetars, clustered around $P \sim 10$s is also indicated after Ref. [58] although their X-ray activity is not driven by rotational energy losses (i.e. they do not belong to the class of RPP) but rather involves accretion or the decay of strong magnetic field. Open squares and crosses denote SGR and AXP, respectively.

field lines. For fully ionized iron ($Z_{26} = 1$) to reach $E_{\rm UHECR} = 10^{20}$eV the pulsar would have to be fastly spinning and highly magnetised (for example $P = 0.001$s and $B_{\rm s} = 10^{12}$G). According to Ref. [72], the region in the $P - B_{\rm s}$ space occupied by pulsars satisfying the requirement $E_{\rm UHECR} = 10^{20}$eV is actually even more constrained than (39) suggests. The magnetic field of the pulsar should not be too strong to make sure that the characteristic time scale of spin-down τ is not too short allowing thus the expanding pre-supernova envelope of mass $M_{\rm env}$ to disperse and become transparent so the iron nuclei would escape it without significant spallation. For $M_{\rm env}$ equal $5M_\odot$ and $50M_\odot$, the magnetic field $B_{\rm s}$ should not exceed $\sim 6 \times 10^{13}$G and $\sim 6 \times 10^{14}$G, respectively (see Fig. 7).
A factor not discussed in Ref. [72] but likely to further constrain the allowed region shown in Fig. 7 to the values of $B_{\rm s}$ not exceeding $\sim 10^{13}$G is the supply of iron nuclei by the neutron star surface: the binding energy $\Delta\varepsilon_c$ of iron increases with increasing $B_{\rm s}$ and for $B_{\rm s} > 10^{13}$G it may exceed 5 keV [77]. Stripping the iron nuclei from the surface at the needed rate ($\dot{N}({\rm Fe}^{26}) \simeq \dot{N}_{\rm GJ}$) would then require a very hot surface (so called "thermionic emission"), with temperature $T_{\rm s} > 30 \cdot \Delta\varepsilon_c/k \simeq 3 \times 10^6$K (according to e.g. Ref. [78,77]; k is Boltzmann's constant) during the first $\sim 10^8$ seconds after the neutron star formation. It is not obvious whether such a requirement is satisfied for $B_{\rm s} > 10^{13}$G, since numerical calculations of the thermal history of magnetised neutron stars are confined to the canonical case of $B_{\rm s} = 10^{12}$G [79].
Let ζ denote a fraction of iron nuclei in the MHD-wind which are subject to acceleration up to $E_{\rm UHECR}$ of (39), and let $\epsilon_{\rm psr}$ denote a fraction of all pulsars in our Galaxy (produced at a rate ~ 0.01 per year) which meet all discussed requirements for producing UHECR. For the flux of UHECR at 10^{20}eV observed by AGASA (at the level of $4 \times 10^{-30}{\rm GeV}^{-1}{\rm cm}^{-2}{\rm s}^{-1}$) to be explained with newly born pulsars within our Galaxy it suffices that the following condition is satisfied: $\epsilon_{\rm psr}\zeta \gtrsim 4 \times 10^{-6}$ (assuming no confinement of the UHECR particles, otherwise the condition is even less stringent). The calculated particle spectrum above 5×10^{19}eV is flat, $N_{\rm Fe} \propto E^{-\gamma}$ with $\gamma = 1$, and therefore should be visible as a distinct component in the CR spectrum, which is much steeper below 5×10^{19}eV with the power-law index $\gamma \simeq 3$. However, the number of events above 5×10^{19}eV detected so far is insufficient for describing them in terms of spectral features [80].

For a pulsar capable of producing UHECR as in Ref. [72], a natural continuation – as it slows down its rotation and becomes a Crab-like object – would be to generate CR particles of lower energy. The hadronic model proposed in Ref. [81] seems to be relevant in this context, since it describes the fate of iron nuclei extracted from the pulsar surface and then subject to acceleration. However, a different acceleration site is considered here: the nuclei are assumed to accelerate within the light cylinder, in outer gaps of Ref. [1] (i.e. the type of accelerator not addressed in this review). Once the nuclei achieve high energy ($\gamma_{\rm Fe} > 10^5$) they disintegrate in the field of soft photons present in the outer gap. Relativistic neutrons extracted in this way decay into protons either inside or outside the surrounding nebula. In the first case, these relativistic protons interact with nebular material via *pp* processes giving rise to VHE photons and neutrinos. In the second case, cosmic ray protons peaking at $\sim 10^{15} - 10^{16}$ eV (close to the knee in the CR spectrum) are produced.

A distinct model of UHECR particles above the GZK cutoff has been recently proposed [82]. It postulates that UHECR are protons accelerated in strong electric fields induced by magnetic field reconnections occuring above the magnetospheres of newly born pulsars formed via accretion-induced collapse of white dwarfs (AIC). The reconnections sites – magnetic null surfaces with sharp reversal of the field across them – are expected to form due to interaction between three magnetic field ingredients: the dipolar magnetic field of the star, magnetic field lines contained in the flow from the inner rim of the accretion disk, and open lines in the coronal wind zone. One surface, called "helmet streamer", is located well off the accretion disk plane. It is accompanied by another null surface, called "reconnection ring", located nearby the inner rim of the disk. Particles (in this particular situation – protons, $Z = 1$) may then reach energies of magnitude

$$E_{\rm UHECR} \simeq Ze\, B_{\rm x}\, \Delta R_{\rm x} \simeq eB_{\rm s} \left(\frac{R_{\rm s}}{R_{\rm x}}\right)^3 \left(\frac{R_{\rm x}}{\Delta R_{\rm x}}\right)^{1/2} \Delta R_{\rm x}$$
$$\simeq Ze\, B_{\rm s} R_{\rm s}^3 \frac{\Omega^{4/3}}{(GM_{\rm s})^{2/3}} \left(\frac{\Delta R_{\rm x}}{R_{\rm x}}\right)^{1/2}, \quad (40)$$

where $B_{\rm x}$ is the magnetic field normal to the particle trajectory at the reconnection site of length $\Delta R_{\rm x}$, G is the gravitational constant, and $R_{\rm x}$ is the radial distance of the site from the star, assumed to be equal to the inner disk radius where the Keplerian disk corotates with the star. Eq. (40) gives, therefore,

$$E_{\rm UHECR} \simeq 3 \cdot 10^{20} Ze\, B_{13}\, P_{\rm ms}^{-4/3} \left(\frac{\Delta R_{\rm x}/R_{\rm x}}{0.1}\right)^{1/2} \,\text{eV} \quad (41)$$

for $M_{\rm s} = 1.4 M_{\odot}$, and $R_{\rm s} = 10^6$cm, while the actual value of $\Delta R_{\rm x}/R_{\rm x}$ is not known. Promising candidates among newly born pulsars would occupy roughly the same region of the $P - \dot{P}$ space as indicated in Fig. 7 for the model of Ref. [72]. Energy losses of the protons due to curvature radiation, as well as due to photopion and $e^{\pm}$ production in the radiation field of the accretion disk, are estimated to be unimportant within the helmet streamer. For the reconnection ring, however, the ambient radiation field is too thick optically for any ultrarelativistic protons to escape from this accelerator. Since the estimated rate of AIC events per galaxy is too low to rely just on our Galaxy, the contribution from all galaxies within the distance of about 50 Mpc (to avoid the GZK effect) should be considered. In order to reach the observed rate of UHECR events it is then necessary that the efficiency of converting magnetic energy into ultrarelativistic particle energy is $\gtrsim 0.1$.

8 Concluding Remarks

High energy astrophysics of neutron stars received an impressive boost from two major satellite missions of the past – ROSAT and CGRO, backed by still ongoing experiments – ASCA, BeppoSAX and RXTE. Theoretical astrophysics of

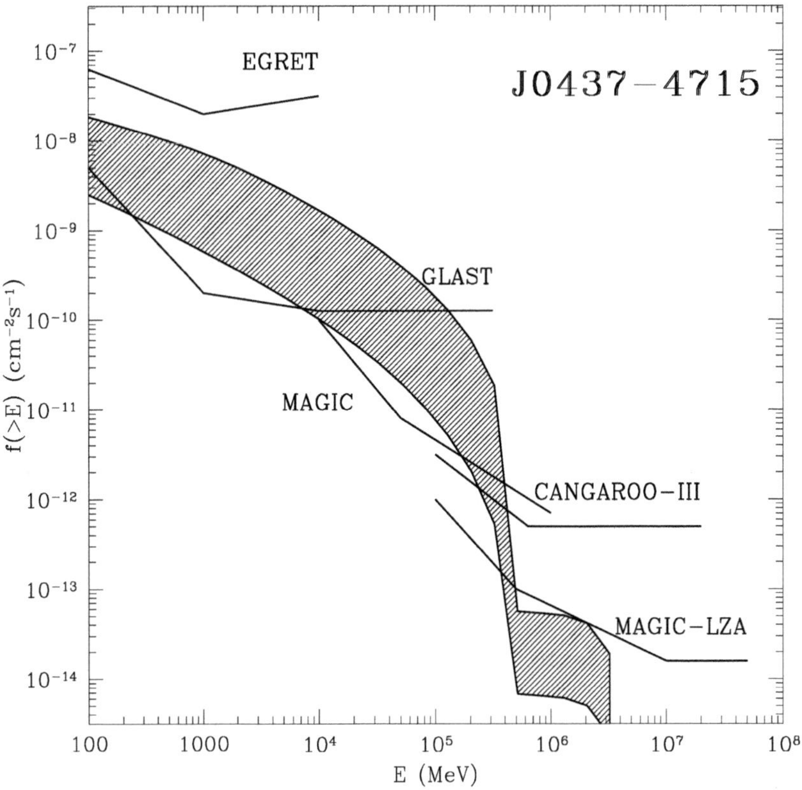

Fig. 8. Cumulative spectral flux of photons expected for the millisecond pulsar J0437-4715 at a distance of 140 pc according to Ref. [83]. The shaded region shows the range of flux levels due to uncertainity in the maximal energy of primary electrons. The main part of the spectrum is due to curvature radiation of the electrons. The additional feature reaching the VHE domain is due to inverse Compton scattering of soft photons from the surface with the temperature 4×10^5K. Sensitivities of EGRET as well as three major HE and VHE future experiments are also indicated. MAGIC–LZA denotes sensitivity of MAGIC in its Large Zenith Angle mode.

RPP was confronted with– and surprised by an unprecedented variety of spectral and temporal properties among the detected sources. Another unexpected challenge came from radio-astronomy, due to the superb performance of The Parkes Multibeam Pulsar Survey, with recent discoveries of radiopulsars, with extremely high magnetic fields in two cases and an extremely long spin period in another case. Numerous modifications (both, minor and major) to the existing models of magnetospheric activity were invented to accomodate at least some of these properties. Several predictions have been presented which would hopefully discriminate between those models. It will be impossible, however, to verify those predictions without achieving higher sensitivities and exploring new energy domains.

A break-through in understanding rotation powered pulsars and their plerionic environments should then come from HE/VHE astronomy of the near future, with its planned satellite and ground-based experiments. Expected sensitivities and energy ranges for some of them are presented in Fig. 8 along with a predicted flux from a nearby millisecond pulsar, overlaid for the sake of comparison. The satellite experiment GLAST [84] will be superior to its predecessor – EGRET on board CGRO – in two aspects. First, its sensitivity at 10 GeV will be three orders of magnitude better than that of EGRET. Second, it will reach the energy of 300 GeV, closing thus for the first time a wide gap in energy between ground-based and satelite experiments. The MAGIC Telescope [85] – a 17 m diameter Imaging Air Čerenkov Telescope (IACT) – is expected to operate with sensitivity about three orders of magnitude higher at 10 GeV than EGRET. Its advanced technology will make it possible to cover the energy range between 10 GeV and 1 TeV, and to reach $\sim 50\,\mathrm{TeV}$ in the Large Zenith Angle mode. Energy ranges of GLAST and MAGIC will overlap over more than one decade in energy. Another proposed IACT, VERITAS [86], will be an array of seven 10 m telescopes, covering the energy range from 50 GeV to 50 TeV with planned sensitivity at 1 TeV about ten times better than MAGIC.

The anticipated progress in exploring the HE/VHE domain will likely decide as well whether the class of rotation powered pulsars and their plerions should be considered as interesting to the UHECR astrophysics.

Acknowledgments

I am grateful to J. Arons, W. Bednarek, T. Bulik, K. S. Cheng, J. Dyks, E. M. de Gouveia Dal Pino, M. Gros, P. Haensel, O. C. de Jager, A. V. Olinto, S. Shibata, T. C. Weekes, and B. Zhang for sharing with me their views and opinions on issues addressed in this review. Financial support by KBN grant 2P03D02117 is acknowledged.

References

1. K. S. Cheng, C. Ho, M. A. Ruderman: Astrophys. J. **300**, 500 (1986)
   K. S. Cheng, C. Ho, M. A. Ruderman: Astrophys. J. **300**, 522 (1986)
2. K. Chen, M. Ruderman: Astrophys. J. **402**, 264 (1993)
3. K. S. Cheng, L. Zhang: Fund. Cosm. Phys. **20**, 177 (1999)
   K. S. Cheng, L. Zhang: Astrophys. J. **515**, 337 (1999)
   L. Zhang, K. S. Cheng: Astrophys. J. **487**, 370 (1997)
4. F. C. Michel: *Theory of Neutron Star Magnetospheres* (The University of Chicago Press, Chicago 1991)
   P. Mészáros: *High-Energy Radiation from Magnetized Neutron Stars* (The University of Chicago Press, Chicago 1992)
5. P. Goldreich, W. H. Julian: Astrophys. J. **157**, 869 (1969)
6. A. K. Harding, I. Contopoulos, D. Kazanas: Astrophys. J. **525**, L125 (1999)
7. J. H. Taylor, R. N. Manchester, A. G. Lyne, F. Camilo: *The Princeton Pulsar Database* (see http://pulsar.princeton.edu) (1995)
   J. H. Taylor, R. N. Manchester, A. G. Lyne: Astrophys. J. Supp. **88**, 529 (1993)

8. F. E. Marshall, E. V. Gotthelf, W. Zhang, J. Middleditch, Q. D. Wang: Astrophys. J. **499** L179 (1998)
9. E. V. Gotthelf, G. Vasisht, M. Boylan-Kolchin, K. Torii: Astrophys. J. **542**, L37 (2000)
10. M. D. Young, R. N. Manchester, S. Johnston: Nature **400**, 848 (1999)
11. W. Becker, J. Trümper: Astron. Astrophys. **326**, 682 (1997)
W. Becker, J. Trümper: Astron. Astrophys. **341** 803 (1999)
12. H. Völk, HEGRA Collaboration: Proc. *GeV-TeV Gamma Ray Astrophysics Workshop : towards a major atmospheric Cherenkov detector VI*, eds. B. L. Dingus, M. H. Salamon, D. B. Kieda: AIP Conference Proceedings, Vol. **515**, p.197 (New York, 2000)
F. Aharonian, et al.: Astron. Astrophys. **346**, 913 (1999)
R. W. Lessard, et al.: Astrophys. J. **531**, L942 (2000)
13. T. Sako, et al.: Astrophys. J. **537**, 422 (2000)
14. T. C. Weekes: Proc. *GeV-TeV Gamma Ray Astrophysics Workshop : towards a major atmospheric Cherenkov detector VI*, eds. B. L. Dingus, M. H. Salamon, D. B. Kieda, AIP Conference Proceedings, Vol. **515**, p.3 (New York, 2000)
M. Catanese, T. C. Weekes: Pub. Astron. Soc. Pac. **111**, 1193 (1999)
15. H. I. Nel, et al.: Astrophys. J. **465**, 898 (1996)
16. L. Kuiper, et al.: Astron. Astrophys. **359**, 615 (2000)
17. H. Ögelman: Proc. *NATO ASI, The Lives of Neutron Stars*, eds. M.A. Alpar, Ü. Kızıloğlu, J. van Paradijs (Dordrecht, Kluwer 1995) pp.101-120
18. A. Possenti, S. Mereghetti, M. Colpi: Astron. Astrophys. **313**, 565 (1996)
19. J. P. Halpern, F. Y.-H. Wang: Astrophys. J. **477**, 905 (1997)
20. P. Slane: Astrophys. J. **437**, 458 (1994)
21. P. Slane, N. Lloyd: Astrophys. J. **452**, L115 (1995)
22. Y. Saito, et al.: Astrophys. J. **477** L37 (1997); Y. Saito, N. Kawai, T. Kamae, S. Shibata: Proc. *Neutron Stars and Pulsars: Thirty Years after the Discovery*, eds. N. Shibazaki, et al. (Universal Academy Press: Tokyo 1998), pp.295-298
23. D. J. Thompson, et al.: Astrophys. J. **516**, 297, (1999)
24. J. M. Fierro, et al.: Astrophys. J. **447**, 807 (1995)
25. B. Rudak, J. Dyks: Month. Not. R. Astron. Soc. **295**, 337 (1998)
26. V. Schönfelder, et al.: Astron. Astrophys. Supp. **143**, 145 (2000)
27. P. C. Schroeder, et al.: Astrophys. J. **450**, 784 (1995)
28. W. N. Johnson et al.: Astrophys. J. Supp. **86**, 693 (1993)
V. Schönfelder, et al.: Astrophys. J. Supp. **86**, 657 (1993)
D. J. Thompson, et al.: Astrophys. J. Supp. **86**, 629 (1993)
29. L. Kuiper, et al.: Astron. Astrophys. **351**, 119 (1999)
30. A. K. Harding, M. G. Baring, P. L. Gonthier: Astrophys. J. **476**, 246 (1997)
M. G. Baring, A. K. Harding: Astrophys. J. **547**, 929 (2001)
31. V. M. Kaspi et al.: Astrophys. J. **528**, 445 (2000)
32. J. K. Daugherty, A. K. Harding: Astrophys. J. **429**, 325 (1994)
S. J. Sturner, C. D. Dermer: Astrophys. J. **420**, L79 (1994)
A. K. Harding, A. G. Muslimov: Astrophys. J. **500**, 862 (1998)
33. M. Takahashi, et al.: Proc. *IAU Coll. 177, Pulsar Astronomy – 2000 and beyond* eds. M. Kramer, N. Wex, R. Wielebinski, ASP Conference Series **202**, pp.353-354 (2000)
34. P. A. Sturrock: Astrophys. J. **164**, 529 (1971)
35. E. T. Scharlemann, J. Arons, W. M. Fawley: Astrophys. J. **222**, 297 (1978)
J. Arons, E. T. Scharlemann: Astrophys. J. **231**, 854 (1979)

36. J. Arons: Proc. *IAU Coll. 177, Pulsar Astronomy – 2000 and beyond* eds. M. Kramer, N. Wex, R. Wielebinski, ASP Conference Series **202**, pp.449-454 (2000)
37. A. G. Muslimov, A. I. Tsygan: Month. Not. R. Astron. Soc. **255**, 61 (1992)
38. A. K. Harding, A. G. Muslimov: Astrophys. J. **508**, 328 (1998)
39. J. Dyks, B. Rudak: Astron. Astrophys. **362**, 1004 (2000)
40. S. J. Sturner, C. D. Dermer, F. C. Michel: Astrophys. J. **445**, 736 (1995)
C. D. Dermer, S. J. Sturner: Astrophys. J. **420**, L75 (1994)
41. X. Y. Xia, G. J. Qiao, X. J. Wu, Y. Q. Hou: Astron. Astrophys. **152**, 93 (1985)
H. K. Chang: Astron. Astrophys. **301**, 456 (1995)
42. S. J. Sturner: Astrophys. J. **446**, 292 (1995)
43. J. K. Daugherty, A. K. Harding: Astrophys. J. **252**, 337 (1982)
44. B. Zhang, A. K. Harding: Astrophys. J. **532**, 1150 (2000)
45. B. Zhang, G. J. Qiao: Astron. Astrophys. **338**, 62 (1998)
46. J. Dyks, B. Rudak, T. Bulik: Proc. *4th INTEGRAL Workshop*, astro-ph/0010301
47. T. Erber: Rev. Mod. Phys. **38**, 626 (1966)
48. J. K. Daugherty, A. K. Harding: Astrophys. J. **273**, 761 (1983)
49. M. S. Strickman, A. K. Harding, O. C. de Jager: Astrophys. J. **524**, 373 (1999)
50. M. S. Strickman, et al.: Astrophys. J. **460**, 735 (1996)
51. D. J. Thompson, A. K. Harding, W. Hermsen, M. P. Ulmer: Proc. *Fourth Compton Symposium*, eds. C. D. Dermer, M. S. Strickman, J. D. Kurfess, AIP (New York, 1997) p.410
52. H. Herold: Phys. Rev. D **19**, 2868 (1979)
53. C. D. Dermer: Astrophys. J. **360**, 197 (1990)
54. J. K. Daugherty, A. K. Harding: Astrophys. J. **309**, 362 (1986)
55. G. G. Pavlov, Y. A. Shibanov, J. Ventura, V. E. Zavlin: Astron. Astrophys. **289**, 837 (1994)
56. J. K. Daugherty, A. K. Harding: Astrophys. J. **336**, 861 (1989)
57. S. L. Adler: Ann. Phys. **67**, 599 (1971)
58. S. Mereghetti: Proc. *NATO ASI The Neutron Star – Black Hole Connection* (1999)
59. F. Camilo, et al.: Astrophys. J. **541**, 367 (2000)
60. M. G. Baring, A. K. Harding: Astrophys. J. **507**, L55 (1998)
61. B. Zhang, A. K. Harding: Astrophys. J. **535**, L51 (2000)
62. R. D. Blandford: Phys. Scripta **T85**, 191 (2000)
63. A. V. Olinto: Phys. Rep. **333-334**, 329 (2000).
64. P. Bhattacharjee, G. Sigl: Phys. Rep. **327**, 109 (2000)
65. A. M. Hillas: Ann. Rev. Astron. Astrophys. **22**, 425 (1984)
66. A. Venkatesan, M. C. Miller, A. V. Olinto: Astrophys. J. **484**, 323 (1997)
67. A. R. Bell: Month. Not. R. Astron. Soc. **257**, 493 (1992)
A. R. Bell, S. G. Lucek: Month. Not. R. Astron. Soc. **283**, 1083 (1996)
68. S. G. Lucek, A. R. Bell: Month. Not. R. Astron. Soc. **268**, 581 (1994)
69. M. J. Rees, J. E. Gunn: Month. Not. R. Astron. Soc. **167**, 1 (1974)
70. C. F. Kennel, F. V. Coroniti: Astrophys. J. **283**, 694 (1984)
C. F. Kennel, F. V. Coroniti: Astrophys. J. **283**, 710 (1984)
71. F. V. Coroniti: Astrophys. J. **349**, 538 (1990)
72. P. Blasi, R. I. Epstein, A. V. Olinto: Astrophys. J. **533**, L123 (2000)
73. R. T. Emmering, R. A. Chevalier: Astrophys. J. **321**, 334 (1987)
74. S. Shibata, N. Kawai, K. Tamura: Proc. *Neutron Stars and Pulsars: Thirty Years after the Discovery*, eds. N. Shibazaki, et al. (Universal Academy Press: Tokyo 1998), pp.457-464

75. J. Arons: Mem. Soc. Ast. Ital. **69**, 989 (1998)
76. M. C. Begelman, Z.-Y. Li: Astrophys. J. **426**, 269 (1994)
77. A. M. Abrahams, S. L. Shapiro: Astrophys. J. **374**, 652 (1991)
V. V. Usov, D. B. Melrose: Aus. J. Phys. **48**, 571 (1995)
78. A. F. Cheng, M. Ruderman: Astrophys. J. **235**, 576 (1980)
79. J. M. Lattimer, K. A. Van Riper, M. Prakash, M. Prakash: Astrophys. J. **425**, 802 (1994)
S. Tsuruta: Phys. Rep. **292**, 1 (1998)
80. J. W. Cronin: Rev. Mod. Phys. **71**, S165 (1999)
81. W. Bednarek, R. J. Protheroe: Phys. Rev. Lett. **79**, 2616 (1997)
82. E. M. de Gouveia Dal Pino, A. Lazarian: Astrophys. J. **535**, L31 (2000)
83. T. Bulik, B. Rudak, J. Dyks: Month. Not. R. Astron. Soc. **317**, 97 (2000)
84. T. Kamae, T. Ohsugi, D. J. Thompson, K. Watanabe: Adv. Sp. Res. **25**, 905 (2000)
85. O. Blanch, et al.: *The MAGIC telescope*, Letter of Intent (1998)
86. T. C. Weekes, et al.: *The Very Energetic Radiation Imaging Telescope Array System (VERITAS)*, preprint (2000)

High-Energy Particles from γ-Ray Bursts*

Eli Waxman

Weizmann Institute of Science, Rehovot 76100, Israel

1 Introduction and Summary

The widely accepted interpretation of the phenomenology of γ-ray bursts (GRBs), bursts of 0.1 MeV–1 MeV photons lasting for a few seconds (see [1] for a review), is that the observable effects are due to the dissipation of the kinetic energy of a relativistically expanding wind, a "fireball," whose primal cause is not yet known (see [2,3] for reviews). The recent detection of "afterglows," delayed low energy (X-ray to radio) emission of GRBs (see [4] for review), confirmed the cosmological origin of the bursts, through the redshift determination of several GRB host-galaxies, and confirmed standard model predictions of afterglows that result from the collision of an expanding fireball with its surrounding medium (see [5] for review). In this review, the production in GRB fireballs of γ-rays, high-energy cosmic-rays and neutrinos is discussed in the light of recent GRB and ultra-high-energy cosmic-ray observations.

The fireball model is described in detail in Sect.2. We do not discuss in this section the issue of GRB progenitors, i.e. the underlying sources producing the relativistic fireballs. At present, the two leading progenitor scenarios are collapses of massive stars [6,7], and mergers of compact objects [8,9]. As explained in Sect.2, the evolution of the fireball and the emission of γ-rays and afterglow radiation (on time scale of a day and longer) are largely independent of the nature of the progenitor. Thus, although present observations provide stringent constraints on the fireball model, the underlying progenitors remain unknown (e.g. [10]; see [4,5] for discussion). In Sect.3, constraints imposed on the fireball model by recent afterglow observations are discussed, which are of importance for high energy particle production.

The association of GRBs and ultra-high energy cosmic-rays (UHECRs) is discussed in Sect.4. Recent afterglow observations strengthen the evidence for GRB and UHECR association, which is based on two key points (see [11] for recent review). First, the constraints imposed on fireball model parameters by recent observations imply that acceleration of protons is possible to energy higher than previously assumed, $\sim 10^{21}$ eV. Second, the inferred local ($z = 0$) GRB energy generation rate of γ-rays, $\sim 10^{44}\mathrm{erg/Mpc^3yr}$, is remarkably similar to the local generation rate of UHECRs implied by cosmic-ray observations.

---

* Based on lectures given at the *ICTP Summer School on Astroparticle Physics and Cosmology* (ICTP, Trieste Italy, June 2000), and at the *VI Gleb Wataghin School on High Energy Phenomenology* (UNICAMP, Campinas Brazil, July 2000).

Martin Lemoine, Günter Sigl (Eds.): LNP 576, pp. 122–154, 2001.

The GRB model for UHECR production makes unique predictions, that may be tested with operating and planned large area UHECR detectors [12,13,14,15]. These predictions are described in Sect.5. In particular, a critical energy is predicted to exist, 10^{20}eV $\leq \epsilon_c < 4 \times 10^{20}$eV, above which a few sources produce most of the UHECR flux, and the observed spectra of these sources is predicted to be narrow, $\Delta\epsilon/\epsilon \sim 1$: the bright sources at high energy should be absent in UHECRs of much lower energy, since particles take longer to arrive the lower their energy. If the sources predicted by this model are detected by planned large area UHECR detectors, this would not only confirm the GRB model for UHECR production, but will also provide constraints on the unknown structure and strength of the inter-galactic magnetic field.

We note, that the AGASA experiment has recently reported the presence of one triplet and 3 doublets of UHECR events above 4×10^{19}eV, with angular separations (within each group) $\leq 2.5°$, roughly consistent with the measurement error [16]. The probability that these multiplets are chance coincidences (as opposed to being produced by point sources) is $\sim 1\%$. Therefore, this observation favors the bursting source model, although more data are needed to confirm it. Testing the predictions of the fireball model for UHECR production would require an exposure 10 times larger than that of present experiments. Such increase is expected to be provided by the HiRes [12] and Auger [13,14] detectors, and by the proposed Telescope Array detector [15].

Predictions for the emission of high energy neutrinos from GRB fireballs are discussed in Sect.6. Implications for planned high energy neutrino telescopes (the IceCube extension of AMANDA, ANTARES NESTOR; see [17] for review) are discussed in detail in Sect.6.4. It is shown that the predicted flux of $\geq 10^{14}$ eV neutrinos may be detectable by Čerenkov neutrino telescopes while the flux above 10^{19} eV may be detectable by large air-shower detectors [18,19,20]. Detection of the predicted neutrino signal will confirm the GRB fireball model for UHECR production and may allow to discriminate between different fireball progenitor scenarios. Moreover, a detection of even a handful of neutrino events correlated with GRBs will allow to test for neutrino properties, e.g. flavor oscillation and coupling to gravity, with accuracy many orders of magnitude better than currently possible.

2 The Fireball Model

2.1 Relativistic Expansion

General phenomenological considerations, based on γ-ray observations, indicate that, regardless of the nature of the underlying sources, GRBs are produced by the dissipation of the kinetic energy of a relativistic expanding fireball. The rapid rise time and short duration, ~ 1 ms, observed in some bursts [21,22] imply that the sources are compact, with a linear scale comparable to a light-ms, $r_0 \sim 10^7$ cm. The high γ-ray luminosity implied by cosmological distances, $L_\gamma \sim 10^{52}$erg s^{-1}, then results in a very high optical depth to pair creation. The energy of observed γ-ray photons is above the threshold for pair production.

The number density of photons at the source n_γ is approximately given by $L_\gamma = 4\pi r_0^2 c n_\gamma \epsilon$, where $\epsilon \simeq 1$MeV is the characteristic photon energy. Using $r_0 \sim 10^7$cm, the optical depth for pair production at the source is

$$\tau_{\gamma\gamma} \sim r_0 n_\gamma \sigma_T \sim \frac{\sigma_T L_\gamma}{4\pi r_0 c \epsilon} \sim 10^{15}, \quad (1)$$

where σ_T is the Thomson cross section.

The high optical depth implies that a thermal plasma of photons, electrons and positrons is created, a "fireball," which then expands and accelerates to relativistic velocities [9,8]. The optical depth is reduced by relativistic expansion of the source: If the source expands with a Lorentz factor Γ, the energy of photons in the source frame is smaller by a factor Γ compared to that in the observer frame, and most photons may therefore be below the pair production threshold.

A lower limit for Γ may be obtained in the following way [23,24]. The GRB photon spectrum is well fitted in the Burst and Transient Source Experiment (BATSE) detectors range, 20 keV to 2 MeV [1], by a combination of two power-laws, $dn_\gamma/d\epsilon_\gamma \propto \epsilon_\gamma^{-\beta}$ with different values of β at low and high energy [25]. Here, $dn_\gamma/d\epsilon_\gamma$ is the number of photons per unit photon energy. The break energy (where β changes) in the observer frame is typically $\epsilon_{\gamma b} \sim 1$MeV, with $\beta \simeq 1$ at energies below the break and $\beta \simeq 2$ above the break. In several cases, the spectrum was observed to extend to energies > 100 MeV [26,1]. Consider then a high energy test photon, with observed energy ϵ_t, trying to escape the relativistically expanding source. Assuming that in the source rest frame the photon distribution is isotropic, and that the spectrum of high energy photons follows $dn_\gamma/d\epsilon_\gamma \propto \epsilon_\gamma^{-2}$, the mean free path for pair production (in the source rest frame) for a photon of energy $\epsilon'_t = \epsilon_t/\gamma$ (in the source rest frame) is

$$l_{\gamma\gamma}^{-1}(\epsilon'_t) = \frac{1}{2}\frac{3}{16}\sigma_T \int d\cos\theta (1-\cos\theta) \int_{\epsilon_{\rm th}(\epsilon'_t,\theta)}^{\infty} d\epsilon \frac{U_\gamma}{2\epsilon^2} = \frac{1}{16}\sigma_T \frac{U_\gamma \epsilon'_t}{(m_e c^2)^2} \,. \quad (2)$$

Here, $\epsilon_{\rm th}(\epsilon'_t, \theta)$ is the minimum energy of photons that may produce pairs interacting with the test photon, given by $\epsilon_{\rm th}\epsilon'_t(1-\cos\theta) \geq 2(m_e c^2)^2$ (θ is the angle between the photons' momentum vectors). U_γ is the photon energy density (in the range corresponding to the observed BATSE range) in the source rest-frame, given by $L_\gamma = 4\pi r^2 \gamma^2 c U_\gamma$. Note , that we have used a constant cross section, $3\sigma_T/16$, above the threshold $\epsilon_{\rm th}$. The cross section drops as $\log(\epsilon)/\epsilon$ for $\epsilon \gg \epsilon_{\rm th}$; however, since the number density of photons drops rapidly with energy, this does not introduce a large correction to $l_{\gamma\gamma}$.

The source size constraint implied by the variability time is modified for a relativistically expanding source. Since in the observer frame almost all photons propagate at a direction making an angle $< 1/\Gamma$ with respect to the expansion direction, radiation seen by a distant observer originates from a conical section of the source around the source-observer line of sight, with opening angle $\sim 1/\Gamma$. Photons which are emitted from the edge of the cone are delayed, compared to those emitted on the line of sight, by $r/2\Gamma^2 c$. Thus, the constraint on source size implied by variability on time scale Δt is

$$r \sim 2\Gamma^2 c\Delta t. \tag{3}$$

The time r/c required for significant source expansion corresponds to comoving time (measured in the source frame) $t_{\rm co.} \simeq r/\Gamma c$. The two-photon collision rate at the source frame is $t_{\gamma\gamma}^{-1} = c/l_{\gamma\gamma}$. Thus, the source optical depth to pair production is $\tau_{\gamma\gamma} = t_{\rm co.}/t_{\gamma\gamma} \simeq r/\Gamma l_{\gamma\gamma}$. Using Eqs. (2) and (3) we have

$$\tau_{\gamma\gamma} = \frac{1}{128\pi}\frac{\sigma_T L_\gamma \epsilon_t}{c^2(m_e c^2)^2\Gamma^6\Delta t}. \tag{4}$$

Requiring $\tau_{\gamma\gamma} < 1$ at ϵ_t we obtain a lower limit for Γ,

$$\Gamma \geq 250\left[L_{\gamma,52}\left(\frac{\epsilon_t}{100\text{MeV}}\right)\Delta t_{-2}^{-1}\right]^{1/6}, \tag{5}$$

where $L_\gamma = 10^{52}L_{\gamma,52}\text{erg/s}$ and $\Delta t = 10^{-2}\Delta t_{-2}$ s.

2.2 Fireball Evolution

As the fireball expands it cools, the photon temperature T_γ in the fireball frame decreases, and most pairs annihilate. Once the pair density is sufficiently low, photons may escape. However, if the observed radiation is due to photons escaping the fireball as it becomes optically thin, two problems arise. First, the photon spectrum is quasi-thermal, in contrast with observations. Second, the source size, $r_0 \sim 10^7$ cm, and the total energy emitted in γ-rays, $\sim 10^{53}$ erg, suggest that the underlying energy source is related to the gravitational collapse of $\sim 1M_\odot$ object. Thus, the plasma is expected to be "loaded" with baryons which may be injected with the radiation or present in the atmosphere surrounding the source. A small baryonic load, $\geq 10^{-8}M_\odot$, increases the optical depth due to Thomson scattering on electrons associated with the "loading" protons, so that most of the radiation energy is converted to kinetic energy of the relativistically expanding baryons before the plasma becomes optically thin [27,28]. To overcome both problems it was proposed [29] that the observed burst is produced once the kinetic energy of the ultra-relativistic ejecta is re-randomized by some dissipation process at large radius, beyond the Thomson photosphere, and then radiated as γ-rays. Collision of the relativistic baryons with the inter-stellar medium [29], and internal collisions within the ejecta itself [30,31,32], were proposed as possible dissipation processes. Most GRBs show variability on time scales much shorter than (typically one hundredth of) the total GRB duration. Such variability is hard to explain in models where the energy dissipation is due to external shocks [33,34]. Thus, it is believed that internal collisions are responsible for the emission of γ-rays.

Let us first consider the case where the energy release from the source is "instantaneous," i.e. on a time scale r_0/c. We assume that most of the energy is released in the form of photons, i.e. that the fraction of energy carried by baryon rest mass M satisfies $\eta^{-1} \equiv Mc^2/E \ll 1$. The initial thickness of the fireball shell is r_0. Since the plasma accelerates to relativistic velocity, all fluid elements move with velocity close to c, and the shell thickness remains constant at r_0 (this breaks at very late time, as discussed below). We are interested in the stage where the optical depth (due to pairs and/or electrons associated with baryons) is high, but only a small fraction of the energy is carried by pairs.

The entropy of a fluid component with zero chemical potential is $S = V(e+p)/T$, where e, p and V are the (rest frame) energy density, pressure and volume. For the photons $p = e/3 \propto T_\gamma^4$. Since initially both the rest mass and thermal energy of baryons is negligible, the entropy is provided by the photons. Conservation of entropy implies

$$r^2\gamma(r)r_0T_\gamma^3(r) = \text{Constant}, \tag{6}$$

and conservation of energy implies

$$r^2\gamma(r)r_0\gamma(r)T_\gamma^4(r) = \text{Constant}\,. \tag{7}$$

Here $\gamma(r)$ is the shell Lorentz factor. Combining Eqs. (6) and (7) we find

$$\gamma(r) \propto r, \quad T_\gamma(r) \propto r^{-1}, \quad n \propto r^{-3}, \tag{8}$$

where n is the rest frame (comoving) baryon number density.

As the shell accelerates the baryon kinetic energy, γMc^2, increases. It becomes comparable to the total fireball energy when $\gamma \sim \eta$, at radius $r_f = \eta r_0$. At this radius most of the energy of the fireball is carried by the baryon kinetic energy, and the shell does not accelerate further. Equation (7) describing energy conservation is replaced with $\gamma =$ Constant. Equation (6), however, still holds. Equation (6) may be written as $T_\gamma^4/nT_\gamma =$ Constant (constant entropy per baryon). This implies that the ratio of radiation energy density to thermal energy density associated with the baryons is r independent. Thus, the thermal energy associated with the baryons may be neglected at all times, and Eq. (6) holds also for the stage where most of the fireball energy is carried by the baryon kinetic energy. Thus, for $r > r_f$ we have

$$\gamma(r) = \Gamma \simeq \eta, \quad T \propto r^{-2/3}, \quad n \propto r^{-2}. \tag{9}$$

Let us consider now the case of extended emission from the source, on time scale $\gg r_0/c$. In this case, the source continuously emits energy at a rate L, and the energy emission is accompanied by mass loss rate $\dot{M} = L/\eta c^2$. For $r < r_f$ the fluid energy density is relativistic, $aT_\gamma^4/nm_pc^2 = \eta r_0/r$, and the speed of sound is $\sim c$. The time it takes the shell at radius r to expand significantly is r/c in the observer frame, corresponding to $t_{\rm co.} \sim r/\gamma c$ in the shell frame. During this time sound waves can travel a distance $cr/\gamma c$ in the shell frame, corresponding to $r/\gamma^2 = r/(r/r_0)^2 = (r_0/r)r_0$ in the observer frame. This implies that at the early stages of evolution, $r \sim r_0$, sound waves had enough time to smooth out spatial fluctuations in the fireball over a scale r_0, but that regions separated by $\Delta r > r_0$ can not interact with each other. Thus, if the emission extends over a time $T_{\rm GRB} \gg r_0/c$, a fireball of thickness $cT_{\rm GRB} \gg r_0$ would be formed, which would expand as a collection of independent, roughly uniform, sub-shells of thickness r_0. Each sub-shell would reach a final Lorentz factor Γ_f, which may vary between sub-shells. This implies that different sub-shells may have velocities differing by $\Delta v \sim c/2\eta^2$, where η is some typical value representative of the entire fireball. Different shells emitted at times differing by Δt, $r_0/c < \Delta t < T_{\rm GRB}$, may therefore collide with each other after a time $t_c \sim c\Delta t/\Delta v$, i.e. at a radius

$$r_i \simeq 2\Gamma^2 c\Delta t = 6 \times 10^{13} \Gamma_{2.5}^2 \Delta t_{-2} \text{ cm}, \tag{10}$$

where $\Gamma = 10^{2.5}\Gamma_{2.5}$. The minimum internal shock radius, $r \sim \Gamma^2 r_0$, is also the radius at which an individual sub-shell may experience significant change in its width r_0, due to Lorentz factor variation across the shell.

2.3 The Allowed Range of Lorentz Factors and Baryon Loading

The acceleration, $\gamma \propto r$, of fireball plasma is driven by radiation pressure. Fireball protons are accelerated through their coupling to the electrons, which are coupled to fireball photons. We have assumed in the analysis presented above, that photons and electrons are coupled throughout the acceleration phase. However, if the baryon loading is too low, radiation may decouple from fireball electrons already at $r < r_f$. The fireball Thomson optical depth is given by the product of comoving expansion time, $r/\gamma(r)c$, and the photon Thomson scattering rate, $n_e c\sigma_T$. The electron and proton comoving number densities are equal, $n_e = n_p$, and are determined by equating the r independent mass flux carried by the wind, $4\pi r^2 c\gamma(r) n_p m_p$, to the mass loss rate from the underlying source, which is related to the rate L at which energy is emitted through $\dot{M} = L/(\eta c^2)$. Thus, during the acceleration phase, $\gamma(r) = r/r_0$, the Thomson optical depth $\tau_T \propto r^{-3}$. τ_T drops below unity at a radius $r < r_f = \eta r_0$ if $\eta > \eta_*$, where

$$\eta_* = \left(\frac{\sigma_T L}{4\pi r_0 m_p c^3}\right)^{1/4} = 1.0 \times 10^3 L_{52}^{1/4} r_{0,7}^{-1/4} . \tag{11}$$

Here $r_0 = 10^7 r_{0,7}$ cm.

If $\eta > \eta_*$ radiation decouples from the fireball plasma at $\gamma = r/r_0 = \eta_*^{4/3}\eta^{-1/3}$. If $\eta \gg \eta_*$, then most of the radiation energy is not converted to kinetic energy prior to radiation decoupling, and most of the fireball energy escapes in the form of thermal radiation. Thus, the baryon load of fireball shells, and the corresponding final Lorentz factors, must be within the range $10^2 \le \Gamma \simeq \eta \le \eta_* \simeq 10^3$ in order to allow the production of the observed non-thermal γ-ray spectrum.

2.4 Fireball Interaction with Surrounding Medium

As the fireball expands, it drives a relativistic shock (blast-wave) into the surrounding gas, e.g. into the inter-stellar medium (ISM) gas if the explosion occurs within a galaxy. In what follows, we refer to the surrounding gas as "ISM gas," although the gas need not necessarily be inter-stellar. At early time, the fireball is little affected by the interaction with the ISM. At late time, most of the fireball energy is transferred to the ISM, and the flow approaches the self-similar blast-wave solution of Blandford & McKee [35]. At this stage a single shock propagates into the ISM, behind which the gas expands with Lorentz factor

$$\Gamma_{BM}(r) = \left(\frac{17E}{16\pi n m_p c^2}\right)^{1/2} r^{-3/2} = 150 \left(\frac{E_{53}}{n_0}\right)^{1/2} r_{17}^{-3/2}, \tag{12}$$

where $E = 10^{53} E_{53}$ erg is the fireball energy, $n = 1 n_0\ \mathrm{cm}^{-3}$ is the ISM number density, and $r = 10^{17} r_{17}$ cm is the shell radius. The characteristic time at which radiation emitted by shocked plasma at radius r is observed by a distant observer is $t \simeq r/4\Gamma_{BM}^2 c$ [36].

The transition to self-similar expansion occurs on a time scale T (measured in the observer frame) comparable to the longer of the two time scales set by the initial conditions: the (observer) GRB duration $T_{\rm GRB}$ and the (observer) time T_Γ at which the self-similar Lorentz factor equals the original ejecta Lorentz factor Γ, $\Gamma_{BM}(t = T_\Gamma) = \Gamma$. Since $t = r/4\Gamma_{BM}^2 c$,

$$T = \max\left[T_{\rm GRB}, 5\left(\frac{E_{53}}{n_0}\right)^{1/3} \Gamma_{2.5}^{-8/3}\ \mathrm{s}\right]. \tag{13}$$

During the transition, plasma shocked by the reverse shocks expands with Lorentz factor close to that given by the self-similar solution,

$$\Gamma_{\rm tr.} \simeq \Gamma_{BM}(t = T) \simeq 245 \left(\frac{E_{53}}{n_0}\right)^{1/8} T_1^{-3/8}, \tag{14}$$

where $T = 10 T_1$ s. The unshocked fireball ejecta propagate at the original expansion Lorentz factor, Γ, and the Lorentz factor of plasma shocked by the reverse shock in the rest frame of the unshocked ejecta is $\simeq \Gamma/\Gamma_{\rm tr.}$. If $T \simeq T_{\rm GRB} \gg T_\Gamma$ then $\Gamma/\Gamma_{\rm tr.} \gg 1$, the reverse shock is relativistic, and the Lorentz factor associated with the random motion of protons in the reverse shock is $\gamma_p^R \simeq \Gamma/\Gamma_{\rm tr.}$.

If, on the other hand, $T \simeq T_\Gamma \gg T_{\rm GRB}$ then $\Gamma/\Gamma_{\rm tr.} \sim 1$, and the reverse shock is not relativistic. Nevertheless, the following argument suggests that the reverse shock speed is not far below c, and that the protons are therefore heated to relativistic energy, $\gamma_p^R - 1 \simeq 1$. The comoving time, measured in the fireball ejecta frame prior to deceleration, is $t_{\rm co.} \simeq r/\Gamma c$. The expansion Lorentz factor is expected to vary across the ejecta, $\Delta\Gamma/\Gamma \sim 1$, due to variability of the underlying GRB source over the duration of its energy release. Such variation would lead to expansion of the ejecta, in the comoving frame, at relativistic speed. Thus, at the deceleration radius, $t_{\rm co.} \simeq \Gamma T$, the ejecta width exceeds $\simeq c t_{\rm co.} \simeq \Gamma c T$.

Since the reverse shock should cross the ejecta over a deceleration time scale, $\simeq \Gamma T$, the reverse shock speed must be close to c. We therefore conclude that the Lorentz factor associated with the random motion of protons in the reverse shock is approximately given by $\gamma_p^R - 1 \simeq \Gamma/\Gamma_{\rm tr.}$ for both $\Gamma/\Gamma_{\rm tr.} \sim 1$ and $\Gamma/\Gamma_{\rm tr.} \gg 1$.

Since $T_{\rm GRB} \sim 10$ s is typically comparable to T_Γ, the reverse shocks are typically expected to be mildly relativistic.

2.5 Fireball Geometry

We have assumed in the discussion so far that the fireball is spherically symmetric. However, a jet-like fireball behaves as if it were a conical section of a spherical fireball as long as the jet opening angle is larger than Γ^{-1}. This is due to the fact that the linear size of causally connected regions, $ct_{\rm co.} \sim r/\Gamma$ in the fireball frame, corresponds to an angular size $ct_{\rm co.}/r \sim \Gamma^{-1}$. Moreover, due to the relativistic beaming of radiation, a distant observer can not distinguish between a spherical fireball and a jet-like fireball, as long as the jet opening angle $\theta > \Gamma^{-1}$. Thus, as long as we are discussing processes that occur when the wind is ultra-relativistic, $\Gamma \sim 300$ (prior to significant fireball deceleration by the surrounding medium), our results apply for both a spherical and a jet-like fireball. In the latter case, L (E) in our equations should be understood as the luminosity (energy) the fireball would have carried had it been spherically symmetric.

2.6 γ-Ray Emission

If the Lorentz factor variability within the wind is significant, internal shocks would reconvert a substantial part of the kinetic energy to internal energy. The internal energy may then be radiated as γ-rays by synchrotron and inverse-Compton emission of shock-accelerated electrons. The internal shocks are expected to be "mildly" relativistic in the fireball rest frame, i.e. characterized by Lorentz factor $\Gamma_i - 1 \sim$ a few. This is due to the fact that the allowed range of shell Lorentz factors is $\sim 10^2$ to $\sim 10^3$ (see Sect.2.3), implying that the Lorentz factors associated with the relative velocities are not very large. Since internal shocks are mildly relativistic, we expect results related to particle acceleration in sub-relativistic shocks (see [37] for review) to be valid for acceleration in internal shocks. In particular, electrons are expected to be accelerated to a power law energy distribution, $dn_e/d\gamma_e \propto \gamma_e^{-p}$ for $\gamma_e > \gamma_m$, with $p \simeq 2$ [38,39,40].

The minimum Lorentz factor γ_m is determined by the following consideration. Protons are heated in internal shocks to random velocities (in the wind frame) $\gamma_p^R - 1 \simeq \Gamma_i - 1 \simeq 1$. If electrons carry a fraction ξ_e of the shock internal energy, then $\gamma_m \simeq \xi_e(m_p/m_e)$. The characteristic frequency of synchrotron emission is determined by γ_m and by the strength of the magnetic field. Assuming that a fraction ξ_B of the internal energy is carried by the magnetic field, $4\pi r_i^2 c\Gamma^2 B^2/8\pi = \xi_B L_{\rm int.}$, the characteristic observed energy of synchrotron photons, $\epsilon_{\gamma b} = \Gamma\hbar\gamma_m^2 eB/m_e c$, is

$$\epsilon_{\gamma b} \simeq 1\xi_B^{1/2}\xi_e^{3/2}\frac{L_{\gamma,52}^{1/2}}{\Gamma_{2.5}^2 \Delta t_{-2}}\text{MeV}. \tag{15}$$

In deriving Eq. (15) we have assumed that the wind luminosity carried by internal plasma energy, $L_{\text{int.}}$, is related to the observed γ-ray luminosity through $L_{\text{int.}} = L_\gamma/\xi_e$. This assumption is justified since the electron synchrotron cooling time is short compared to the wind expansion time (unless the equipartition fraction ξ_B is many orders of magnitude smaller than unity), and hence electrons lose all their energy radiatively. Fast electron cooling also results in a synchrotron spectrum $dn_\gamma/d\epsilon_\gamma \propto \epsilon_\gamma^{-1-p/2} = \epsilon_\gamma^{-2}$ at $\epsilon_\gamma > \epsilon_{\gamma b}$, consistent with observed GRB spectra [25].

At present, there is no theory that allows the determination of the values of the equipartition fractions ξ_e and ξ_B. Equation (15) implies that fractions not far below unity are required to account for the observed γ-ray emission. We note, that build up of magnetic field to near equipartition by electro-magnetic instabilities is expected to be a generic characteristic of collisionless shocks (see discussion in ref. [37] and references therein), and is inferred to occur in other systems, e.g. in supernova remnant shocks (e.g. [41,42]).

The γ-ray break energy $\epsilon_{\gamma b}$ of most GRBs observed by BATSE detectors is in the range of 100 keV to 300 keV [43]. It may appear from Eq. (15) that the clustering of break energies in this narrow energy range requires fine tuning of fireball model parameters, which should naturally produce a much wider range of break energies. This is, however, not the case [44]. Consider the dependence of $\epsilon_{\gamma b}$ on Γ. The strong Γ dependence of the pair-production optical depth, Eq. (4), implies that if the value of Γ is smaller than the minimum value allowed by Eq. (5), for which $\tau_{\gamma\gamma}(\epsilon_\gamma = 100\text{MeV}) \simeq 1$, most of the high energy photons in the power-law distribution produced by synchrotron emission, $dn_\gamma/d\epsilon_\gamma \propto \epsilon_\gamma^{-2}$, would be converted to pairs. This would lead to high optical depth due to Thomson scattering on $e^\pm$, and hence to strong suppression of the emitted flux [44]. For fireball parameters such that $\tau_{\gamma\gamma}(\epsilon_\gamma = 100\text{MeV}) \simeq 1$, the break energy implied by Eqs. (15) and (5) is

$$\epsilon_{\gamma b} \simeq 1\xi_B^{1/2}\xi_e^{3/2}\frac{L_{\gamma,52}^{1/6}}{\Delta t_{-2}^{2/3}}\text{MeV}. \tag{16}$$

As explained in Sect.2.3, shell Lorentz factors can not exceed $\eta_* \simeq 10^3$, for which break energies in the X-ray range, $\epsilon_{\gamma b} \sim 10$ keV, may be obtained. We note, however, that the radiative flux would be strongly suppressed in this case too [44]. If the typical Γ of radiation emitting shells is close to η_*, then the range of Lorentz factors of wind shells is narrow, which implies that only a small fraction of wind kinetic energy would be converted to internal energy which can be radiated from the fireball.

Thus, the clustering of break energies at ~ 1 MeV is naturally accounted for, provided that the variability time scale satisfies $\Delta t \le 10^{-2}$ s, which implies an upper limit on the source size, since $\Delta t \geq r_0/c$. We note, that a large fraction of bursts detected by BATSE show variability on the shortest resolved time scale, ~ 10 ms [33]. In addition, a natural consequence of the model is the existence of low luminosity bursts with low, 1 keV to 10 keV, break energies [44]. Such "X-ray bursts" may have recently been identified [45].

For internal collisions, the observed γ-ray variability time, $\sim r_i/\Gamma^2 c \simeq \Delta t$, reflects the variability time of the underlying source, and the GRB duration reflects the duration over which energy is emitted from the source. Since the wind Lorentz factor is expected to fluctuate on time scales ranging from the shortest variability time r_0/c to the wind duration $T_{\rm GRB}$, internal collisions will take place over a range of radii, $r \sim \Gamma^2 r_0$ to $r \sim \Gamma^2 c T_{\rm GRB}$.

2.7 Afterglow Emission

Let us consider the radiation emitted from the reverse shocks, during the transition to self-similar expansion. The characteristic electron Lorentz factor (in the plasma rest frame) is $\gamma_m \simeq \xi_e(\Gamma/\Gamma_{\rm tr.})m_p/m_e$, where the internal energy per proton in the shocked ejecta is $\simeq (\Gamma/\Gamma_{\rm tr.})m_p c^2$. The energy density U is $E \simeq 4\pi r^2 c T \Gamma_{\rm tr.}^2 U$, and the number of radiating electrons is $N_e \simeq E/\Gamma m_p c^2$. Using Eq. (14) and $r = 4\Gamma_{\rm tr.}^2 cT$, the characteristic (or peak) energy of synchrotron photons (in the observer frame) is [46]

$$\epsilon_{\gamma m} \simeq \hbar \Gamma_{\rm tr.} \gamma_m^2 \frac{eB}{m_e c} = 2\xi_{e,-1}^2 \xi_{B,-1}^{1/2} n_0^{1/2} \Gamma_{2.5}^2 \, {\rm eV}, \tag{17}$$

and the specific luminosity, $L_\epsilon = dL/d\epsilon_\gamma$, at $\epsilon_{\gamma m}$ is

$$L_m \simeq \frac{1}{2\pi\hbar} \Gamma_{\rm tr.} \frac{e^3 B}{m_e c^2} N_e \simeq 10^{61} \xi_{B,-1}^{1/2} E_{53}^{5/4} T_1^{-3/4} \Gamma_{2.5}^{-1} n_0^{1/4} \, {\rm s}^{-1}, \tag{18}$$

where $\xi_e = 0.1\xi_{e,-1}$, and $\xi_B = 0.1\xi_{B,-1}$.

Here too, we expect a power law energy distribution, $dN_e/d\gamma_e \propto \gamma_e^{-p}$ for $\gamma_e > \gamma_m$, with $p \simeq 2$. Since the radiative cooling time of electrons in the reverse shock is long compared to the ejecta expansion time, the specific luminosity extends in this case to energy $\epsilon_\gamma > \epsilon_{\gamma m}$ as $L_\epsilon = L_m(\epsilon_\gamma/\epsilon_{\gamma m})^{-1/2}$, up to photon energy $\epsilon_{\gamma c}$. Here $\epsilon_{\gamma c}$ is the characteristic synchrotron frequency of electrons for which the synchrotron cooling time, $6\pi m_e c/\sigma_T \gamma_e B^2$, is comparable to the ejecta (rest frame) expansion time, $\sim \Gamma_{\rm tr.} T$. At energy $\epsilon_\gamma > \epsilon_{\gamma c}$,

$$\epsilon_{\gamma c} \simeq 0.1 \xi_{B,-1}^{-3/2} n_0^{-1} E_{53}^{-1/2} T_1^{-1/2} \, {\rm keV}, \tag{19}$$

the spectrum steepens to $L_\epsilon \propto \epsilon_\gamma^{-1}$.

The shock driven into the ISM continuously heats new gas, and produces relativistic electrons that may produce the delayed afterglow radiation observed on time scales $t \gg T$, typically of order days to months. As the shock-wave decelerates, the emission shifts to lower frequency with time. Since we are interested in proton acceleration to high energy and in the production of high energy neutrinos, which take place primarily in the internal and reverse shocks (see Sect.4, Sect.6), we do not discuss in detail the theory of late-time afterglow emission.

3 Some Implications of Afterglow Observations

Afterglow observations lead to the confirmation, as mentioned in the Introduction, of the cosmological origin of GRBs [4], and confirmed [47,48] standard model predictions [49,50,51,52] of afterglow that results from synchrotron emission of electrons accelerated to high energy in the highly relativistic shock driven by the fireball into its surrounding gas. Since we are interested mainly in the earlier, internal collision phase of fireball evolution, we do not discuss afterglow observations in detail. We note, however, several implications of afterglow observations which are of importance for the discussion of UHECR production.

The following point should be clarified in the context of afterglow observations. The distribution of GRB durations is bimodal, with broad peaks at $T_{\rm GRB} = 0.2$ s and $T_{\rm GRB} = 20$ s [1]. The majority of bursts belong to the long duration, $T_{\rm GRB} \sim 20$ s, class. The detection of afterglow emission was made possible thanks to the accurate GRB positions provided on hour time scale by the *BeppoSAX* satellite [53]. Since the detectors on board this satellite trigger only on long bursts, afterglow observations are not available for the sub-population of short, $T_{\rm GRB} \sim 0.2$ s, bursts. Thus, while the discussion of the fireball model presented in Sect.2, based on γ-ray observations and on simple phenomenological arguments, applies to both long and short duration bursts, the discussion below of afterglow observations applies to long duration bursts only. It should therefore be kept in mind that short duration bursts may constitute a different class of GRBs, which, for example, may be produced by a different class of progenitors and may have a different redshift distribution than the long duration bursts.

Prior to the detection of afterglows, it was commonly assumed that the farthest observed GRBs lie at redshift $z \sim 1$. Following the detection of afterglows and the determination of GRB redshifts, it is now clear that most GRB sources lie within the redshift range $z \sim 0.5$ to $z \sim 2$, with some bursts observed at $z > 3$. For the average GRB γ-ray fluence, 1.2×10^{-5}erg/cm^2 in the 20 keV to 2 MeV band, this implies characteristic isotropic γ-ray energy and luminosity $E_\gamma \sim 10^{53}$ erg and $L_\gamma \sim 10^{52}$ erg/s (in the 20 keV to 2 MeV band), about an order of magnitude higher than the values assumed prior to afterglow detection (Here, and throughout the paper we assume a flat universe with $\Omega = 0.3$, $\Lambda = 0.7$, and $H_0 = 65$km/s Mpc). These estimates are consistent with more detailed analyses of the GRB luminosity function and redshift distribution. Mao & Mo, e.g., find, for the cosmological parameters we use, a median GRB energy of $\simeq 0.6 \times 10^{53}$erg in the 50keV to 300keV band [54], corresponding to a median GRB energy of $\simeq 2 \times 10^{53}$erg in the 20 keV to 2 MeV band.

The determination of GRB redshifts also lead to a modification of GRB rate estimates. Since most observed GRB sources lie within the redshift range $z \sim 0.5$ to $z \sim 2$, observations essentially determine the GRB rate per unit volume at

$z \sim 1$. The observed rate of 10^3/yr implies $R_{\rm GRB}(z=1) \simeq 3/{\rm Gpc}^3{\rm yr}$. The present, $z=0$, rate is less well constrained, since available data are consistent with both no evolution of GRB rate with redshift, and with strong evolution (following, e.g., star formation rate), in which $R_{\rm GRB}(z=1)/R_{\rm GRB}(z=0) \sim 10$ [55,56]. Detailed analyses, assuming $R_{\rm GRB}$ is proportional to star formation rate, lead to $R_{\rm GRB}(z=0) \sim 0.5/{\rm Gpc}^3{\rm yr}$ [4]. The implied local ($z=0$) γ-ray energy generation rate by GRBs in the 20 keV to 2 MeV band is therefore

$$\dot{\varepsilon}_\gamma(z=0) = 10^{44}\zeta \,{\rm erg/Mpc}^3{\rm yr}, \tag{20}$$

with ζ in the range of $\sim 10^{-0.5}$ to $\sim 10^{0.5}$. Note, that $\dot{\varepsilon}_\gamma$ is independent of the fireball geometry. If fireballs are conical jets of solid angle $\Delta\Omega$, then the total energy released by each burst is smaller by a factor $\Delta\Omega/4\pi$ than the isotropic energy, and the GRB rate is larger by the same factor.

Due to present technical limitations of the experiments, afterglow radiation is observed in most cases only on time scale $\gg 10$ s. At this stage, radiation is produced by the external shock driven into the surrounding gas, and afterglow observations therefore do not provide direct constraints on plasma parameters at the internal and reverse shocks, where protons are accelerated to ultra-high energy. In one case, however, that of GRB 990123, optical emission has been detected on ~ 10 s time scale [57]. The most natural explanation of the observed optical radiation is synchrotron emission from electrons accelerated to high energy in the reverse shocks driven into fireball ejecta at the onset of interaction with the surrounding medium [58,59], as explained in Sect.2.7. This observation provides therefore direct constraints on the fireball ejecta plasma. First, it provides strong support for one of the underlying assumptions of the dissipative fireball scenario described in Sect.2.2, that the energy is carried from the underlying source in the form of proton kinetic energy. This is due to the fact that the observed radiation is well accounted for in a model where a shock propagates into fireball plasma composed of protons and electrons (rather than, e.g., pair plasma). Second, comparison of the observed flux with model predictions, Eqs. (17) and (18), implies $\xi_e \sim \xi_B \sim 10^{-1}$.

Afterglow observations imply that a significant fraction of the energy initially carried by the fireball is converted into γ-rays, i.e. that the observed γ-ray energy provides a rough estimate of the total fireball energy. This has been demonstrated for one case, that of GRB970508, by a comparison of the total fireball energy derived from long term radio observations with the energy emitted in γ-rays [60,61], and for a large number of bursts by a comparison of observed γ-ray energy with the total fireball energy estimate based on X-ray afterglow data [62]. In the context of the fireball model described in Sect.2, the inferred high radiative efficiency implies that a significant fraction of the wind kinetic energy must be converted to internal energy in internal shocks, and that electrons must carry a significant fraction of the internal energy, i.e. that ξ_e should be close to unity. We have already shown, see Sect.2.6, that ξ_e values not far below unity are required to account for the observed γ-ray emission. Conversion in internal shocks of a large fraction of fireball kinetic energy to internal energy is possibly provided the variance in the Lorentz factors of fireball shells is large [44].

In accordance with the implications of afterglow observations, we assume in the discussion below of UHECR and neutrino production in GRB fireballs, that the fraction of fireball energy converted to internal energy carried by electrons, and hence to γ-rays, is large. For discussion of UHECR production under the assumption that only a negligible fraction, $\sim m_e/m_p$, of the fireball energy is converted to radiation see ref. [63].

4 UHECRs from $\gamma-$Ray Burst Fireballs

4.1 Fermi Acceleration in $\gamma-$Ray Bursts

In the fireball model, the observed radiation is produced, both during the GRB and the afterglow, by synchrotron emission of shock accelerated electrons. In the region where electrons are accelerated, protons are also expected to be shock accelerated. This is similar to what is thought to occur in supernovae remnant shocks, where synchrotron radiation of accelerated electrons is the likely source of non-thermal X-rays (recent ASCA observations give evidence for acceleration of electrons in the remnant of SN1006 to 10^{14}eV [64]), and where shock acceleration of protons is believed to produce cosmic rays with energy extending to $\sim 10^{15}$eV (see, e.g., [37] for review). Thus, it is likely that protons, as well as electrons, are accelerated to high energy within GRB fireballs. Let us consider the constraints that should be satisfied by the fireball parameters in order to allow acceleration of protons to $\sim 10^{20}$ eV.

We consider proton Fermi acceleration in fireball internal shocks, which take place as the fireball expands over a range of radii, $r \sim \Gamma^2 r_0$ to $r \sim \Gamma^2 c T_{\rm GRB}$, and at the reverse shocks driven into fireball ejecta due to interaction with surrounding medium at $r \sim \Gamma^2 cT \sim \Gamma^2 c T_{\rm GRB}$ (see Sect.2.2, Sect.2.4). Both internal and reverse shocks are, in the wind rest-frame, mildly relativistic, i.e. characterized by Lorentz factors $\Gamma_i - 1 \sim 1$. Moreover, since reverse shocks do not cause strong deceleration of fireball plasma, see Sect.2.4, the expansion Lorentz factor $\Gamma_{\rm tr.}$ of fireball plasma shocked by reverse shocks is similar to the fireball Lorentz factor Γ prior to interaction with the surrounding medium. Thus, plasma parameters, e.g. energy and number density, in the reverse shocks are similar to those obtained in internal shocks due to variability on time scale $\Delta t \sim T$. Results obtained below for internal shocks are therefore valid also for reverse shocks, provided Δt is replaced with T.

Since the shocks we are interested in are mildly relativistic, we expect results related to particle acceleration in sub-relativistic shocks (see [37] for review) to be valid for our scenario. The predicted energy distribution of accelerated protons is therefore $dn_p/d\epsilon_p \propto \epsilon_p^{-2}$ [38,39,40], similar to the predicted electron energy spectrum, which is consistent with the observed photon spectrum (see Sect.2.6).

The most restrictive requirement, which rules out the possibility of accelerating particles to energy $\sim 10^{20}$ eV in most astrophysical objects, is that the particle Larmor radius r_L should be smaller than the system size [65]. In our scenario we must apply a more stringent requirement, namely that r_L should be smaller than the largest scale l over which the magnetic field fluctuates, since otherwise Fermi acceleration may not be efficient. We may estimate l as follows. The comoving time, i.e. the time measured in the wind rest frame, is $t = r/\Gamma c$. Thus, regions separated by a comoving distance larger than r/Γ are causally disconnected, and the wind properties fluctuate over comoving length scales up to $l \sim r/\Gamma$. We must therefore require $r_L < r/\Gamma$. A somewhat more stringent requirement is related to the wind expansion. Due to expansion the internal energy is decreasing and therefore available for proton acceleration (as well as for γ-ray production) only over a comoving time $t \sim r/\Gamma c$. The typical Fermi acceleration time is $t_a = f r_L/c\beta^2$ [65,37], where βc is the Alfvén velocity and $f \sim 1$ [66]. In our scenario $\beta \simeq 1$ leading to the requirement $f r_L < r/\Gamma$. This condition sets a lower limit to the required comoving magnetic field strength. Using the relations $r_L = \epsilon'_p/eB = \epsilon_p/\Gamma eB$, where $\epsilon'_p = \epsilon_p/\Gamma$ is the proton energy measured in the fireball frame, and $4\pi r^2 c\Gamma^2 B^2/8\pi = \xi_B L_\gamma/\xi_e$, the constraint $f r_L < r/\Gamma$ may be written as [67],

$$\frac{\xi_B}{\xi_e} > 0.02 f^2 \Gamma_{2.5}^2 \epsilon_{p,20}^2 L_{\gamma,52}^{-1}, \tag{21}$$

where $\epsilon_p = 10^{20}\epsilon_{p,20}$ eV is the accelerated proton energy. Note, that this constraint is independent of r, the internal collision radius.

The accelerated proton energy is also limited by energy loss due to synchrotron radiation and interaction with fireball photons. As discussed in Sect.6, the dominant energy loss process is synchrotron cooling. The condition that the synchrotron loss time, $t_{sy} = (6\pi m_p^4 c^3/\sigma_T m_e^2)\epsilon_p^{-1} B^{-2}$, should be smaller than the acceleration time sets an upper limit to the magnetic field strength. Since the equipartition field decreases with radius, $B_{e.p.} \propto r^{-2}$, the upper limit on the magnetic field may be satisfied simultaneously with Eq. (21) provided that the internal collisions occur at large enough radius [67],

$$r > 10^{12} f^2 \Gamma_{2.5}^{-2} \epsilon_{p,20}^3 \mathrm{cm}. \tag{22}$$

Since collisions occur at radius $r \simeq \Gamma^2 c\Delta t$, the condition Eq. (22) is equivalent to a lower limit on Γ

$$\Gamma > 130 f^{1/2} \epsilon_{p,20}^{3/4} \Delta t_{-2}^{-1/4}. \tag{23}$$

From Eqs. (21) and (23), we infer that a dissipative ultra-relativistic wind, with luminosity and variability time implied by GRB observations, satisfies the constraints necessary to allow the acceleration of protons to energy $> 10^{20}$ eV, provided that the wind bulk Lorentz factor is large enough, $\Gamma > 100$, and that the magnetic field is close to equipartition with electrons. The former constraint, $\Gamma > 100$, is remarkably similar to that inferred based on the γ-ray spectrum, and $\Gamma \sim 300$ is the "canonical" value assumed in the fireball model. The latter constraint, magnetic field close to equipartition, must be satisfied to account for both γ-ray emission (see Sect.2.6) and afterglow observations (see Sect.3).

Finally, two points should be clarified. First, it has recently been claimed that ultra-high energy protons would lose most of their energy adiabatically, i.e.

due to expansion, before they escape the fireball [68]. This claim is based on the assumptions that internal shocks, and therefore proton acceleration, occur at $r \sim \Gamma^2 r_0$ only, and that subsequently the fireball expands adiabatically. Under these assumptions, protons would lose most their energy by the time they escape. However, as emphasized both in this section and in Sect.2.2, internal shocks are expected to occur over a wide range of radii, and in particular at $r \sim \Gamma^2 cT$ during the transition to self-similar expansion. Thus, proton acceleration to ultra-high energy is expected to operate over a wide range of radii, from $r \sim \Gamma^2 r_0$ up to $r \sim \Gamma^2 cT$, where ultra-high energy particles escape.

Second, it has recently been claimed in [69] that the conditions at the external shock driven by the fireball into the ambient gas are not likely to allow proton acceleration to ultra-high energy. Regardless of the validity of this claim, it is irrelevant for the acceleration in internal shocks, the scenario considered for UHECR production in GRBs in both [67] and [70]. Moreover, it is not at all clear that UHECRs can not be produced at the external shock, since the magnetic field may be amplified ahead of the shock by the streaming of high energy particles. For discussion of high energy proton production in the external shock and its possible implications see ref. [71].

4.2 UHECR Flux and Spectrum

Fly's Eye [72,73] and AGASA [74,75,76] results confirm the flattening of the cosmic-ray spectrum at $\sim 10^{19}$ eV, evidence for which existed in previous experiments with weaker statistics [77]. Fly's Eye data is well fitted in the energy range $10^{17.6}$ eV to $10^{19.6}$ eV by a sum of two power laws: A steeper component, with differential number spectrum $J \propto E^{-3.50}$, dominating at lower energy, and a shallower component, $J \propto E^{-2.61}$, dominating at higher energy, $E > 10^{19}$ eV. The flattening of the spectrum, combined with the lack of anisotropy and the evidence for a change in composition from heavy nuclei at low energy to light nuclei (protons) at high energy [77,72,73,78,79], suggest that a Galactic component of heavy nuclei, $J \propto E^{-3.50}$, dominates the cosmic-ray flux at low energy, while an extra-Galactic component of protons, $J \propto E^{-2.61}$, dominates the flux at high energy, $> 10^{19}$ eV.

The GRB energy observed in γ-rays reflects the fireball energy in accelerated electrons. If accelerated electrons and protons carry similar energy, as indicated by afterglow observations [62] (see Sect.3), then the GRB cosmic-ray production rate is [see Eq. (20)]

$$\epsilon_p^2 \frac{d\dot{n}_p}{d\epsilon_p}\bigg|_{z=0} \simeq 10^{44} \mathrm{erg/Mpc^3 yr}. \tag{24}$$

In Fig. 1 we compare the UHECR spectrum, reported by the Fly's Eye [73], the Yakutsk [80], and the AGASA experiments [81], with that expected from a homogeneous cosmological distribution of sources, each generating a power law differential spectrum of high energy protons $dn/d\epsilon_p \propto \epsilon_p^{-2}$. The absolute flux measured at 3×10^{18} eV differs between the various experiments, corresponding

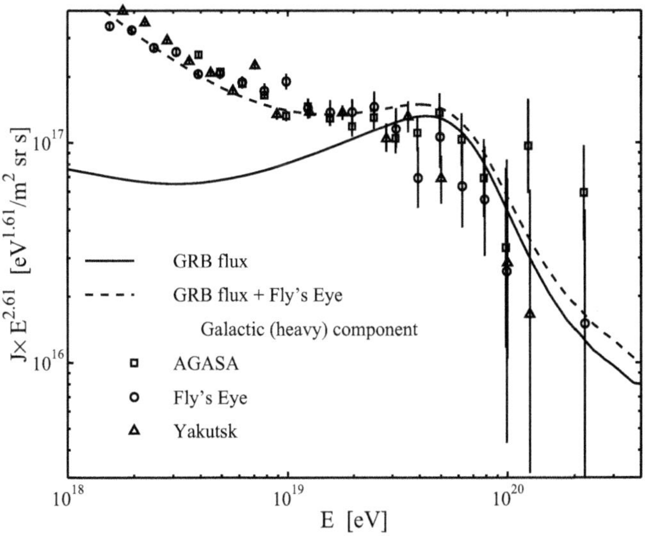

Fig. 1. The UHECR flux expected in a cosmological model, where high-energy protons are produced at a rate $(\epsilon_p^2 d\dot{n}_p/d\epsilon_p)_{z=0} = 0.8 \times 10^{44}\mathrm{erg/Mpc^3yr}$ as predicted by the GRB model [Eq. (24)], solid line, compared to the Fly's Eye [73], Yakutsk [80] and AGASA [81] data. 1σ flux error bars are shown. The highest energy points are derived assuming the detected events represent a uniform flux over the energy range 10^{20} eV–3×10^{20} eV. The dashed line is the sum of the GRB model flux and the Fly's Eye fit to the Galactic heavy nuclei component, $J \propto \epsilon^{-3.5}$ [73] (with normalization increased by 25%).

to a systematic $\simeq 10\%$ ($\simeq 20\%$) over-estimate of event energies in the AGASA (Yakutsk) experiment compared to the Fly's Eye experiment (see also [75]). In Fig. 1, the Yakutsk energy normalization is used. For the model calculation, a flat universe, $\Omega = 0.3$, $\Lambda = 0.7$ and $H_0 = 65$km/Mpc s were assumed. The calculation is similar to that described in [82]. The generation rate of cosmic-rays (per unit comoving volume) was assumed to evolve rapidly with redshift following the luminosity density evolution of QSOs [83], which is also similar to that describing the evolution of star formation rate [84,85]: $\dot{n}_{CR}(z) \propto (1+z)^\alpha$ with $\alpha \simeq 3$ [86] at low redshift, $z < 1.9$, $\dot{n}_{CR}(z) =$ Const. for $1.9 < z < 2.7$, and an exponential decay at $z > 2.7$ [87]. The cosmic-ray spectrum at energy $> 10^{19}$ eV is little affected by modifications of the cosmological parameters or of the redshift evolution of cosmic-ray generation rate. This is due to the fact that cosmic-rays at this energy originate from distances shorter than several hundred Mpc. The spectrum and flux at $\epsilon_p > 10^{19}$ eV is mainly determined by the present ($z = 0$) generation rate and spectrum, which in the model shown in Fig. 1 is $\epsilon_p^2(d\dot{n}_p/d\epsilon)_{z=0} = 0.8 \times 10^{44}\mathrm{erg/Mpc^3yr}$.

The suppression of model flux above $10^{19.7}$ eV is due to energy loss of high energy protons in interaction with the microwave background, i.e. to the "GZK cutoff" [88,89]. The available data do not allow to determine the existence (or absence) of the "cutoff" with high confidence. The AGASA results show an excess (at a $\sim 2.5\sigma$ confidence level) of events compared to model predictions above 10^{20}eV. This excess is not confirmed, however, by the other experiments. Moreover, since the 10^{20}eV flux is dominated by sources at distances < 100 Mpc, over which the distribution of known astrophysical systems (e.g. galaxies, clusters of galaxies) is inhomogeneous, significant deviations from model predictions presented in Fig. 1 for a uniform source distribution are expected at this energy [82]. Clustering of cosmic-ray sources leads to a standard deviation, σ, in the expected number, N, of events above 10^{20} eV, given by $\sigma/N = 0.9(d_0/10\mathrm{Mpc})^{0.9}$ [90], where d_0 is the unknown scale length of the source correlation function and $d_0 \sim 10$ Mpc for field galaxies.

Thus, GRB fireballs would produce UHECR flux and spectrum consistent with that observed, provided the efficiency with which the wind kinetic energy is converted to γ-rays, and therefore to electron energy, is similar to the efficiency with which it is converted to proton energy, i.e. to UHECRs [67]. There is, however, one additional point which requires consideration [67]. The energy of the most energetic cosmic ray detected by the Fly's Eye experiment is in excess of 2×10^{20}eV, and that of the most energetic AGASA event is $\sim 2 \times 10^{20}$eV. On a cosmological scale, the distance traveled by such energetic particles is small: < 100Mpc (50Mpc) for the AGASA (Fly's Eye) event (e.g., [91]). Thus, the detection of these events over a ~ 5yr period can be reconciled with the rate of nearby GRBs, ~ 1 per 100 yr to ~ 1 per 1000 yr out to 100Mpc, only if there is a large dispersion, ≥ 100yr, in the arrival time of protons produced in a single burst (This implies that if a direct correlation between high energy CR events and GRBs, as suggested in [92], is observed on a ~ 10yr time scale, it would be strong evidence *against* a cosmological GRB origin of UHECRs).

The required dispersion is likely to occur due to the combined effects of deflection by random magnetic fields and energy dispersion of the particles [67]. Consider a proton of energy ϵ_p propagating through a magnetic field of strength B and correlation length λ. As it travels a distance λ, the proton is typically deflected by an angle $\alpha \sim \lambda/r_L$, where $r_L = \epsilon_p/eB$ is the Larmor radius. The typical deflection angle for propagation over a distance d is $\theta_s \simeq (2d/9\lambda)^{1/2}\lambda/r_L$. This deflection results in a time delay, compared to propagation along a straight line,

$$\tau(\epsilon_p, d) \simeq \theta_s^2 d/4c \simeq 10^7 \epsilon_{p,20}^{-2} d_{100}^2 \lambda_{\mathrm{Mpc}} B_{-8}^2 \quad \mathrm{yr}, \tag{25}$$

where $d = 100 d_{100}$Mpc, $\lambda = 1\lambda_{\mathrm{Mpc}}$ Mpc and $B = 10^{-8} B_{-8}$ G (see also Sect. in this volume). Here, we have chosen numerical values corresponding to the current upper bound on the inter-galactic magnetic field, $B\lambda^{1/2} \leq 10^{-8}\mathrm{G\ Mpc}^{1/2}$ [93,94]. The upper bound on the (systematic increase with redshift of the) Faraday rotation measure of distant, $z \leq 2.5$, radio sources, $RM < 5\mathrm{rad/m}^2$, implies an upper bound $B \leq 10^{-11}(h/0.75)(\Omega_b h^2)^{-1}$ G on an inter-galactic field coherent over cosmological scales [94]. Here, h is the Hubble constant in units of 100km/s Mpc and Ω_b is the baryon density in units of the closure density. For a magnetic field coherent on scales $\sim \lambda$, this implies $B\lambda^{1/2} \leq 10^{-8}(h/0.65)^{1/2}(\Omega_b h^2/0.04)^{-1}$G Mpc$^{1/2}$.

The random energy loss UHECRs suffer as they propagate, owing to the production of pions, implies that at any distance from the observer there is some finite spread in the energies of UHECRs that are observed with a given fixed energy. For protons with energies $> 10^{20}$eV the fractional RMS energy spread is of order unity over propagation distances in the range 10 – 100Mpc (e.g. [91]). Since the time delay is sensitive to the particle energy, this implies that the spread in arrival time of UHECRs with given observed energy is comparable to the average time delay at that energy $\tau(\epsilon_p, d)$ (This result has been confirmed by numerical calculations in [95]). Thus, the required time spread, $\tau > 100$ yr, is consistent with the upper bound, $\tau < 10^7$ yr, implied by the present upper bound to the inter-galactic magnetic field.

5 γ–Ray Burst Model Predictions for UHECR Experiments

5.1 The Number and Spectra of Bright Sources

The initial proton energy, necessary to have an observed energy ϵ_p, increases with source distance due to propagation energy losses. The rapid increase of the initial energy after it exceeds, due to electron-positron production, the threshold for pion production effectively introduces a cutoff distance, $d_c(\epsilon_p)$, beyond which sources do not contribute to the flux above ϵ_p. The function $d_c(\epsilon_p)$ is shown in Fig. 3 (adapted from [96]). Since $d_c(\epsilon_p)$ is a decreasing function of ϵ_p, for a given number density of sources there is a critical energy ϵ_c, above which only one source (on average) contributes to the flux. In the GRB model ϵ_c depends on the product of the burst rate $R_{\rm GRB}$ and the time delay. The number of sources contributing, on average, to the flux at energy ϵ_p is [96]

$$N(\epsilon_p) = \frac{4\pi}{5} R_{\rm GRB} d_c(\epsilon_p)^3 \tau\left[\epsilon_p, d_c(\epsilon_p)\right] \quad , \tag{26}$$

and the average intensity resulting from all sources is

$$J(\epsilon_p) = \frac{1}{4\pi} R_{\rm GRB} \frac{dn_p}{d\epsilon_p} d_c(\epsilon_p) \quad , \tag{27}$$

where $dn_p/d\epsilon_p$ is the number per unit energy of protons produced on average by a single burst (this is the formal definition of $d_c(\epsilon_p)$). The critical energy ϵ_c is given by

$$\frac{4\pi}{5} R_{\rm GRB} d_c(\epsilon_c)^3 \tau\left[\epsilon_c, d_c(\epsilon_c)\right] = 1 \quad . \tag{28}$$

ϵ_c, the energy beyond which a single source contributes on average to the flux, depends on the unknown properties of the inter-galactic magnetic field, $\tau \propto B^2\lambda$. However, the rapid decrease of $d_c(\epsilon_p)$ with energy near 10^{20}eV implies that ϵ_c is only weakly dependent on the value of $B^2\lambda$, as shown in Fig. 2. In The

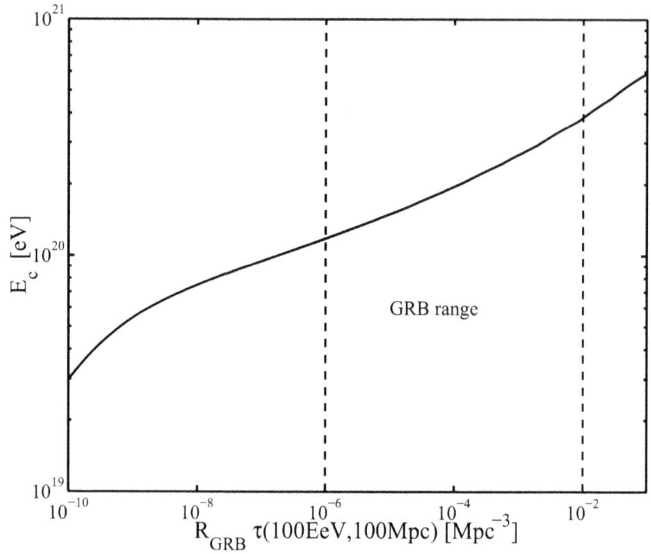

Fig. 2. ϵ_c, the energy beyond which a single GRB contributes on average to the UHECR flux, as a function of the product of GRB rate, $R_{\rm GRB} \simeq 1/{\rm Gpc}^3$, and the time delay of a 10^{20} eV proton originating at 100 Mpc distance. The time delay depends on the unknown inter-galactic field, $\tau \propto B^2\lambda$. Dashed lines show the allowed range of $B^2\lambda$: The lower limit is set by the requirement that at least a few GRB sources be present at $d < 100$ Mpc, and the upper limit by the Faraday rotation bound $B\lambda^{1/2} \leq 10^{-8}{\rm G\ Mpc}^{1/2}$ [94], see Eq. (25) and the discussion the follows it.

GRB model, the product $R_{\rm GRB}\tau(d = 100{\rm Mpc}, \epsilon_p = 10^{20}{\rm eV})$ is approximately limited to the range 10^{-6} Mpc^{-3} to 10^{-2} Mpc^{-3} [The lower limit is set by the requirement that at least a few GRB sources be present at $d < 100$ Mpc, and the upper limit by the Faraday rotation bound $B\lambda^{1/2} \leq 10^{-8}{\rm G\ Mpc}^{1/2}$ [94], see Eq. (25), and $R_{\rm GRB} \leq 1/\ {\rm Gpc}^3{\rm yr}$]. The corresponding range of values of ϵ_c is $10^{20}{\rm eV} \leq \epsilon_c < 4 \times 10^{20}{\rm eV}$.

Fig. 3 presents the flux obtained in one realization of a Monte-Carlo simulation described by Miralda-Escudé & Waxman [96] of the total number of UHECRs received from GRBs at some fixed time. For each realization the positions (distances from Earth) and times at which cosmological GRBs occurred were randomly drawn, assuming an intrinsic proton generation spectrum $dn_p/d\epsilon_p \propto \epsilon_p^{-2}$, and $\epsilon_c = 1.4 \times 10^{20}{\rm eV}$. Most of the realizations gave an overall spectrum similar to that obtained in the realization of Fig. 3 when the brightest source of this realization (dominating at 10^{20}eV) is not included. At $\epsilon_p < \epsilon_c$, the number of sources contributing to the flux is very large, and the overall UHECR flux received at any given time is near the average (the average flux is that obtained when the UHECR emissivity is spatially uniform and time independent). At $\epsilon_p > \epsilon_c$, the flux will generally be much lower than the average, because there will be no burst within a distance $d_c(\epsilon_p)$ having taken place sufficiently recently.

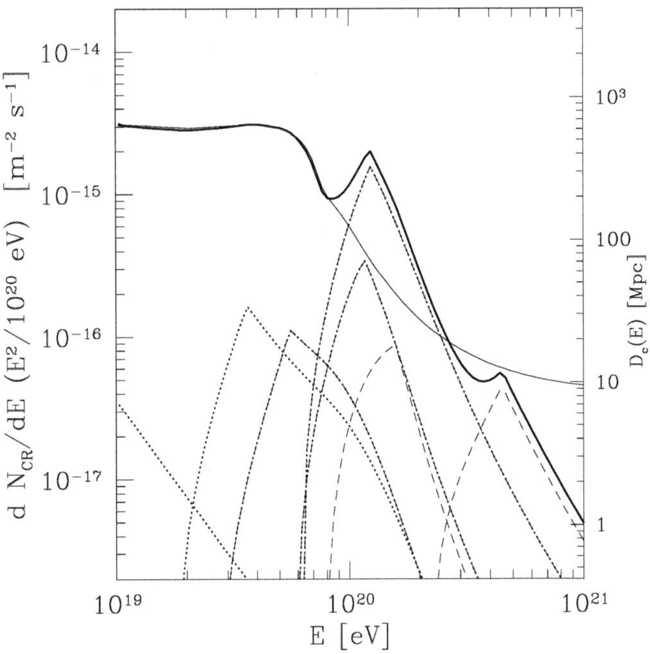

Fig. 3. Results of a Monte-Carlo realization of the bursting sources model, with $\epsilon_c = 1.4 \times 10^{20}$ eV: Thick solid line- overall spectrum in the realization; Thin solid line- average spectrum, this curve also gives $d_c(\epsilon_p)$; Dotted lines- spectra of brightest sources at different energies.

There is, however, a significant probability to observe one source with a flux higher than the average. A source similar to the brightest one in Fig. 3 appears $\sim 5\%$ of the time.

At any fixed time a given burst is observed in UHECRs only over a narrow range of energy, because if a burst is currently observed at some energy ϵ_p then UHECRs of much lower energy from this burst have not yet arrived, while higher energy UHECRs reached us mostly in the past. As mentioned above, for energies above the pion production threshold, $\epsilon_p \sim 5 \times 10^{19}$eV, the dispersion in arrival times of UHECRs with fixed observed energy is comparable to the average delay at that energy. This implies that the spectral width $\Delta\epsilon_p$ of the source at a given time is of order the average observed energy, $\Delta\epsilon_p \sim \epsilon_p$. Thus, bursting UHECR sources should have narrowly peaked energy spectra, and the brightest sources should be different at different energies. For steady state sources, on the other hand, the brightest source at high energies should also be the brightest one at low energies, its fractional contribution to the overall flux decreasing to low energy only as $d_c(\epsilon_p)^{-1}$. A detailed numerical analysis of the time dependent energy spectrum of bursting sources is given in [97,98] (see also the contribution by G. Sigl on propagation issues and interactions of UHECRs with extra-galactic magnetic fields).

5.2 Spectra of Sources at $\epsilon_p < 4 \times 10^{19}$eV

The detection of UHECRs above 10^{20}eV imply that the brightest sources must lie at distances smaller than 100Mpc. UHECRs with $\epsilon_p \leq 4 \times 10^{19}$eV from such bright sources will suffer energy loss only by pair production, because at $\epsilon_p < 5 \times 10^{19}$ eV the mean-free-path for pion production interaction (in which the fractional energy loss is $\sim 10\%$) is larger than 1Gpc. Furthermore, the energy loss due to pair production over 100Mpc propagation is only $\sim 5\%$.

In the case where the typical displacement of the UHECRs due to deflections by inter-galactic magnetic fields is much smaller than the correlation length, $\lambda \gg d\theta_s(d, \epsilon_p) \simeq d(d/\lambda)^{1/2}\lambda/r_L$, all the UHECRs that arrive at the observer are essentially deflected by the same magnetic field structures, and the absence of random energy loss during propagation implies that all rays with a fixed observed energy would reach the observer with exactly the same direction and time delay. At a fixed time, therefore, the source would appear mono-energetic and point-like. In reality, energy loss due to pair production results in a finite but small spectral and angular width, $\Delta\epsilon_p/\epsilon_p \sim \delta\theta/\theta_s \leq 1\%$ [99].

In the case where the typical displacement of the UHECRs is much larger than the correlation length, $\lambda \ll d\theta_s(\epsilon_p, d)$, the deflection of different UHECRs arriving at the observer are essentially independent. Even in the absence of any energy loss there are many paths from the source to the observer for UHECRs of fixed energy ϵ_p that are emitted from the source at an angle $\theta \leq \theta_s$ relative to the source-observer line of sight. Along each of the paths, UHECRs are deflected by independent magnetic field structures. Thus, the source angular size would be of order θ_s and the spread in arrival times would be comparable to the characteristic delay τ, leading to $\Delta\epsilon_p/\epsilon_p \sim 1$ even when there are no random energy losses. The observed spectral shape of a nearby ($d < 100$Mpc) bursting source of UHECRs at $\epsilon_p < 4 \times 10^{19}$eV was derived for the case $\lambda \ll d\theta_s(d, \epsilon_p)$ in [99], and is given by

$$\frac{dN}{d\epsilon_p} \propto \sum_{n=1}^{\infty}(-1)^{n+1}\, n^2 \exp\left[-\frac{2n^2\pi^2\epsilon^2}{\epsilon_0^2(t,d)}\right] \quad , \tag{29}$$

where $\epsilon_0(t, d) = de(2B^2\lambda/3ct)^{1/2}$. For this spectrum, the ratio of the RMS UHECR energy spread to the average energy is 30%

Figure 4 shows the line $\theta_s d = \lambda$ in the $B - \lambda$ plane, for a source at a distance $d = 30$Mpc observed at energy $\epsilon_p \simeq 10^{19}$eV. Since the $\theta_s d = \lambda$ line divides the allowed region in the plane at $\lambda \sim 1$Mpc, measuring the spectral width of bright sources would allow to determine if the field correlation length is much larger, much smaller, or comparable to 1Mpc.

6 High-Energy Neutrinos

6.1 Internal Shock (γ−Ray Burst) Neutrinos

Neutrinos at Energies $\sim 10^{14}$ eV. Protons accelerated in the fireball to high energy lose energy through photo-meson interaction with fireball photons.

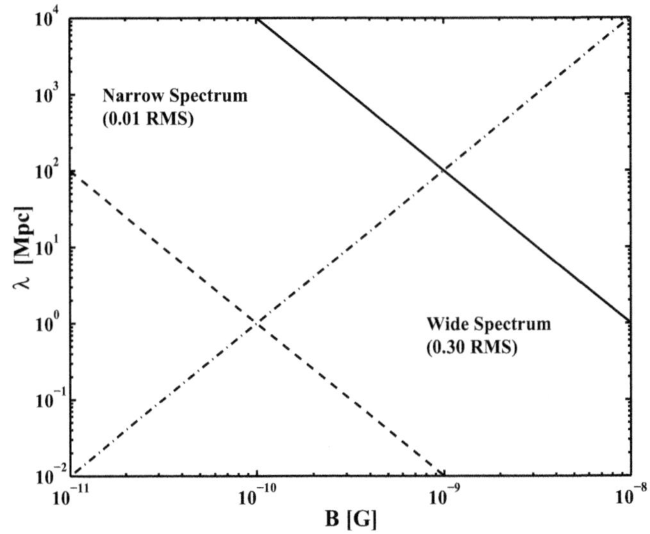

Fig. 4. The line $\theta_s d = \lambda$ for a source at 30Mpc distance observed at energy $\epsilon_p \simeq 10^{19}$eV (dot-dash line), shown with the Faraday rotation upper limit $B\lambda^{1/2} \le 10^{-8}$G Mpc$^{1/2}$ (solid line), and with the lower limit $B\lambda^{1/2} \ge 10^{-10}$G Mpc$^{1/2}$ required in the GRB model [see Eq. (25)].

The decay of charged pions produced in this interaction, $\pi^+ \to \mu^+ + \nu_\mu \to e^+ + \nu_e + \overline{\nu}_\mu + \nu_\mu$, results in the production of high energy neutrinos. The key relation is between the observed photon energy, ϵ_γ, and the accelerated proton's energy, ϵ_p, at the threshold of the Δ-resonance. In the observer frame,

$$\epsilon_\gamma \, \epsilon_p = 0.2 \, \mathrm{GeV}^2 \, \Gamma^2 \, . \tag{30}$$

For $\Gamma \simeq 300$ and $\epsilon_\gamma = 1$ MeV, we see that characteristic proton energies $\sim 10^{16}$ eV are required to produce pions. Since neutrinos produced by pion decay typically carry 5% of the proton energy (see below), production of $\sim 10^{14}$ eV neutrinos is expected [100,101].

The fractional energy loss rate of a proton with energy $\epsilon'_p = \epsilon_p/\Gamma$ measured in the wind rest frame due to pion production is

$$\begin{aligned} t_\pi^{-1}(\epsilon'_p) &\equiv -\frac{1}{\epsilon'_p}\frac{d\epsilon'_p}{dt} \\ &= \frac{1}{2\gamma_p^2} c \int_{\epsilon_0}^{\infty} d\epsilon \, \sigma_\pi(\epsilon)\xi(\epsilon)\epsilon \int_{\epsilon/2\Gamma_p}^{\infty} dx \, x^{-2} \frac{dn_\gamma}{d\epsilon_\gamma}(\epsilon_\gamma = x) \, , \end{aligned} \tag{31}$$

where $\gamma_p = \epsilon'_p/m_p c^2$, $\sigma_\pi(\epsilon)$ is the cross section for pion production for a photon with energy ϵ in the proton rest frame, $\xi(\epsilon)$ is the average fraction of energy lost to the pion, $\epsilon_0 = 0.15$GeV is the threshold energy, and $dn_\gamma/d\epsilon_\gamma$ is the photon density per unit photon energy in the wind rest frame. In deriving Eq. (31) we have assumed that the photon distribution in the wind rest frame is isotropic. The GRB photon spectrum is well fitted in the BATSE detector range, 20 keV to 2 MeV, by a combination of two power-laws, $dn_\gamma/d\epsilon_\gamma \propto \epsilon_\gamma^{-\beta}$, with $\beta \simeq 1$ at $\epsilon_\gamma < \epsilon_{\gamma b}$, $\beta \simeq 2$ at $\epsilon_\gamma > \epsilon_{\gamma b}$ and $\epsilon_{\gamma b} \sim 1$MeV [25]. Thus, the second integral in Eq. (31) may be approximated by

$$\int_\epsilon^\infty \mathrm{d}x\, x^{-2}\frac{dn_\gamma}{d\epsilon_\gamma}(\epsilon_\gamma = x) \simeq \frac{1}{1+\beta}\frac{U_\gamma}{2\epsilon'^3_{\gamma b}}\left(\frac{\epsilon}{\epsilon'_{\gamma b}}\right)^{-(1+\beta)}, \tag{32}$$

where U_γ is the photon energy density (in the range corresponding to the observed BATSE range) in the wind rest-frame, $\beta = 1$ for $\epsilon < \epsilon'_{\gamma b}$ and $\beta = 2$ for $\epsilon > \epsilon'_{\gamma b}$. $\epsilon'_{\gamma b}$ is the break energy measured in the wind frame, $\epsilon'_{\gamma b} = \epsilon_{\gamma b}/\Gamma$. The main contribution to the first integral in Eq. (31) is from photon energies $\epsilon \sim \epsilon_{\rm peak} = 0.3$GeV, where the cross section peaks due to the Δ resonance. Approximating the integral by the contribution from the resonance we obtain

$$t_\pi^{-1}(\epsilon'_p) \simeq \frac{U_\gamma}{2\epsilon'_{\gamma b}}c\sigma_{\rm peak}\xi_{\rm peak}\frac{\Delta\epsilon}{\epsilon_{\rm peak}}\min(1, 2\gamma_p\epsilon'_{\gamma b}/\epsilon_{\rm peak})\,. \tag{33}$$

Here, $\sigma_{\rm peak} \simeq 5\times10^{-28}{\rm cm}^2$ and $\xi_{\rm peak} \simeq 0.2$ are the values of σ and ξ at $\epsilon = \epsilon_{\rm peak}$, and $\Delta\epsilon \simeq 0.2$GeV is the peak width.

The energy loss of protons due to pion production is small during the acceleration process. Once accelerated, the time available for proton energy loss by pion production is comparable to the wind expansion time as measured in the wind rest frame, $t_{\rm co} \sim r/\Gamma c$. Thus, the fraction of energy lost by protons to pions is $f_\pi \simeq r/\Gamma c t_\pi$. The energy density in the BATSE range, U_γ, is related to the luminosity L_γ by $L_\gamma = 4\pi r^2\Gamma^2 c U_\gamma$. Using this relation in Eq. (33), and $r = 2\Gamma^2 c\Delta t$, f_π is given by [100]

$$f_\pi(\epsilon_p) \simeq 0.1\frac{L_{\gamma,52}}{\epsilon_{\gamma b,\rm MeV}\Gamma_{2.5}^4\Delta t_{-2}} \times \begin{cases} 1, & \text{if } \epsilon_p > \epsilon_{pb}; \\ \epsilon_p/\epsilon_{pb}, & \text{otherwise.} \end{cases} \tag{34}$$

The proton break energy is

$$\epsilon_{pb} \simeq 10^{16}\Gamma_{2.5}^2(\epsilon_{\gamma b,\rm MeV})^{-1}\,\mathrm{eV}\,. \tag{35}$$

The value of f_π, Eq. (34), is strongly dependent on Γ. It has recently been pointed out [102] that if the Lorentz factor Γ varies significantly between bursts, with burst to burst variations $\Delta\Gamma/\Gamma \sim 1$, then the resulting neutrino flux will be dominated by a few neutrino bright bursts, and may significantly exceed the flux implied by Eq. (34), derived for typical burst parameters. This may strongly enhance the detectability of GRB neutrinos by planned neutrino telescopes [103]. However, as explained in Sect.2.6, the Lorentz factors of fireballs producing observed GRBs can not differ significantly from the minimum value allowed by Eq. (5), $\Gamma \simeq 250$, for which the fireball pair production optical depth, Eq. (4), is $\simeq 1$ for $\epsilon_\gamma = 100$ MeV: Lower Lorentz factors lead to optically thick fireballs, while higher Lorentz factors lead to low luminosity X-ray bursts (which may have already been identified). Thus, for Lorentz factors consistent with observed GRB spectra, for which $\tau_{\gamma\gamma}(\epsilon_\gamma = 100\text{MeV}) \simeq 1$), we find

$$f_\pi(\epsilon_p) \simeq 0.2 \frac{L_{\gamma,52}^{1/3}}{\epsilon_{\gamma b,\text{MeV}} \Delta t_{-2}^{1/3}} \times \begin{cases} 1, & \text{if } \epsilon_p > \epsilon_{pb}; \\ \epsilon_p/\epsilon_{pb}, & \text{otherwise.} \end{cases} \tag{36}$$

A detailed analysis, using Monte-Carlo simulations of the internal shock model, confirms that for fireball parameter range consistent with observed GRB characteristics, f_π at $\epsilon_p > \epsilon_{pb}$ is limited to the range of $\sim 10\%$ to 30% [104].

Thus, for parameters typical of a GRB producing wind, a significant fraction of the energy of protons accelerated to energies larger than the break energy, $\sim 10^{16}$eV, would be lost to pion production. Roughly half of the energy lost by protons goes into π^0 's and the other half to π^+ 's. Neutrinos are produced by the decay of π^+'s, $\pi^+ \rightarrow \mu^+ + \nu_\mu \rightarrow e^+ + \nu_e + \overline{\nu}_\mu + \nu_\mu$ [the large optical depth for high energy γ's from π^0 decay, Eq. (4), would not allow these photons to escape the wind]. The mean pion energy is 20% of the energy of the proton producing the pion. This energy is roughly evenly distributed between the π^+ decay products. Thus, approximately half the energy lost by protons of energy ϵ_p is converted to neutrinos with energy $\sim 0.05\epsilon_p$. Equation (34) then implies that the spectrum of neutrinos above $\epsilon_{\nu b} = 0.05\epsilon_{pb}$ follows the proton spectrum, and is harder (by one power of the energy) at lower energy.

If GRBs are the sources of UHECRS, then using Eq. (24) the expected GRB neutrino flux is [101]

$$\begin{aligned} \epsilon_\nu^2 \Phi_{\nu_\mu} &\simeq \epsilon_\nu^2 \Phi_{\bar{\nu}_\mu} \simeq \epsilon_\nu^2 \Phi_{\nu_e} \simeq \frac{c}{4\pi} \frac{f_\pi}{8} \epsilon_p^2 (d\dot{n}_p/d\epsilon_p) t_H \\ &\simeq 1.5 \times 10^{-9} \frac{f_\pi(\epsilon_{pb})}{0.2} \min\{1, \epsilon_\nu/\epsilon_{\nu b}\} \text{GeV}\,\text{cm}^{-2}\text{s}^{-1}\text{sr}^{-1}, \end{aligned} \tag{37}$$

where $t_H \simeq 10^{10}$ yr is the Hubble time. The factor of 1/8 is due to the fact that charged pions and neutral pions are produced with roughly equal probabilities (and each neutrino carries $\sim 1/4$ of the pion energy).

The GRB neutrino flux can also be estimated directly from the observed gamma-ray fluence. The BATSE detectors measure the GRB fluence F_γ over two decades of photon energy, ~ 0.02MeV to ~ 2MeV, corresponding to a decade of radiating electron energy (the electron synchrotron frequency is proportional to the square of the electron Lorentz factor). If electrons carry a fraction ξ_e of the energy carried by protons, then the muon neutrino fluence of a single burst is $\epsilon_\nu^2 dN_\nu/d\epsilon_\nu \simeq 0.25(f_\pi/\xi_e)F_\gamma/\ln(10)$. The average neutrino flux per unit time and solid angle is obtained by multiplying the single burst fluence with the GRB rate per solid angle, $\simeq 10^3$ bursts per year over 4π sr. Using the average burst fluence $F_\gamma = 10^{-5}\text{erg/cm}^2$, we obtain a muon neutrino flux $\epsilon_\nu^2 \Phi_\nu \simeq 3\times 10^{-9}(f_\pi/\xi_e)\text{GeV/cm}^2\text{s sr}$. Thus, the neutrino flux estimated directly from the gamma-ray fluence agrees with the estimate Eq. (37) based on the cosmic-ray production rate.

Neutrinos at Energy > 10^{16} eV. The neutrino spectrum Eq. (37) is modified at high energy, where neutrinos are produced by the decay of muons and pions whose life time $\tau_{\mu,\pi}$ exceeds the characteristic time for energy loss due to adiabatic expansion and synchrotron emission [100,105,101]. The synchrotron loss time is determined by the energy density of the magnetic field in the wind rest frame. For the characteristic parameters of a GRB wind, the muon energy for which the adiabatic energy loss time equals the muon life time, ϵ_μ^a, is comparable to the energy ϵ_μ^s at which the life time equals the synchrotron loss time, τ_μ^s. For pions, $\epsilon_\pi^a > \epsilon_\pi^s$. This, and the fact that the adiabatic loss time is independent of energy and the synchrotron loss time is inversely proportional to energy, imply that synchrotron losses are the dominant effect suppressing the flux at high energy. The energy above which synchrotron losses suppress the neutrino flux is

$$\frac{\epsilon^s_{\nu_\mu(\bar\nu_\mu,\nu_e)}}{\epsilon_{\nu b}} \simeq \left(\frac{\xi_B}{\xi_e}L_{\gamma,52}\right)^{-1/2} \Gamma_{2.5}^2 \Delta t_{-2}\epsilon_{\gamma b,\mathrm{MeV}} \times \begin{cases} 10, & \text{for } \bar\nu_\mu,\ \nu_e; \\ 100, & \text{for } \nu_\mu\ . \end{cases} \tag{38}$$

The efficiency of neutrino production in internal collisions decreases with Δt, $f_\pi \propto \Delta t^{-1}$ [see Eq. (34)], since the radiation energy density is lower at larger collision radii. However, at larger radii synchrotron losses cut off the spectrum at higher energy, $\epsilon^s(\Delta t) \propto \Delta t$ [see Eq. (38)]. Collisions at large radii therefore result in extension of the neutrino spectrum of Eq. (37) to higher energy, beyond the cutoff energy Eq. (38),

$$\epsilon_\nu^2 \Phi_\nu \propto \epsilon_\nu^{-1}, \quad \epsilon_\nu > \epsilon_\nu^s. \tag{39}$$

Comparison with Other Authors We note, that the results presented above were derived using the Δ-approximation, i.e. assuming that photo-meson interactions are dominated by the contribution of the Δ-resonance. It has recently been shown [106], that for photon spectra harder than $dn_\gamma/d\epsilon_\gamma \propto \epsilon_\gamma^{-2}$, the contribution of non-resonant interactions may be important. Since in order to interact with the hard part of the photon spectrum, $\epsilon_\gamma < \epsilon_{\gamma b}$, the proton energy must exceed the energy at which neutrinos of energy $\epsilon_{\nu b}$ are produced, significant modification of the Δ-approximation results is expected only for $\epsilon_\nu \gg \epsilon_{\nu b}$, where the neutrino flux is strongly suppressed by synchrotron losses.

The neutrino flux from GRBs is small above 10^{19}eV, and a neutrino flux comparable to the γ-ray flux is expected only below $\sim 10^{17}$eV, in agreement with the results of Ref. [105]. Our result is not in agreement, however, with that of ref. [107], where a much higher flux at $\sim 10^{19}$eV is obtained based on the equations of ref. [100], which are the same equations as used here[1].

[1] The parameters chosen in [107] are $L_\gamma = 10^{50}$erg/s, $\Delta t = 10$s, and $\Gamma = 100$. Using equation (4) of ref. [100], which is the same as Eq. (34) of the present paper, we obtain for these parameters $f_\pi = 1.6 \times 10^{-4}$, while the author of [107] obtains, using the same equation, $f_\pi = 0.03$.

6.2 Reverse Shock (Afterglow) Neutrinos, $\sim 10^{18}$ eV

Let us now consider neutrino emission from photo-meson interactions of protons accelerated to high energies in the reverse shocks driven into the fireball ejecta at the initial stage of interaction of the fireball with its surrounding gas, which occurs on time scale $T \sim 10$ s, comparable to the duration of the GRB itself (see Sect.2.4). Optical–UV photons are radiated by electrons accelerated in shocks propagating backward into the ejecta (see Sect.2.7), and may interact with accelerated protons. The interaction of these low energy, 10 eV–1 keV, photons and high energy protons produces a burst of duration $\sim T$ of ultra-high energy, 10^{17}–10^{19} eV, neutrinos [as indicated by Eq. (30)] via photo-meson interactions [110].

Afterglows have been detected in several cases; reverse shock emission has only been identified for GRB 990123 [57]. Both the detections and the non-detections are consistent with shocks occurring with typical model parameters [46,59,58], suggesting that reverse shock emission may be common. The predicted neutrino emission depends, however, upon parameters of the surrounding medium that can only be estimated once more observations of the prompt optical afterglow emission are available. We first consider the case where the density of gas surrounding the fireball is $n \sim 1\mathrm{cm}^{-3}$, a value typical to the inter-stellar medium and consistent with GRB 990123 observations.

The photon density in Eq. (31) is related to the observed specific luminosity by $dn_\gamma/d\epsilon_\gamma(x) = L_\epsilon(\Gamma x)/(4\pi r^2 c\Gamma x)$. For proton Lorentz factor $\epsilon_0/2\epsilon'_{\gamma c} \ll \gamma_p < \epsilon_0/2\epsilon'_{\gamma m}$, where primed energies denote rest frame energies (e.g. $\epsilon'_{\gamma m} = \epsilon_{\gamma m}/\Gamma_{\rm tr.}$), photo-meson production is dominated by interaction with photons in the energy range $\epsilon_{\gamma m} < \epsilon_\gamma \ll \epsilon_{\gamma c}$, where $L_\epsilon \propto \epsilon_\gamma^{-1/2}$ (see Sect.2.7). For this photon spectrum, the contribution to the first integral of Eq. (31) from photons at the Δ-resonance is comparable to that of photons of higher energy, and we obtain

$$t_\pi^{-1}(\epsilon'_p) \simeq \frac{2^{5/2}}{2.5}\frac{L_m}{4\pi r^2\Gamma_{\rm tr.}}\left(\frac{\epsilon_{\rm peak}}{\gamma_p\epsilon'_{\gamma m}}\right)^{-1/2}\frac{\sigma_{\rm peak}\xi_{\rm peak}\Delta\epsilon}{\epsilon_{\rm peak}}. \quad (40)$$

$\Gamma_{\rm tr.}$ is the expansion Lorentz factor of plasma shocked by the reverse shocks, given by Eq. (14). The time available for proton energy loss by pion production is comparable to the expansion time as measured in the wind rest frame, $\sim r/\Gamma_{\rm tr.}c$. Thus, the fraction of energy lost by protons to pions is

$$f_\pi(\epsilon_p) \simeq 0.1\left(\frac{L_m}{10^{61}\mathrm{s}^{-1}}\right)\left(\frac{\Gamma_{\rm tr.}}{250}\right)^{-5} T_1^{-1} \times (\epsilon_{\gamma m,\mathrm{eV}}\epsilon_{p,20})^{1/2}. \quad (41)$$

Equation (41) is valid for protons in the energy range

$$4\times 10^{18}\left(\frac{\Gamma_{\rm tr.}}{250}\right)^2(\epsilon_{\gamma c,\mathrm{keV}})^{-1}\mathrm{eV} < \epsilon_p < 4\times 10^{21}\left(\frac{\Gamma_{\rm tr.}}{250}\right)^2(\epsilon_{\gamma m,\mathrm{eV}})^{-1}\mathrm{eV}\,. \quad (42)$$

Such protons interact with photons in the energy range $\epsilon_{\gamma m}$ to $\epsilon_{\gamma c}$, where the photon spectrum $L_\epsilon \propto \epsilon_{\gamma c}^{1/2}$ and the number of photons above interaction threshold is $\propto \epsilon_p^{1/2}$. At lower energy, protons interact with photons of energy $\epsilon_\gamma > \epsilon_{\gamma c}$, where $L_\epsilon \propto \epsilon^{-1}$ rather then $L_\epsilon \propto \epsilon^{-1/2}$. At these energies therefore $f_\pi \propto \epsilon_p$.

Since approximately half the energy lost by protons of energy ϵ_p is converted to neutrinos with energy $\sim 0.05\epsilon_p$, Eq. (42) implies that the spectrum of neutrinos below $\epsilon_{\nu b} \simeq 10^{17}(\Gamma_{\rm tr.}/250)^2(\epsilon_{\gamma c,{\rm keV}})^{-1}$eV is harder by one power of the energy then the proton spectrum, and by half a power of the energy at higher energy. For a power law differential spectrum of accelerated protons $dn_p/d\epsilon_p \propto \epsilon_p^{-2}$, the differential neutrino spectrum is $dn_\nu/d\epsilon_\nu \propto \epsilon_\nu^{-\alpha}$ with $\alpha = 1$ below the break and $\alpha = 3/2$ above the break. Assuming that GRBs are indeed the sources of ultra-high energy cosmic rays, then Eqs. (41,42) and (24) imply that the expected neutrino intensity is

$$\begin{aligned}\epsilon_\nu^2 \Phi_{\nu_\mu} &\simeq \epsilon_\nu^2 \Phi_{\bar\nu_\mu} \simeq \epsilon_\nu^2 \Phi_{\nu_e} \\ &\simeq 10^{-10}\frac{f_\pi^{[19]}}{0.1}\left(\frac{\epsilon_\nu}{10^{17}{\rm eV}}\right)^\beta {\rm GeV\,cm^{-2}s^{-1}sr^{-1}},\end{aligned} \tag{43}$$

where $f_\pi^{[19]} \equiv f_\pi(\epsilon_{p,20} = 2)$ and $\beta = 1/2$ for $\epsilon_\nu > 10^{17}$eV and $\beta = 1$ for $\epsilon_\nu < 10^{17}$eV.

Some GRBs may result from the collapse of a massive star, in which case the fireball is expected to expand into a pre-existing wind (e.g. [108,109]). For typical wind parameters, the transition to self-similar behavior takes place at a radius where the wind density is $n \simeq 10^4{\rm cm}^{-3} \gg 1{\rm cm}^{-3}$. The higher density implies a lower Lorenz factor of the expanding plasma during the transition stage, and a larger fraction of proton energy lost to pion production. Protons of energy $\epsilon_p \geq 10^{18}$ eV lose all their energy to pion production in this case. If most GRBs result from the collapse of massive stars, then the predicted neutrino flux is [110,111]

$$\epsilon_\nu^2 \Phi_\nu \simeq 10^{-8} \min\{1, \epsilon_\nu^{\rm ob.}/10^{17}{\rm eV}\}{\rm GeV\,cm^{-2}s^{-1}sr^{-1}}. \tag{44}$$

The neutrino flux is expected to be strongly suppressed at energy $\epsilon_\nu > 10^{19}$ eV, since protons are not expected to be accelerated to energy $\epsilon_p \gg 10^{20}$ eV. If protons are accelerated to much higher energy, the ν_μ ($\bar\nu_\mu$, ν_e) flux may extend to $\sim 10^{21} n_0^{-1/2}\xi_{B,-1}^{-1/2}$ eV ($\sim 10^{20} n_0^{-1/2}\xi_{B,-1}^{-1/2}$ eV). At higher energy, synchrotron losses of pions (muons) will suppress the neutrino flux.

6.3 Inelastic *p*-*n* Collisions

The acceleration, $\gamma \propto r$, of fireball plasma emitted from the source of radius r_0 (see Sect.2.2) is driven by radiation pressure. Fireball protons are accelerated through their coupling to the electrons, which are coupled to fireball photons. Fireball neutrons, which are expected to exist in most progenitor scenarios, are coupled to protons by nuclear scattering as long as the comoving *p*-*n* scattering time is shorter than the comoving wind expansion time $r/\gamma c = r_0/c$. As the fireball plasma expands and accelerates, the proton density decreases, $n_p \propto r^{-2}\gamma^{-1}$, and neutrons may become decoupled. For $\eta > \eta_{pn}$, where

$$\eta_{pn} \simeq 400 L_{52}^{1/4} r_{0,7}^{-1/4}, \tag{45}$$

neutrons decouple from the accelerating plasma prior to saturation, $\gamma = \eta$, at $\Gamma = \eta_{pn}^{4/3}\eta^{-1/3}$ [112,113]. In this case, relativistic relative velocities between protons and neutrons arise, which lead to pion production through inelastic nuclear collisions. Since decoupling occurs at a radius where the collision time is similar to wind expansion time, each n leads on average to one pair of $\nu\bar{\nu}$. The typical comoving neutrino energy, ~ 50 MeV, implies an observed energy ~ 10 GeV. A typical burst, $E = 10^{53}$ erg at $z = 1$, with significant neutron to proton ratio and $\eta > 400$ will therefore produce a fluence $F(\nu_e + \bar{\nu}_e) \sim 0.5F(\nu_\mu + \bar{\nu}_\mu) \sim 10^{-4}\mathrm{cm}^{-2}$ of ~ 10 GeV neutrinos.

Relativistic relative p-n velocities, leading to neutrino production through inelastic collisions, may also result from diffusion of neutrons between regions of the fireball wind with large difference in Γ [114]. If, for example, plasma expanding with very high Lorentz factor, $\Gamma > 100$, is confined to a narrow jet surrounded by a slower, $\Gamma \sim 10$ wind, internal collisions within the slower wind can heat neutrons to relativistic temperature, leading to significant diffusion of neutrons from the slower wind into the faster jet. Such process may operate for winds with $\eta < 400$ as well as for $\eta > 400$, and may lead, for certain (reasonable) wind parameter values, to ~ 10 GeV neutrino flux similar to that due to p-n decoupling in a $\eta > 400$ wind.

6.4 Implications

The high energy neutrinos predicted in the dissipative wind model of GRBs may be observed by detecting the Cerenkov light emitted by high energy muons produced by neutrino interactions below a detector on the surface of the Earth (see [115] for a recent review). The probability $P_{\nu\mu}$ that a neutrino would produce a high energy muon in the detector is approximately given by the ratio of the high energy muon range to the neutrino mean free path. For the neutrinos produced in internal shocks, $\epsilon_\nu \sim 10^{14}$ eV, $P_{\nu\mu} \simeq 1.3 \times 10^{-6}(\epsilon_\nu/1\mathrm{TeV})$ [115]. Using Eq. (37), the expected flux of neutrino induced muons is

$$J_\mu \simeq 10\frac{f_\pi(\epsilon_{pb})}{0.2}\mathrm{km}^{-2}\mathrm{yr}^{-1}\,. \tag{46}$$

The rate is almost independent of $\epsilon_{\nu b}$, due to the increase of $P_{\nu\mu}$ with energy. The rate Eq. (46) is comparable to the background expected due to atmospheric neutrinos [115]. However, neutrino bursts should be easily detected above the background, since the neutrinos would be correlated, both in time and angle, with the GRB γ-rays. A km^2 neutrino detector should detect each year ~ 10 neutrinos correlated with GRBs. Note, that at the high energies considered, knowledge of burst direction and time will allow to discriminate the neutrino signal from the background by looking not only for upward moving neutrino induced muons, but also by looking for down-going muons.

The predicted flux of $\sim 10^{17}$ eV neutrinos, produced by photo-meson interactions during the onset of fireball interaction with its surrounding medium,

Eqs. (43,44), may be more difficult to detect. For the energy range of afterglow neutrinos, the probability $P_{\nu\mu}$ that a neutrino would produce a high energy muon with the currently required long path within the detector is $P_{\nu\mu} \simeq 3\times10^{-3}(\epsilon_\nu/10^{17}\mathrm{eV})^{1/2}$ [115,116]. This implies, using Eq. (43), an expected detection rate of muon neutrinos of $\sim 0.06/\mathrm{km}^2\mathrm{yr}$ (over 2π sr), assuming fireballs explode in and expand into typical inter-stellar medium gas. If, on the other hand, most GRB progenitors are massive stars and fireballs expand into a pre-existing stellar wind, Eq. (44) implies a detection of several muon induced neutrinos per year in a $1\mathrm{km}^3$ detector. We note, that GRB neutrino detection rates may be significantly higher than derived based on the above simple arguments, because the knowledge of neutrino direction and arrival time may relax the requirement for long muon path within the detector.

Air-showers could be used to detect ultra-high energy neutrinos (see the contribution by P. Billoir in this volume). The neutrino acceptance of the planned Auger detector, $\sim 10^4\mathrm{km}^3\mathrm{sr}$ [18], seems too low. The effective area of proposed space detectors [19,20] may exceed $\sim 10^6\mathrm{km}^2$ at $\epsilon_\nu > 2\times10^{19}$ eV, detecting several tens of GRB correlated events per year, provided that the neutrino flux extends to $\epsilon_\nu > 2\times10^{19}$ eV. Since, however, the GRB neutrino flux is not expected to extend well above $\epsilon_\nu \sim 10^{19}$ eV, and since the acceptance of space detectors decrease rapidly below $\sim 10^{19}$ eV, the detection rate of space detectors would depend sensitively on their low energy threshold.

Detection of high energy neutrinos will test the shock acceleration mechanism and the suggestion that GRBs are the sources of ultra-high energy protons, since $\geq 10^{14}$ eV ($\geq 10^{18}$ eV) neutrino production requires protons of energy $\geq 10^{16}$ eV ($\geq 10^{19}$ eV). The dependence of $\sim 10^{17}$ eV neutrino flux on fireball environment imply that the detection of high energy neutrinos will also provide constraints on GRB progenitors.

Inelastic *p*-*n* collisions may produce ~ 10 GeV neutrinos with a fluence of $\sim 10^{-4}\mathrm{cm}^{-2}$ per burst, due to either *p*-*n* decoupling in a wind with high neutron fraction and high, > 400, Lorentz factor [112,113], or to neutron diffusion in a wind with, e.g., strong deviation from spherical symmetry [114]. The predicted number of events in a $1\mathrm{km}^3$ neutrino telescope is $\sim 10\mathrm{yr}^{-1}$. Such events may be detectable in a suitably densely spaced detector. Detection of ~ 10 GeV neutrinos will constrain the fireball neutron fraction, and hence the GRB progenitor.

Detection of neutrinos from GRBs could be used to test the simultaneity of neutrino and photon arrival to an accuracy of ~ 1 s (~ 1 ms for short bursts), checking the assumption of special relativity that photons and neutrinos have the same limiting speed. These observations would also test the weak equivalence principle, according to which photons and neutrinos should suffer the same time delay as they pass through a gravitational potential. With 1 s accuracy, a burst at 100 Mpc would reveal a fractional difference in limiting speed of 10^{-16}, and a fractional difference in gravitational time delay of order 10^{-6} (considering the Galactic potential alone). Previous applications of these ideas to supernova 1987A (see [117] for review), where simultaneity could be checked only to an accuracy of order several hours, yielded much weaker upper limits: of order

10^{-8} and 10^{-2} for fractional differences in the limiting speed and time delay respectively.

The model discussed above predicts the production of high energy muon and electron neutrinos. However, if the atmospheric neutrino anomaly has the explanation it is usually given, oscillation to ν_τ's with mass ~ 0.1 eV [118,119,120], then one should detect equal numbers of ν_μ's and ν_τ's. Up-going τ's, rather than μ's, would be a distinctive signature of such oscillations. Since ν_τ's are not expected to be produced in the fireball, looking for τ's would be an "appearance experiment." To allow flavor change, the difference in squared neutrino masses, Δm^2, should exceed a minimum value proportional to the ratio of source distance and neutrino energy [117]. A burst at 100 Mpc producing 10^{14}eV neutrinos can test for $\Delta m^2 \geq 10^{-16}\text{eV}^2$, 5 orders of magnitude more sensitive than solar neutrinos.

Acknowledgments

This work was supported in part by grants from the Israel-US BSF (BSF-9800343), MINERVA, and AEC (AEC-38/99). EW is the Incumbent of the Beracha foundation career development chair.

References

1. G. J. Fishman, C. A. Meegan: Ann. Rev. Astron. Astrophys. **33**, 415 (1995)
2. P. Mészáros: Proc. *17th Texas Symposium on Relativistic Astrophysics and Cosmology*, Annals Ny. Acad. Sci. No. 759 (NY Acad. Sci., NY 1995)
3. T. Piran: Phys. Rep. **333**, 529 (2000)
4. S. R. Kulkarni, et al.: Proc. *5th Huntsville Gamma-Ray Burst Symposium*, to appear, astro-ph/0002168
5. P. Mészáros: Astron. Astrophys. Supp. **138**, 533 (1999)
6. C. L. Fryer, S. E. Woosley: Astrophys. J. **502**, L9 (1998)
7. B. Paczynski: Proc. *Gamma Ray Bursts: 4th Huntsville Symp.*, eds. C. A. Meegan, R. D. Pierce, & T. M. Koshut (Woodbury: AIP), 783 (1998)
8. J. Goodman: Astrophys. J. **308**, L47 (1986)
9. B. Paczyński: Astrophys. J. **308**, L43 (1986)
10. M. Livio, E. Waxman: Astrophys. J. **538**, 187 (2000)
11. E. Waxman: Nucl. Phys. B (Proc. Suppl.) **87B**, 345 (2000)
12. S. C. Corbató, et al.: Nuc. Phys. **28B** (Proc. Suppl.), 36 (1992); see also http://www.physics.adelaide.edu.au/astrophysics/FlysEye.html
13. J. W. Cronin: Nucl. Phys. B (Proc. Suppl.) **28B**, 213 (1992)
14. A. A. Watson: Proc. *Nuclear and Particle Phys. 1993*, eds. MacGregor, I. J. D. & Doyle, A. T. (Inst. Phys. Conf. Ser. **133**) (Bristol:IoP), 135; see also http://www.auger.org/
15. M. Teshima, et al.: Nuc. Phys. **28B** (Proc. Suppl.), 169 (1992); see also http://www-ta.icrr.u-tokyo.ac.jp/
16. M. Takeda, et al.: Astrophys. J. **522**, 225 (1999)
17. F. Halzen: Proc. *17th International Workshop on Weak Interactions and Neutrinos* (Cape Town, South Africa, January 1999), p.123, astro-ph/9904216

18. K. S. Capelle, J. W. Cronin, G. Parente, E. Zas: Astropart. Phys. **8**, 321 (1998)
19. J. Linsley: Proc. *19th International Cosmic Ray Conference*, 3, 438 (1985)
20. Y. Takahashi: Proc. *24th International Cosmic Ray Conference*, 3, 595 (1995)
21. P. N. Bhat, et al.: Nature **359**, 217 (1992)
22. G. J. Fishman, et al.: Astrophys. J. Supp. **92**, 229 (1994)
23. J. H. Krolik, E. A. Pier: Astrophys. J. **373**, 277 (1991)
24. M. Baring: Astrophys. J. **418**, 391 (1993)
25. D. Band, et al.: Astrophys. J. **413**, 281 (1993)
26. M. Sommer, et al.: Astrophys. J. **422**, L63 (1994)
27. B. Paczyński: Astrophys. J. **363**, 218 (1990)
28. A. Shemi, T. Piran: Astrophys. J. **365**, L55 (1990)
29. M. Rees, P. Mészáros: Month. Not. R. Astron. Soc. **258**, 41P (1992)
30. R. Narayan, B. Paczyńsky, T. Piran: Astrophys. J. **395**, L83 (1992)
31. B. Paczyński, G. Xu: Astrophys. J. **427**, 708 (1994)
32. P. Mészáros, M. Rees: Month. Not. R. Astron. Soc. **269**, 41P (1994)
33. E. Woods, A. Loeb: Astrophys. J. **453**, 583 (1995)
34. R. Sari, T. Piran: Astrophys. J. **485**, 270 (1997)
35. R. D. Blandford, C. F. Mckee: Phys. Fluids **19**, 1130 (1976)
36. E. Waxman: Astrophys. J. **491**, L19 (1997)
37. R. Blandford, D. Eichler: Phys. Rep. **154**, 1 (1987)
38. W. I. Axford, E. Leer, G. Skadron, G.: Proc. *15th International Cosmic Ray Conference*, Plovdiv (Bulgaria) (1977)
39. A. R. Bell: Month. Not. R. Astron. Soc. **182**, 147 (1978)
40. R. D. Blandford, J. P. Ostriker: Astrophys. J. **221**, L229 (1978)
41. D. J. Helfand, R. H. Becker: Astrophys. J., **314**, 203 (1987)
42. P. J. Cargill, K. Papadopoulos K.: Astrophys. J., **329**, L29 (1988)
43. J. J. Brainerd, et al.: Proc. *19th Texas Symposium Relativistic Astrophysics and Cosmology*, eds. E. Aubourg et al., Nuc. Phys. B (Proc. Supp.) **80B** (2000)
44. D. Guetta, M. Spada, E. Waxman: Astrophys. J. **557**, 399 (2001)
45. R. M. Kippen: Proc. *Second Rome Workshop: Gamma-Ray Bursts in the Aferglow Era* (Rome, 2000), to appear, astro-ph/0201277
46. E. Waxman, B. T. Draine: Astrophys. J. **537**, 796 (2000)
47. E. Waxman: Astrophys. J. **485**, L5 (1997)
48. A. M. J. Wijers, M. J. Rees, P. Mészáros: Month. Not. R. Astron. Soc. **288**, L51 (1997)
49. B. Paczyński, J. Rhoads: Astrophys. J. **418**, L5 (1993)
50. J. I. Katz: Astrophys. J. **432**, L107 (1994)
51. P. Mészáros, M. Rees: Astrophys. J. **476**, 232 (1997)
52. M. Vietri: Astrophys. J. **478**, L9 (1997)
53. E. Costa, et al.: Nature **387**, 783 (1997)
54. S. Mao, H. J. Mo: Astron. Astrophys. **339**, L1 (1998)
55. M. Krumholtz, S. E. Thorsett, F. A. Harrison: Astrophys. J. **506**, L81 (1998)
56. D. W. Hogg, A. S. Fruchter: Astrophys. J. **520**, 54 (1999)
57. C. W. Akerlof, et al.: Nature, **398**, 400 (1999)
58. P. Mészáros, M. Rees: Month. Not. R. Astron. Soc. **306**, L39 (1999)
59. R. Sari, T. Piran: Astrophys. J. **517**, L109 (1999)
60. E. Waxman, S. Kulkarni, D. Frail, D.: Astrophys. J. **497**, 288 (1998)
61. D. Frail, E. Waxman, S. Kulkarni: Astrophys. J. **537**, 191 (2000)
62. D. L. Freedman, E. Waxman: Astrophys. J. **547**, 922 (2001)
63. T. Totani: Astron. Astrophys. Supp. **142**, 443 (2000)

64. K. Koyama, et al.: Nature **378**, 255 (1995)
65. A. M. Hillas: Ann. Rev. Astron. Astrophys. **22**, 425 (1984)
66. R. M. Kulsrud: Proc. *Particle Acceleration Mechanisms in Astrophysics*, eds. J. Arons, C. Max, and C. McKee (AIP 56, NY 1979)
67. E. Waxman: Phys. Rev. Lett. **75**, 386 (1995)
68. J. P. Rachen, P. Mészáros, P.: Proc. *4th Huntsville Symposium on GRBs*, AIP Conf. Proc. **428**, Gamma Ray Bursts, p. 776 (1998)
69. Y. A. Gallant, A. Achterberg: Month. Not. R. Astron. Soc. **305**, L6 (1999)
70. M. Vietri: Astrophys. J. **453**, 883 (1995)
71. C. D. Dermer: Astrophys. J., submitted, astro-ph/0005440
72. D. J. Bird, et al.: Phys. Rev. Lett. **71**, 3401 (1993)
73. D. J. Bird, et al.: Astrophys. J. **424**, 491 (1994)
74. N. Hayashida, et al.: Phys. Rev. Lett. **73**, 3491 (1994)
75. S. Yoshida, et al.: Astropart. Phys. **3**, 151 (1995)
76. M. Takeda, et al.: Phys. Rev. Lett. **81**, 1163 (1998)
77. A. A. Watson: Nucl. Phys. B (Proc. Suppl.) **22B**, 116 (1991)
78. T. K. Gaisser, et al.: Phys. Rev. D **47**, 1919 (1993)
79. B. R. Dawson, R. Meyhandan, K. M. Simpson: Astropart. Phys. **9**, 331 (1998)
80. N. N. Efimov, et al.: Proc. *International Symposium on Astrophysical Aspects of the Most Energetic Cosmic-Rays*, eds. M. Nagano and F. Takahara (World Scientific, Singapore, 1991), p. 20
81. N. Hayashida, et al.: Astrophys. J. **522**, 225 (1999) (and appendix astro-ph/0008102)
82. E. Waxman: Astrophys. J. **452**, L1 (1995)
83. B. J. Boyle, R. J. Terlevich: Month. Not. R. Astron. Soc. **293**, L49 (1998)
84. S. J. Lilly, O. Le Fevre, F. Hammer, D. Crampton: Astrophys. J. **460**, L1 (1996)
85. P. Madau, et al.: Month. Not. R. Astron. Soc. **283**, 1388 (1996)
86. P. C. Hewett, C. B. Foltz, F. Chaffee, F.: Astrophys. J. **406**, 43 (1993)
87. M. Schmidt, D. P. Schneider, J. E. Gunn: Astron. J. **110**, 68 (1995)
88. K. Greisen: Phys. Rev. Lett. **16**, 748 (1966)
89. G. T. Zatsepin, V. A. Kuzmin: JETP Lett. **4**, 78 (1966)
90. J. N. Bahcall, E. Waxman: Astrophys. J. **542**, 542 (2000)
91. F. A. Aharonian, J. W. Cronin: Phys. Rev. D **50**, 1892 (1994)
92. M. Milgrom, V. Usov: Astrophys. J. **449**, L37 (1995)
93. P. P. Kronberg, M. Reinhardt, M. Simard-Normandin: Astron. Astrophys. **61**, 771 (1977)
94. P. P. Kronberg: Rep. Prog. Phys. **57**, 325 (1994)
95. E. Waxman, P. Coppi: Astrophys. J. **464**, L75 (1996)
96. J. Miralda-Escudé, E. Waxman: Astrophys. J. **462**, L59 (1996)
97. G. Sigl, M. Lemoine, A. V. Olinto: Phys. Rev. D **56**, 4470 (1997)
98. M. Lemoine, G. Sigl, A. V. Olinto, D. N. Schramm: Astrophys. J. **486**, L115 (1997)
99. E. Waxman, J. Miralda-Escudé: Astrophys. J. **472**, L89 (1996)
100. E. Waxman, J. N. Bahcall: Phys. Rev. Lett. **78**, 2292 (1997)
101. E. Waxman, J. N. Bahcall: Phys. Rev. D **59**, 023002 (1999)
102. F. Halzen, D. W. Hooper: Astrophys. J. **527**, L93 (1999)
103. J. Alvarez-Muniz, F. Halzen, D. W. Hooper: Phys. Rev. D **62**, 093015 (2000)
104. D. Guetta, M. Spada, E. Waxman: Astrophys. J. **559**, 101 (2001)
105. J. P. Rachen, P. Mészáros: Phys. Rev. D **58**, 123005 (1998)
106. A. Mücke, et al.: Pub. Astron. Soc. Aust. **16**, 160 (1999)

107. M. Vietri: Phys. Rev. Lett. **80**, 3690 (1998)
108. R. A. Chevalier, Z. Li: Astrophys. J., **536**, 195 (2000)
109. S. E. Woosley, A. I. MacFadyen: Astron. Astrophys. Supp. **138**, 499 (1999)
110. E. Waxman, J. N. Bahcall: Astrophys. J. **541**, 707 (2000)
111. Z. G. Dai, T. Lu: Astrophys. J. **551**, 249 (2001)
112. E. V. Derishev, V. V. Kocharovsky, Vl. V. Kocharovsky: Astrophys. J. **521**, 640 (1999)
113. J. N. Bahcall, P. Mészáros: Phys. Rev. Lett., 85, 1362 (2000)
114. P. Mészáros, M. Rees: Astrophys. J. **541**, L5 (2000)
115. T. K. Gaisser, F. Halzen, T. Stanev: Phys. Rep. **258**, 173 (1995)
116. R. Gandhi, C. Quigg, M. Reno, I. Sarcevic: Phys. Rev. D **58**, 093009 (1998)
117. J. N. Bahcall: *Neutrino Astrophysics*, Cambridge University Press (New-York, 1989), pp. 438–460
118. D. Casper, et al.: Phys. Rev. Lett. **66**, 2561 (1991)
119. Y. Fukuda, et al.: Phys. Lett. **B335**, 237 (1994)
120. G. L. Fogli, E. Lisi: Phys. Rev. D **52**, 2775 (1995)

Cosmic Magnetic Fields from the Perspective of Ultra-High-Energy Cosmic Rays Propagation

Gustavo Medina Tanco

Insituto Astronômico e Geofísico, Universidade de São Paulo, Av. Miguel Stéfano 4200, CEP 04301-904 SP, Brazil

Abstract. The Ultra High Energy Cosmic Ray (UHECR) and Gamma Ray Burst (GRB) fields are very similar in many respects. Both have gone through a burst of activity in source modeling, both have a history of "repeaters" and, for both, the discovery of isotropy in arrival directions mean a fundamental change of current views and ideas. As in the case of gamma ray bursts (GRB), one can expect that the next significant step in our understanding of the UHECR problem will come when some source is finally identified as data accumulates from the new large exposure experiments under construction. However, UHECR are most likely charged particles, and so there is probably not such a thing as an easily identifiable optical counterpart, as there turned out to be with GRBs. Intervening galactic and intergalactic magnetic fields can affect propagation at energies of hundreds of EeV in a very significant way. This coupling means that both, cosmic magnetic fields and UHECR will have to be tackled together. UHECR will be, at the same time, the object of study and an invaluable diagnostic tool for magnetic fields inside the 100 Mpc sphere defined by the Greisen-Zatsepin-Kuzmin cut-off. The prospects for the future will be discussed.

1 General Considerations

Cosmic rays (CR) span an enormous interval in energy: more than 11 orders of magnitude. Several are the production, acceleration, and propagation mechanisms, as well as the experimental techniques, involved over this huge energy range encompassing a plummeting flux that goes down from 10^4 m^{-2} s^{-1} at ~ 1 GeV to 1 (100 km^2)$^{-1}$ yr^{-1} at ~ 100 EeV, the highest energies ever detected; see the introduction by P. Biermann and G. Sigl in this volume.

The spectrum is remarkably regular over this energy interval, and can be described by a succession of power laws separated by a few breaks: the knee at a few PeV, the second knee at ~ 0.75 EeV and the ankle at $\sim 5-10$ EeV. Of particular relevance to the study UHECR is the tale-telling GZK cut-off, long ago postulated at $\gtrsim 40$ EeV [1,2] but apparently not present according to current data.

Figure 1 shows an economical way of getting an insight into the general characteristics of CR over their whole energy range. There, the gyroradius of charged nuclei (p-O-Fe) as a function of energy for two fiducial magnetic fields values, 10^{-6} and 10^{-9} Gauss respectively, representative of the galactic magnetic field (GMF) and intergalactic magnetic field (IGMF), are given.

It can be seen that, at energies below 10^{18} eV, CR particles have gyroradii much smaller than the thickness of the galactic disk. At these relatively low

Martin Lemoine, Günter Sigl (Eds.): LNP 576, pp. 155–180, 2001.

energies CR propagate diffusively and are effectively confined inside the disk. However, at energies above $\sim 10^{18}$ eV, the gyroradii are much larger than the typical Galaxy thickness, and the diffusive approximation breaks down. Furthermore, for an average IGMF of $\sim 10^{-9}$ Gauss [3] the gyroradius of the lightest nuclei in the intergalactic medium are in excess of 100 Mpc at $E > 10^{20}$ eV.

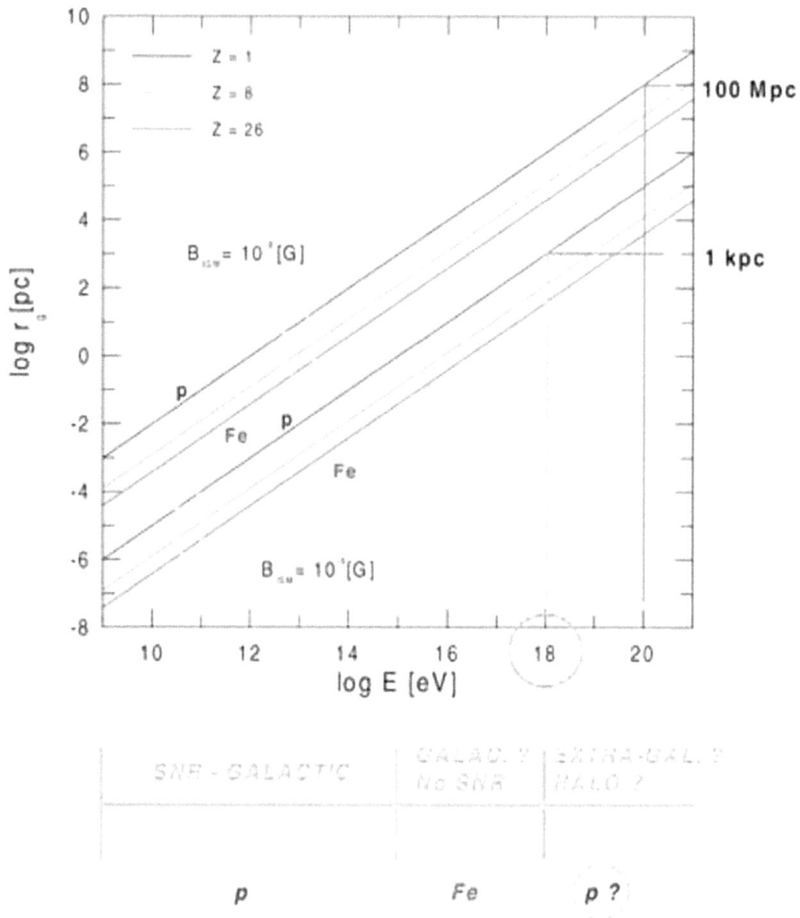

Fig. 1. Gyroradii of CR nuclei as a function of energy for typical GMF and IGMF values

Up to energies around the knee, CR are undisputedly galactic and, very likely, accelerated by first order Fermi processes at galactic supernova remnant (SNR) shock waves [4,5,6,7,8]. At energies between the knee and the ankle the origin of the particles is more uncertain though still galactic. SNR propagating through the wind of its progenitor star could play an important role in accelerating these nuclei, explaining also the observed gradual change to a heavier component [9].

Beyond the ankle, the increase in the maximum depth of air showers points to a change back to a lighter (proton dominated?) component [10]. At the same time, the arrival directions of UHECR seem remarkably isotropic [11,12], showing no evidence of correlation with either the galactic or the Supergalactic planes [13]. As light nuclei don't propagate diffusively inside the galactic plane at these energies, the previous observations point strongly to particle sources outside the galactic disk, either located in an extended galactic halo or extra-galactic.

As for the identity of the UHECR particles and their sources, a wide spectrum of possibilities is allowed by the scarce available data. Deeply penetrating primaries seem to be ruled out by Fly's Eye [14] and AGASA [15] data with an upper limit (at the 90% CL) 10 times smaller than the observed UHECR flux above $10^{19.5}$ eV [10]. There is no indication either in the AGASA, Haverah Park [16] or Fly's Eye [17] data set, that the primaries above 10^{20} eV are gamma-rays, as generally favored by top-down UHECR production mechanisms. Therefore, so far, observations leave hadrons as the most likely primaries.

Neutrons, even at the highest energies detected, have a Lorentz factor $\Gamma \sim 10^{11}$ and, therefore, decay into protons after a free fly of only ~ 1 Mpc. Thus, UHECR are probably charged hadrons. Heavy nuclei, on the other hand, interact strongly with the infrared (IR) background and photodisintegrate, which severely limit their travel distance. Although definite conclusions are critically dependent on the IR background flux assumed [18,19,20,21], it is likely that UHECR are mainly light nuclei, possibly protons.

Protons also interact with the all-pervading cosmic microwave background radiation (CMBR) via photo-pion production interactions. The interaction length is ~ 6 Mpc at $E \sim 10^{19.6}$ eV, with an inelasticity $\Delta E \sim 20$ % per interaction [22,23,24]. Therefore, unless the particles have unreasonably high energy at the sources, the flux at Earth should originate in a relatively small volume of nearby universe, the GZK-sphere, with a radius of the order of $50 - 100$ Mpc [25,26]. Furthermore, the energy spectrum resulting from a homogeneous distribution of sources should present a distinguishable GZK-cut-off at $E > 10^{19.5}$ eV [22,23,27]. Since the latter cut-off is apparently not present in the available data [28,29] there is probably some kind of clustering or even a local enhancement in the UHECR source density distribution.

Beyond these "facts", rather few things can be confidently ascertain about UHECR, and the way is open even for new physics to come into play at some level. The problem looks, in many respects, very similar to the GRB problem before the Beppo-Sax-era: few data, an essentially unknown distance scale, unknown particle source, unknown powering mechanism and possibly even unknown particles involved. Chances are, therefore, that further advancement in the UHECR field will come when particle sources are identified (or not!) with astrophysical counterparts. There is, however, a potentially fundamental difference at this point: we may be dealing with charged particles in the case of UHECR. This means that the particle trajectories are coupled, in principle, to an unknown extent to the intervening magnetic fields, both galactic and intergalactic.

Unfortunately, our knowledge of the topology and intensity of the large scale magnetic fields in the local universe and galactic halo is deficient; it resumes, mainly, to upper limits or rotation measurement constraints to the product of the magnetic field intensity times the square root of the coherence length averaged along some few lines of sight [3]. The implicit assumption of a particular field model with rather low IGMF intensities, combined with the high energy of particles beyond the GZK cut-off, has led to the notion of a random walk propagation scenario [30,25,31,32] where small particle deflections,

$$\theta(d,E) \simeq 0.025^o \left(\frac{d}{\lambda}\right)^{1/2} \left(\frac{\lambda}{10\text{Mpc}}\right) \left(\frac{B}{10^{-11}\text{G}}\right) \left(\frac{E}{10^{20}\text{eV}}\right) , \qquad (1)$$

allow for an easy identification of the sources. Here, d is the distance to the source, and λ and B are coherence length and strength of the magnetic field (see also Sect. 5 of the contribution by Sigl in this volume). Nevertheless, the propagation character of UHECR can be strongly dependent on the model assumed for the magnetic fields inside the propagation region, the GZK-sphere.

2 Magnetic Scenarios

Figure 2 shows, schematically, the UHECR propagation region. The main regions to distinguish are: (a) the source and its immediate, probably magnetized, environment, (b) the intergalactic medium, (c) the galactic halo, (d) the galactic disk, (e) the heliosphere and (f) the Earth magnetosphere.

The sources and their environments are essentially unknown. Magnetic fields, if present, are probably relevant to the acceleration mechanism and/or escape probability of the particles. However, we can always redefine the size of the source to include this environment and, consequently, we will not make such a distinction in what follows.

The heliosphere and Earth magnetosphere, despite their relatively large coherent magnetic fields, have length scales small enough to produce only minute deflections and, from the point of view of propagation can be neglected. The latter does not mean, however, that these magnetic fields cannot play a relevant role in other phenomena related to UHECR [33,34].

The main components of the problem are then, at least from the point of view of the magnetic field, the IGMF and GMF (both halo and disk field).

2.1 The Galactic Magnetic Field

The grand design of the Galactic magnetic field is difficult to observe from our position inside the system. However, a global picture exists, particularly reliable within 2 – 3 kpc from the Sun [35], based on Zeeman splitting of radio and maser lines, and Faraday rotation measurements from pulsars and extragalactic sources, complemented with frequency dependent time delay measurements of pulsar signals. Several parametrizations are available [36,37] for the regular

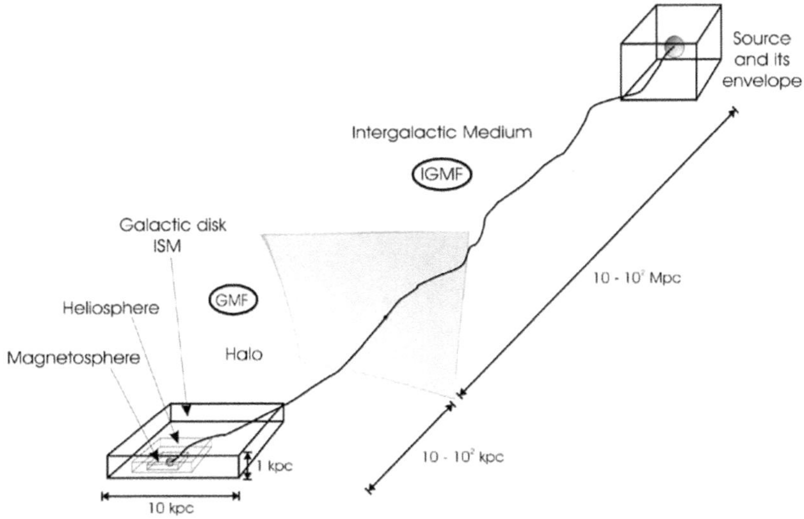

Fig. 2. Schematical representation of the UHECR propagation region

component of the GMF. Radio polarization observations (see Fig. 3) of other phase-on and edge-on spiral galaxies confirm the existence of such a large scale ordered magnetic component, encompassing the visible spiral disk structure, and possibly extending into the galaxy halo. These polarization observations of other spiral galaxies represent an important test, since rotation measures of the GMF sample mainly the warm component of the interstellar medium, which has a filling factor of only ~ 20 %, rising doubts about the global relevance of the latter results.

Furthermore, the large scale magnetic field structures in the few tens of spiral galaxies studied so far, present a narrow spectrum of topologies. Basically, in the plane of the galaxy the field is either axisymmetric or bisymmetric while in the direction perpendicular to the galaxy plane the horizontal component has either quadrupolar (horizontal field even in z and perpendicular field odd in z) or dipolar (horizontal field odd in z and perpendicular field even in z) symmetry, and mode mixing can also occur [35]. It is not clear which of these symmetries actually applies to our own galaxy. Some authors favor an axisymmetric model [40,41], while others prefer a bisymmetric one [42,37]. The symmetry of the horizontal component with respect to the midplane is somehow masked by local fluctuations in the vicinity of the Sun, therefore, claims exist for both, odd and even symmetries, although the latter probably agrees better with the observations. The perpendicular component is very small near the midplane and is difficult to disentangle from local inhomogeneities. At least one reversal in the field direction exist internal to the solar circle at $0.2 - 0.3$ kpc in the direction of the Sagitarius arm and there is widespread controversy regarding the existence of either one or two additional inner and one outer reversals. In any case, the Sun is too close to a reversal and, therefore, the local GMF intensity may not

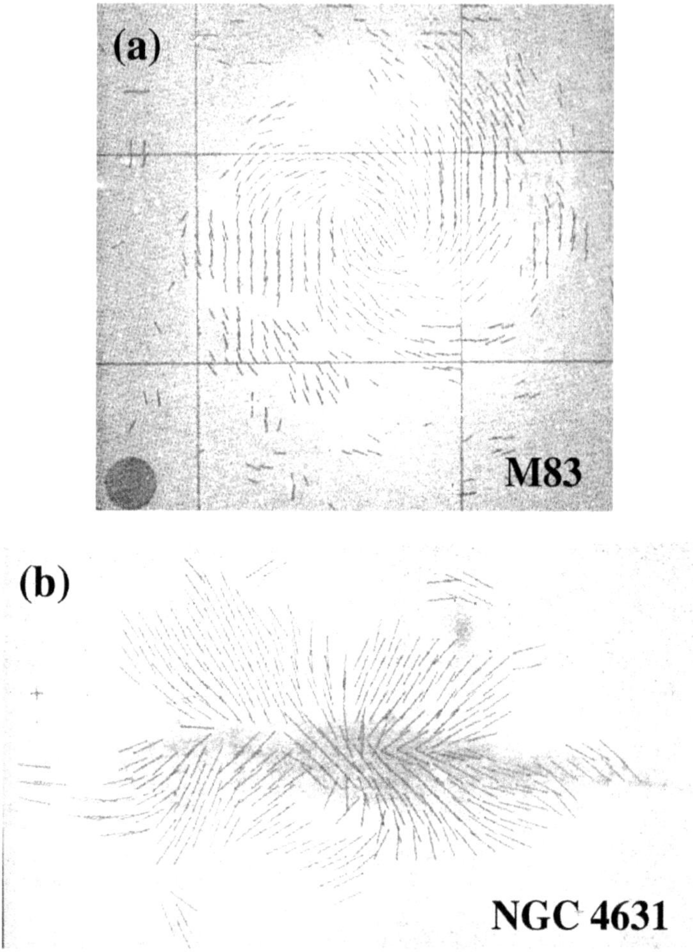

Fig. 3. The regular component of the magnetic field in spiral galaxies [38,39]

be representative of the Galaxy as a whole, where typical values might run as high as 4–6 μG.

Superimposed on the regular component is a random component of comparable, or even larger intensity. Its origin and maintenance mechanism are poorly known. The strength of the rms random magnetic field, however, seems to reflect equipartition with the kinetic energy in the turbulence. A comprehensive study of Faraday rotation measurement differences between pairs of pulsars points to an rms amplitude of $4 - 6$ μG and coherence length between 10 and 100 pc [43]. A plausible mechanism to produce the random field component is the small scale dynamo, which would produce magnetic flux ropes with a length of about 50–100 pc, thickness of $\sim$ 5–10 pc and a filling factor of $\sim 1\%$ embedded in the warm phase of the ISM [44]. Some kind of consensus seems to exist around a model in which the strength of the regular field is 1.8 ± 0.3 μG and the total local field is ~ 5 μG probably increasing towards the Galactic center [45].

Therefore, even if the present uncertainties preclude the exact knowledge of the GMF, a limited spectrum of possible configurations exists which can be used to check different UHECR propagation scenarios. Furthermore, in the future, whenever better statistics on UHECR are available, these data could be used to further improve our knowledge on the GMF.

Given a parameterization for the GMF, particles arriving at Earth from any direction in the sky can be back-tracked through the galactic disk and halo indexmagnetic field!galactic halo to the border of the halo where the IGMF starts. This procedure defines a mapping between the arrival directions at the border of the halo and the arrival directions at the position of the Earth inside the Galaxy. In this way, the intrinsic angular error boxes due to the existence of a GMF can be estimated as a function of arrival direction. A couple of examples of such calculation are given, in galactic coordinates, in Fig. 4 [46,47] for an axisymmetric, quadrupolar, without a z (perpendicular to the galactic plane) component (ASS-S) model of GMF [48]. Two maps are shown for incoming (a) protons and (b) iron nuclei at 2.5×10^{20} eV. It can be seen that, for protons in this particular GMF model, the intrinsic error boxes due to the GMF are $< 1^o$ for most of the sky; the exception are those lines of sight that cross the central region of the Galaxy and first quadrant at low galactic latitudes, where the angular uncertainty can grow rapidly beyond 4°. It must be kept in mind, however, that it is in the central regions of the Galaxy where our knowledge of the GMF is poorer. Iron nuclei, on the other hand, suffer very large deflections over most of the observable sky even at energies as high as 2.5×10^{20} eV.

In a recent series of papers, Harari, Mollerach and Roulet [49,50,51] have studied the propagation of UHECR with a particular emphasis on lensing effects. Both, more "standard" GMF models as the ones described above, and a GMF model including a galactic magnetized wind (as in Ahn et al. [52]) were analyzed. They show that multiple images from a single source can occur quite frequently, complicating the unambiguous identification of individual UHECR sources and allowing the formation of spurious clusters of events. If the exposures are large enough to allow the arrival to the detector of several events coming from a single source, then flux magnification effects can also be expected in the vicinity of caustic curves. These phenomena are, in general, energy dependent and should also distort the observed energy spectrum.

2.2 The Intergalactic Magnetic Field

Luminous matter, as traced by galaxies, as well as dark matter, as traced by galaxies and clusters large scale velocity fields, is distributed inhomogeneously in the universe. Groups, clusters, superclusters, walls, filaments and voids are known to exist at all observed distances and are very well mapped in the local universe. Hence, the distribution of matter inside the GZK-sphere is highly inhomogeneous and so is, very likely, the distribution of UHECR sources.

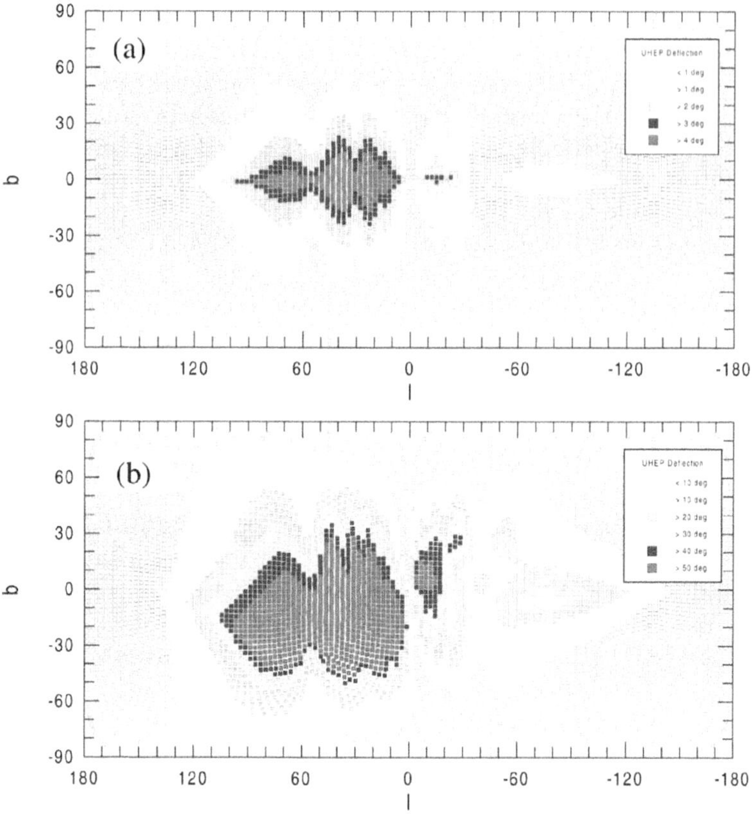

Fig. 4. Angular error box as a function of arrival direction in galactic coordinates for UHECR of $E = 2.5 \times 10^{20}$ eV. The incoming particles are (a) protons and (b) Fe nuclei. ASS-S GMF model

Synchrotron emission and multi-wavelength radio polarization measurements show that magnetic fields are widespread in the Universe. But how do they encompass the structure seen in the distribution of matter we do not yet know [3].

The available limits on the IGMF come from Faraday rotation measurements in clusters of galaxies and suggest that $B_{\rm IGM} \times \lambda^{1/2} < 10^{-9}$ G $\times$ Mpc$^{1/2}$ [3], where λ is the field reversal scale. Note, however, that this kind of measurement doesn't set an actual limit to the intensity of the magnetic field unless the reversal scale is known along a particular line of sight. The latter means that, depending on the structure of the IGMF, substantially different scenarios can be envisioned that are able to satisfy the rotation measurement constraints.

To complicate things further, the rotation measure is not just a measure of the integral of the component of $B_{\rm IGM}$ projected along the line of sight. In fact,

$$RM \propto \int_0^s n_e(s) \boldsymbol{B}_{\rm IGM} . d\boldsymbol{s} \qquad (2)$$

and the electron density, n_e, along the line of sight plays as important a role as the magnetic field in defining the observed value of the Faraday rotation measure. In the past, oversimplified models of the electron density have been used, usually assuming a homogeneous distribution scaled with redshift. It has been shown recently [53], that those estimates can be up to one order of magnitude smaller than models assuming a inhomogeneous electron density distribution which follows the observed Lyman-α forest distribution, i.e., $\sim 10^{-8}$ vs. 10^{-9} G. Furthermore, the scatter among different lines of sight seems to be too large to determine a significant limit.

Unfortunately, we do not know what is the actual large scale structure of the IGMF. Nevertheless, we can imagine two extreme scenarios that are likely to bound the true IGMF structure. In Fig. 5 calculations of large scale structure formation by Ryu and co-workers [54] have been modified by hand to exemplify these scenarios. The top frame displays Ryu's IGMF simulation results in the background showing how, by $z = 0$, the magnetic field has been convected together with the accretion flows into walls, filaments and clusters, depleting the voids from field. According to these calculations, the magnetic field is confined in high density, small filling factor regions, bounded by a rather thin skin of rapidly decreasing intensity, surrounded by large volumes of negligible IGMF. As suggested by the free-hand lines on top of Fig. 5, the IGMF inside structures is highly correlated in scales of up to tens of pc. Furthermore, in order to comply with the rotation measurement constraints mentioned before, the intensity of the magnetic field inside the density structures should be correspondingly higher. Therefore, a plausible upper limit could range from 0.1 to 1 μG, i.e., from comparable to, to several times larger than, a field in equipartition with the thermal energy of the plasma in filaments and sheets. Notably, this is also comparable with GMF values within the interstellar medium. We will call the latter scenario *laminar-structure* .

The second model, that we will call *cellular-structure* , is depicted in the bottom panel of Fig. 5. We imagine the space divide into adjacent cells, each one with a uniform magnetic field randomly oriented. We identify the size of a cell with the magnitude of the local reversal scale. Furthermore, one can assume that the intensity of the magnetic field scales as some power of the local matter (electron) density and, consequently, the rotation measurement constraint $B_{\rm IGM} \times \lambda^{1/2} < 10^{-9}$ G $\times$ Mpc$^{1/2}$ tells how the reversal scale, i.e., the size of the cells, should be scaled. A convenient reference, such as the IGMF in the Virgo [55] or Coma [56] cluster can be used for normalization. The cellular-structure scenario leads to a more widespread IGMF, filling even the voids. The observational constraints imply then that the IGMF varies much more smoothly, from 10^{-10}G inside voids to a few times 10^{-9}–10^{-8}G inside walls and filaments, only reaching high values, $0.1 - 1$ μG, inside and around clusters of galaxies.

Observations are not enough at present to distinguish between these two scenarios, but we can still try to asses what are their implications for UHECR propagation.

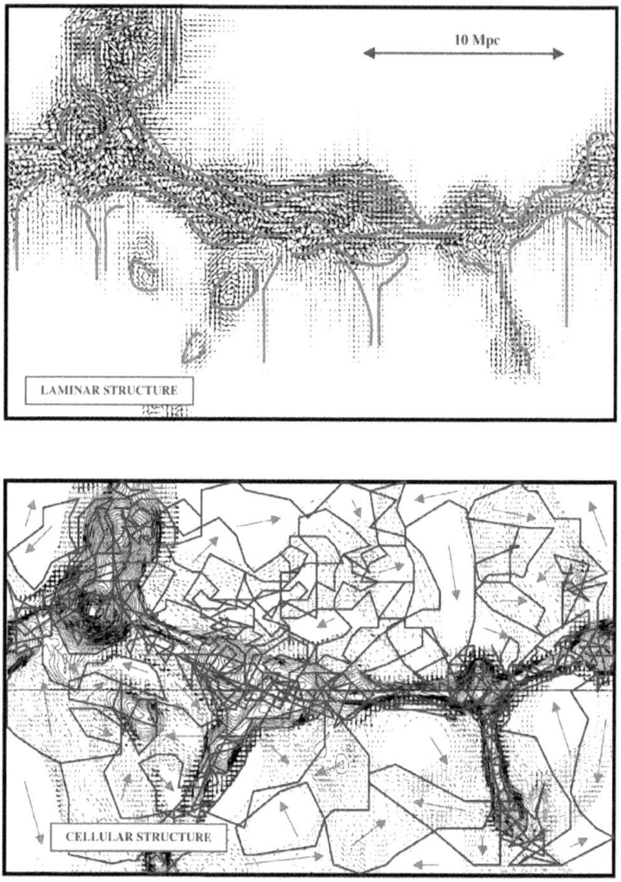

Fig. 5. two possible extreme models for the IGMF structure: (top) laminar structure and (bottom) cellular structure. These schematic plots are adaptations by hand made on top of IGMF and density calculations by Ryu, Kang, and Biermann [54]

2.3 UHECR Propagation in a Laminar IGMF

This is the most difficult scenario to dealt with because it does not accept a statistical treatment and results are very dependent on details about the exact magnetic field configuration inside the GZK-sphere, which is beyond our present knowledge.

A simpler approach is to study the UHECR emissivity of a single wall surrounded by a void [58,57]. Figure 6 shows the corresponding model for a wall immersed in a void; the magnetic field inside the wall has two components, o uniform field along the z-axis of intensity 0.1 μG and a random component with a Kolmogorov power law spectrum of amplitude equal to 30% of the regular component. One hundred UHECR sources are included inside the wall, and each one of them injects protons at the same rate and with the same power low energy spectrum, $dN_{\rm inj}/dE \propto E^{-2}$. Pair production and photo-pion production losses

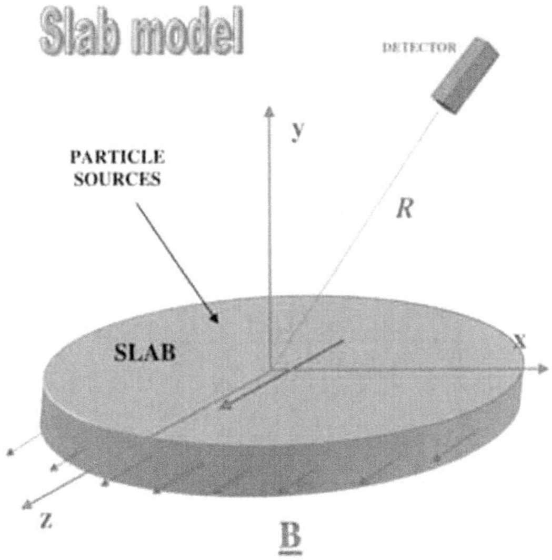

Fig. 6. Simplified model of a wall, or slab, containing UHECR sources and surrounded by a void. The magnetic field configuration is representative of the laminar model [57]

in interactions with the IR and microwave backgrounds respectively are also included. The wall has a radius of 20 Mpc, a thickness of 5 Mpc and is sandwiched by a transition layer 5 Mpc in thickness where the magnetic field decreases exponentially up to negligible values inside the surrounding void. Once the system reaches steady state, a detector can be shifted around the wall to simulate observers at arbitrary positions with respect to the wall. In a real situation, this system could be representative, for example, of the Supergalactic plane (SGP); in that case the Milky Way, i.e. we, the observers, should be located at some point on the x-z plane (but we don't know at what angle with respect to the z-axis). The simulations show that the UHECR flux measured can vary by three orders of magnitude depending of the relative orientation between the wall, the field and the observer. At the same time, almost all directional information is lost, and the strength of the GZK-cut-off would vary considerably as a function of orientation [57].

The previous effects can be intuitively understood by looking at Fig. 7, which shows a cross section of the wall in Fig. 6 at the plane $z = 0$. Several particles trajectories are shown for proton injection at $E = 100$ EeV, with different azimuthal angles and a slight elevation with respect to the x-y plane. It can be seen that there is nothing like a random walk: particles tend to be trapped inside the wall and move in a systematic way. Most of the particles drift perpendicularly to the regular field while their guiding centers bounce along the field. It can also be seen how the gyroradii decrease as the particles lose energy in interactions with the radiation backgrounds. Even the few particles that do escape from the wall, do so in an non-isotropic manner (e.g., predominantly to the right for $y > 0$).

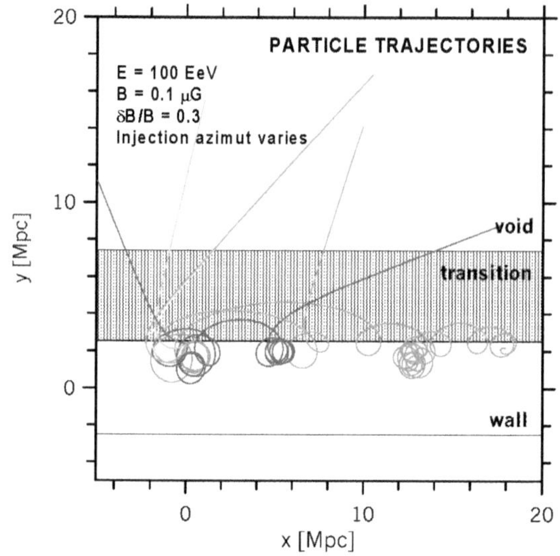

Fig. 7. Cross section of the wall in Fig. 6 at the plane $z = 0$. Several particles trajectories are shown for proton injection at $E = 100$ EeV

The laminar IGMF model is, actually, the worst scenario for doing some kind of astronomy with UHECR. It would be very difficult to interpret the UHECR angular data and to identify individual particle sources. Furthermore, the significance of any statistical analysis would be greatly impaired due to systematics.

Further studies on this model have also been performed by Sigl, Lemoine, and Biermann [59,60] and will be discussed in the contribution by G. Sigl in this volume.

2.4 UHECR Propagation in a Cellular IGMF

The cellular model is the easiest scenario to deal with numerically and, by far, the most promising from the point of view of the astrophysics of UHECR. This is also the IGMF model that has been used probably more frequently in the literature [25,27,30,31,32,61,62,63,64,65,66,67,68,69].

The main assumption here, is that the intensity of the magnetic field scales with density. Indeed, for those spatial scales where measurements are available, the intensity of the magnetic fields seem to correlate remarkably well with the density of thermal gas in the medium. This is valid at least at galactic and smaller scales [70,71]. It is apparent that B can be reasonably well fitted by a single power law over ~ 14 orders of magnitude in thermal gas density at sub-galactic scales. A power law correlation, though with a different power law index, is also suggested at very large scales (c.f. Fig. 5 in [71]), from galactic halos indexmagnetic field!galactic halo to the environments outside galaxy clusters, over ~ 4 orders of magnitude in thermal gas density. This view [70] is, however, still controversial [3]. In fact, magnetic fields in galaxy clusters are roughly ~ 1 μG, which is of the order of interstellar magnetic fields; furthermore, supracluster emission around the Coma cluster suggests μG fields in extended regions beyond cluster cores. The latter could indicate that the IGMF cares little about the density of the associated thermal gas density, having everywhere an intensity close to the microwave background-equivalent magnetic field strength, $B_{\rm BGE} \simeq 3 \times 10^{-6}$ G.

Taking the view that a power law scaling exist, a model can be devised in which the IGMF correlates with the distribution of matter as traced, for example, by the distribution of galaxies. A high degree of non-homogeneity should then be expected, with relatively high values of $B_{\rm IGMF}$ over small regions (< 1 Mpc) of high matter density. These systems should be immersed in vast low density/low $B_{\rm IGMF}$ regions with $B_{\rm IGMF} < 10^{-9}$ G. Furthermore, in accordance with rotation measurements, the topology of the field should be such that it is structured coherently on scales of the order of the coherence length λ which, in turn, scales with IGMF intensity: $\lambda \propto B^{-2}_{\rm IGMF}(r)$. $\boldsymbol{B}_{\rm IGMF}$ should be independently oriented at distances $> \lambda$. Therefore, a 3D ensemble of cells can be constructed, with cell size given by the coherence length, λ, and such that: $\lambda \propto B^{-2}_{\rm IGMF}(r)$, while $B_{\rm IGMF} \propto \rho^{0.35}_{\rm gal}(r)$ [70] or $\propto \rho^{2/3}_{\rm gal}(r)$ (for frozen-in field compression), where $\rho_{\rm gal}$ is the galaxy density, and the IGMF is uniform inside cells of size λ and randomly oriented with respect to adjacent cells [65,69]. The observed IGMF value at some given point, like the Virgo cluster, can be used as the normalization condition for the magnetic field intensity. The density of galaxies, $\rho_{\rm gal}$, is estimated using either redshift catalogs [like the CfA Redshift Catalogue [66,27] or the PSCz [72]], or large scale structure formation simulations [69]. The latter is a convenient way to cope with, or at least to assess the importance of, the several biases involved in the use of galaxy redshift surveys to sample the true spatial distribution of matter in 3D space.

The relevant energy losses for UHECR during propagation are: pair production via γ–γ with CMB for photons, redshift, pair production and photopion production in interactions with the CMB for nucleons and, for heavy nuclei, also photodisintegration in interactions with the IR background. All of these can be appropriately included [73,74,75,76]. The spatial distribution of the sources of UHECR is tightly linked to the nature of the main particle acceleration/production mechanism involved. However, in most models, particles will either be accelerated at astrophysical sites that are related to baryonic matter, or produced via decay of dark matter particles. In both cases the distribution of galaxies (luminous matter) should be an acceptable, if certainly not optimal, tracer of the sources.

Once the previously described scenario is built, test particles can be injected at the sources and propagated through the intergalactic medium and intervening IGMF to the detector at Earth.

Figure 8a-b show the arrival probability distribution of UHECR protons as a function of galactic coordinates for a distribution of sources following the

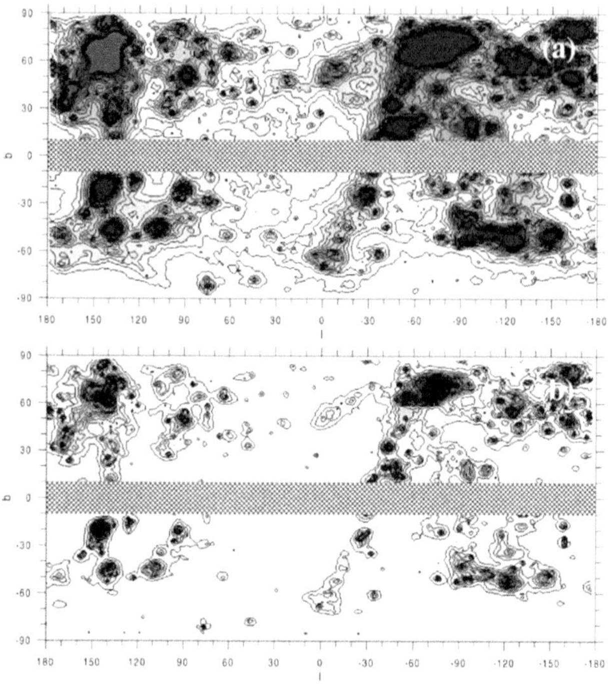

Fig. 8. Arrival probability distribution of protons (linear scale) as a function of galactic coordinates for a distribution of sources following the distribution of luminous matter inside 100 Mpc. $dN_{\rm inj}/dE \propto E^{-2}$, with (a) $E_{\rm inj} > 4 \times 10^{19}$ eV and (b) $E_{\rm inj} > 10^{20}$ eV. The fiducial IGMF of Sect. 2.4 is used

distribution of luminous matter inside 100 Mpc (CfA2 catalog). A power law injection energy spectrum at the sources is assumed, $dN_{\rm inj}/dE \propto E^{-2}$, with (a) $E_{\rm inj} > 4 \times 10^{19}$ eV and (b) $E_{\rm inj} > 10^{20}$ eV respectively.

It can be seen that, in contrast to the laminar IGMF case, in this scenario information regarding the large scale distribution of the sources inside the GZK-sphere can be easily recoverable. The SGP and the Virgo cluster, in particular, are clearly visible between $l \simeq 0$–-100. It can also be appreciated the increase in resolution as the energy reaches the 100 EeV range and the gyroradii of UHECR protons become comparable to the size of the GZK-sphere. It is also in the cellular model that the deflection angle of the incoming particle with respect to the true angular position of the source (see Fig. 9) is small enough for an UHECR astronomy to develop at the largest energies.

3 Theory versus Data

From the astrophysical point of view, there are at present two elementary observations that can be used to constraint the production and propagation models

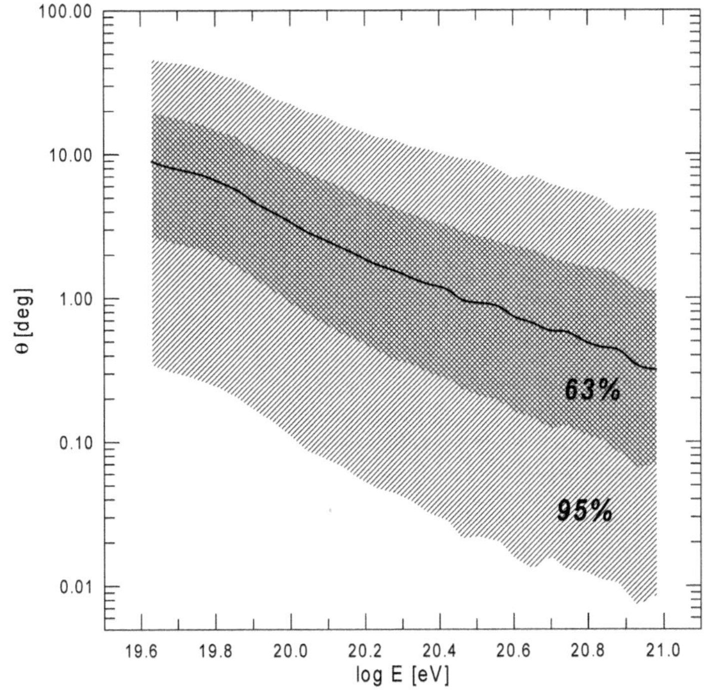

Fig. 9. Median and 63% and 95% C.L. for the deflection angle of an incoming UHECR proton with respect to the true angular position of the source for the example in Fig. 8. All sky average

of UHECR: arrival energy and arrival direction of the individual particles, which are statistically encoded as spectral energy distribution and angular distribution respectively.

The extension of the UHECR spectrum beyond the GZK cut-off has long been hinted by the extreme high energy events of Volcano Ranch [77,78], Haverah Park [79,80], Fly's Eye [81] and AGASA [82], and recently confirmed by the latter experiment [28,83].

It is not clear, however, whether the available data (461 events for $E > 10^{19}$ eV, and only 6 events for $E > 10^{20}$ eV) is sufficient to support any conjecture about the actual shape of the spectrum above 10^{20} eV. Furthermore, it is the nearby sources that are expected to be responsible for this region of the spectrum and their distribution is far from isotropic or homogeneous. Therefore, it is not clear either what is the influence that the differential exposure in declination, peculiar to the AGASA experiment, has on the deduced spectral shape at the highest energies.

Figure 10 [27] try to assess both, the statistical significance of the AGASA result [28] at the very end of the energy spectrum, and the degree to which it is compatible with a non-homogeneous distribution of sources that follows closely

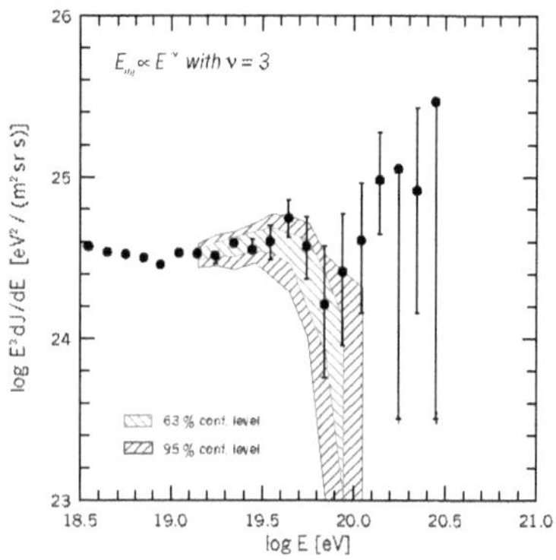

Fig. 10. Statistical significance of the observed UHECR energy spectrum and the distribution of the particle sources: (a) homogeneous, (b) CfA2 catalog [27]. The fiducial IGMF of Sect. 2.4 is used

the spatial distribution of luminous matter in the nearby Universe (as given by the CfA2 galaxy redshift catalog).

In Fig. 10a (left), the observed spectrum is compared with the energy spectrum resulting from a homogeneous distribution of sources from $z = 0$ to $z = 0.1$ in a Friedmann-Robertson-Walker metric. The fiducial cellular IGMF model is used (see above). Individual sources were treated as standard candles supplying the same luminosity in UHECR protons above 10^{19} eV. The injected spectrum was a power law, $dN/dE \propto E^{-\nu}$, with $\nu = 3$ above the latter threshold. The shown confidence levels take into account the dependence in declination of the quoted exposure of AGASA, an energy-independent Gaussian distribution of the errors in the determination of the arrival energy of (assumed) protons and the bin and sample size above 10^{19} used in Ref. [28]. It can be seen from Fig. 10a that this model is able to fit the observed AGASA spectrum quite well up to $\sim 10^{20}$ eV. At higher energies, however, AGASA observations seem unaccountable by the homogeneous approximation, even when the quoted errors are considered. This result is statistically significant despite the low number of UHECR events available.

Figure 10b (right), compares the observed spectrum with the energy spectrum resulting from a inhomogeneous distribution, that of the galaxies in the CfA2 catalog. Everything else is as in the simulation in Fig. 10a. It can be seen that, in this case, it is not so clear that the GZK cut-off is not present in the data. Therefore, caution must be exercised in the interpretation of the UHECR energy spectrum until more data is available.

It must also be noted, that the use of redshift catalogs to characterize the absolute spatial density distribution of baryonic matter is not free of problems. Catalogs are limited by a series of biases (flux limits, selection effects, absorption, galactic obscuration, sky coverage, surface brightness, etc., just to mention a few) whose real magnitude is difficult, if not impossible, to assess. These problems

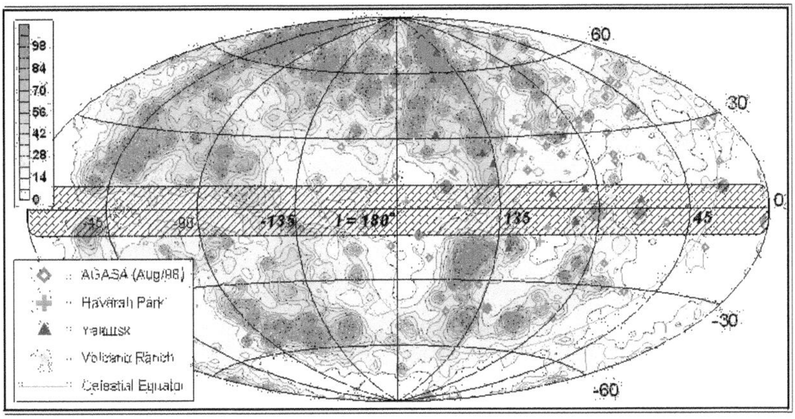

Fig. 11. Expected arrival probability distribution (contour lines in the background) for the same UHECR source and IGMF cellular model as above compared with the actual published data above 4×10^{19} eV from AGASA, Haverah Park, Yakutsk, and Volcano Ranch. The fiducial IGMF of Sect. 2.4 is used

can be specially severe when different redshift surveys are combined in a single catalog. Blanton, Blasi, and Olinto [72], for example, taking into account flux limits for an older version of the CfA2 catalog (only 17% sky coverage) and the IRAS PSCz survey [84], argue for a lower local over density, which makes even worst the disagreement between observations and expectations. Face to these inherent uncertainties, we stress that the result expressed in Fig. 10b must be interpreted more as a word of caution than as a closed result.

Figure 11 shows an Aitoff (equal area) projection, in galactic coordinates, of the expected arrival probability distribution (contour lines in the background) for the same UHECR source and IGMF cellular model as above compared with the actual published data above 4×10^{19} eV from AGASA (47 events [11]), Haverah Park (27 [85]), Yakutsk (24 [86]) and Volcano Ranch [87]). The mask covers the plane of the galaxy, where the actual distribution of galaxies is not well known due to obscuration by dust. The curved, thick line is the celestial equator.

The arrival probability contours trace quite well the local large scale structures. The SGP, in particular, can be easily distinguished running from North to South approximately along the $l = 135$ galactic meridian. It is apparent from Fig. 11 that, despite some conspicuous clusters in the vicinity of the SGP, the actual observed distribution of UHECR is much more isotropic than what one would expect if their sources aggregate like the luminous matter. Unfortunately, given the non-uniform exposure in declination of the various experiments and the low number statistics involved, it is not trivial to quantify this statement.

The view of the AGASA group [11] is that supra-GZK events arrive isotropically at Earth. However, this picture is further complicated by the detection of three doublets and a triplet within a separation angle of 2.5^o.

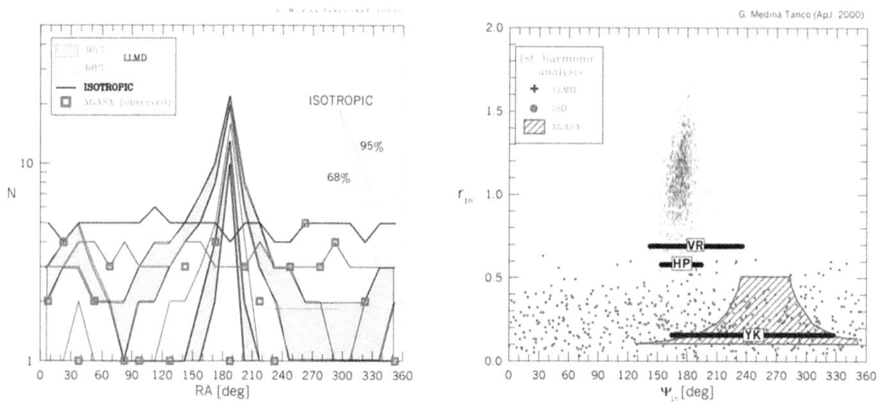

Fig. 12. (a) Arrival directions (right ascension) of UHECR above 4×10^{19} eV for the LLMD (shaded regions) and homogeneous models (lines), as well as the observed AGASA data (squares). 68% and 95% confidence levels are shown for both models. Confidence levels are calculated for each individual bin after 1000 independent experiments (with 47 events each - same number as the AGASA sample) were performed. (b) Amplitude and phase of the first harmonic calculated for 10^3 samples drawn form the isotropic (circles) and anisotropic (LLMD - crosses) distributions. The size of individual samples is 47 protons, as in AGASA. The hatched region is the (1σ) error box calculated from AGASA observations, while the thick horizontal bars are the 1σ error bars for the phases of Volcano Ranch, Haverah Park and Yakutsk experiments. In both cases, (a) and (b), samples are selected with the same declination distribution expected for the AGASA experiment [88]

Further insight can be gained, as in the case of the energy spectrum, by comparing the data with what should be expected from a homogeneous and from an inhomogeneous distribution of sources that follows closely the local luminous matter spatial distribution (LLMD).

The most elemental analysis that can be made regarding isotropy is one-dimensional, in right ascension (RA), where other complicating factors like non-uniform exposure in declination and low number statistics are more easily dealt with. Figures 12a and 12b show different forms of visualizing the distribution of events in RA. The shaded bands (Fig. 12a) in the background correspond to the 68% and 95% confidence levels of the expected distribution of events in RA for a sample of size 47 protons originated in the LLMD scenario. Despite the small size of the samples some features are clearly seen. The largest peak is the signal from the Virgo-Coma line of sight towards the North galactic pole. The opposite half of the SGP (towards the second quadrant in galactic latitude) is responsible for the smaller peak around 30°. The deep depressions surrounding the Virgo peak correspond to the Orion (left) and Local (right) voids, the most prominent structures in our immediate neighborhood, combined with the spurious effect of obscuration of the galaxy distribution due to the galactic plane.

The thick continuous lines in Fig. 12a correspond to the 68% and 95% confidence levels of the distribution in RA when the incident UHECR flux is isotropic. We can see that, even with so few events, both limits should be distinguishable.

The heavy squares represent the AGASA data (same bin size as for the models above) and are consistent with an isotropic distribution. No signature is seen from the Virgo peak and, furthermore, the most populated bins fall in a region corresponding to the Local Void.

A more quantitative treatment to characterize the anisotropy in RA is the first harmonic analysis [89]. Thus, given a data sample, the amplitude $r_{\rm 1h} = \sqrt{a_{\rm 1h}^2 + b_{\rm 1h}^2}$ and phase $\Psi_{\rm 1h} = \tan^{-1}(b_{\rm 1h}/a_{\rm 1h})$ are calculated, where $a_{\rm 1h} = \frac{2}{N}\sum_{i=1}^{N}\cos\alpha_i$, $b_{\rm 1h} = \frac{2}{N}\sum_{i=1}^{N}\sin\alpha_i$ and α_i is the right ascension of an individual event.

$r_{\rm 1h}$ and $\Psi_{\rm 1h}$ are calculated for 10^3 samples drawn form the isotropic and anisotropic (LLMD) distributions and the results are shown in Fig. 12b with small dots and crosses respectively. Both cases are very well discriminated in the $r_{\rm 1h}$-$\Psi_{\rm 1h}$ plane. The error box for the first harmonic of AGASA data (calculated by [90]) is also displayed (hatched region), and is completely consistent with an isotropic UHECR flux. Moreover, the AGASA result by itself, seems completely inconsistent with the LLMD scenario. However, when the phase and amplitudes obtained from other major experiments are considered (large, thick horizontal bars in Fig. 12b for Haverah Park -HP- Volcano Ranch -VR- Yakutsk -YK-; see [90]) the picture looks suggestively different, since all the phase observations are clustered inside the same quadrant in RA, covering the right wing of the Virgo peak. That is, despite the fact that every isolated measurement is consistent with isotropy, the observed phases seem to show a systematic enhancement in the direction of the interface between the SGP and the large adjacent Local void. It must also be noted that Haverah Park and Volcano Ranch data behave more like a transition between the isotropic and LLMD scenarios. Three out of four first harmonic phases (HP, YK and VR) include the North galactic pole within one S.D. level, while the forth (AGASA) include it within two S.D.. The exclusion of the observed UHECR events inside the obscuration band, $b < 10^o$, changes the phase of the AGASA result by only 6° (from 258° to 252°) and, therefore, previous conclusions are unchanged by this effect.

Clearly, a two-dimensional analysis of the data would be highly desirably in order to answer questions as simple as whether the data is isotropic or unimodal. One way of doing this, given the small number of events involved and the non-uniformity of the distribution of events in declination due to experimental limitations, is to analyze the normalized eigenvalues τ_1, τ_2 and τ_3 of the orientation matrix $\mathbf{T}$ of the data. Defining $\mathbf{T_{i,j}} = \mathbf{\Sigma_{k=1}^{N} v_i^k v_j^k}$, where $\mathbf{v^k}$ are the N unit vectors representing the data over the celestial sphere and assuming $0 \le \tau_1 \le \tau_2 \le \tau_3 \le 1$, the shape, $\gamma = \log_{10}(\tau_3/\tau_2)/\log_{10}(\tau_2/\tau_1)$, and the strength parameter, $\zeta = \log_{10}(\tau_3/\tau_1)$, can be built [91]. The shape criterion γ is useful in discriminating girdle-type distributions from clustered distributions. The larger the value of γ the more clustered is the distribution. Uniform, nearly isotropic, distributions have $\zeta \sim 0$. Because of the nature of the experimental

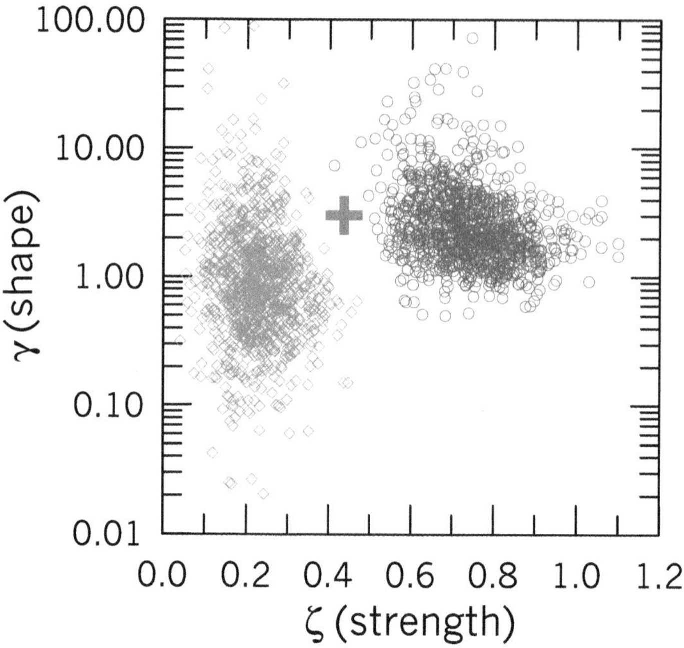

Fig. 13. Two-dimensional eigenvector analysis. Heavy cross is the AGASA observation. Rhombuses and circles correspond to homogeneous and LLMD cases [88]

setup, the observed distribution of UHECR is girdle in nature, regardless of the isotropicity of the UHECR flux. Therefore, in Fig. 13 we compare the results for 10^3 isotropic (rhombes) and LLMD (circles) samples respectively with the AGASA sample in the γ-ζ plane. It can be seen that the isotropic and LLMD scenarios should be very well separated with the available data, albeit its smallness. AGASA data (thick cross), on the other hand, does not fit either of these scenarios, being an intermediate case.

While a first order interpretation of the AGASA data certainly points to an isotropic flux of UHECR, consideration of the first harmonic analysis of other data sets and of two-dimensional tests over the AGASA data itself, as well as expected numbers of doublets for isotropic and anisotropic samples [88], point to a more complicated picture with a certain degree of mixture of both limiting cases.

At least three scenarios in which such a result could be obtained:

1. The sources involve bottom-up mechanisms associated with luminous matter but some of the events are scattered in the intergalactic medium such that we observe the combination of a diffuse and a direct component (see the scheme in Fig. 14). Under plausible conditions [69], suitable scattering centers could be radio ghosts [92], plasma fossils of galactic nuclear activity.

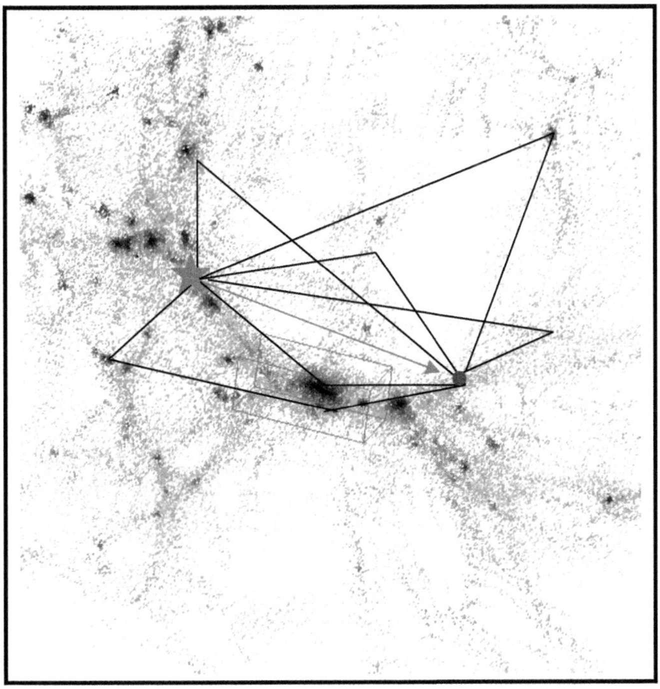

Fig. 14. Radio ghosts could act as scattering centers of UHECR in the intergalactic medium, giving rise to a diffuse particle flux besides the direct, anisotropic, flux component. Thus, the observer (dot) would see the source (star) immersed in a CR background. For details see Ref. [69]

2. The sources involve bottom-up mechanisms associated with luminous matter but there is a large local magnetic structure, like a magnetized Galactic wind, which isotropize the UHECR flux upon traversing the galactic halo [52,93]. As the energy of the particles increases, and as long as they all have the same mass, the degree of isotropization should decrease making the galactic pole visible. Figure 16 shows schematically how such a focusing effect could take place inside an hypothetical Galactic wind, in the context of our nearby universe. Figure 15 shows the backtracking of the real data above 10^{20} eV to the border of the Galactic halo (from Ahn et al. [52]) when no random field component is included in the wind model. Although not shown, reasonable levels of MHD turbulence do not change the qualitative nature of the result. Clustering could also be a natural consequence of the Galactic magnetized wind scenario due to lensing effects [51]. In any case, this scenario should produce a strong asymmetry between the observable UHECR flux at the Northern and Southern Earth hemispheres. Consequently, this represents an ideal theoretical ground to be tested by the next generation Pierre Auger observatory in Malargüe, Argentina.

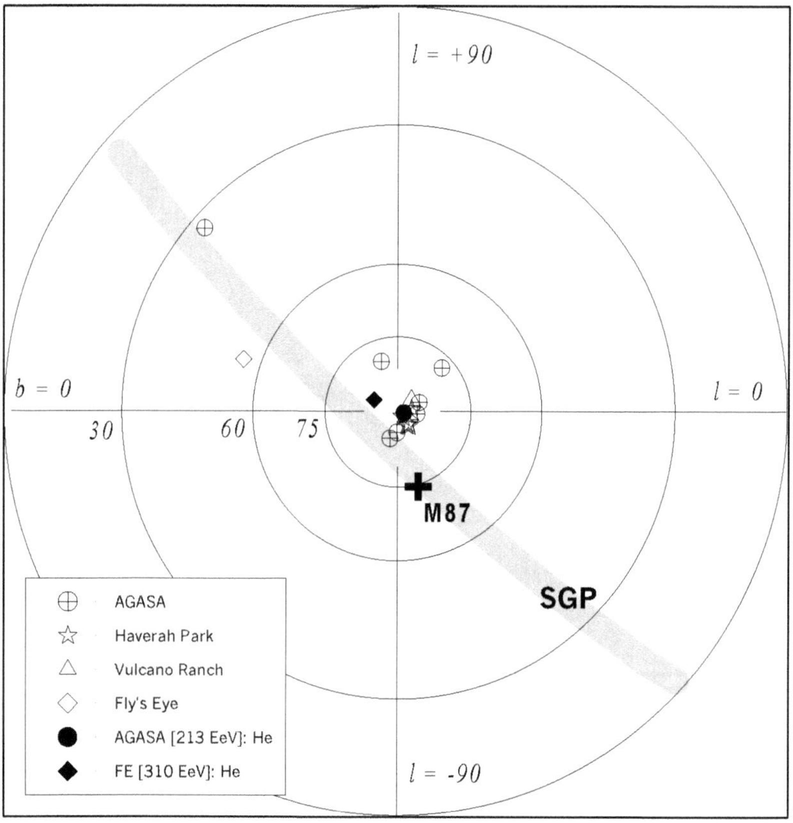

Fig. 15. Mapping of UHECR data onto the border of a magnetized Galactic wind [52]. The thick band represents the position of the Supergalactic plane, the most prominent large scale structure

3. The sources involve top-down mechanisms associated with dark matter whose distribution roughly associates with the LLMD. In this case, the observed flux is the combination of an extragalactic component, whose signature is not very different from that of the LLMD, and a component originated in the halo of our own galaxy [90,94,95]. It has been claimed [96] that, under general conditions, the halo component should dominate the extragalactic flux by at least two orders of magnitude. This is only true, however, in the unrealistic case of dark matter uniformly distributed in intergalactic space. Nevertheless, dark matter aggregates strongly and tends to be overabundant, by factors of $\sim 10^2$, in the center of galaxy clusters when compared to its abundance in the halos of isolated galaxies. It can therefore be shown that, in a sample of 50 events, and assuming Virgo as the only source of extragalactic events, 3–7 events should originate in Virgo and arrive inside a solid angle of approximately the size of the cluster. This could give rise to a slight anisotropy that correlates with the SGP when

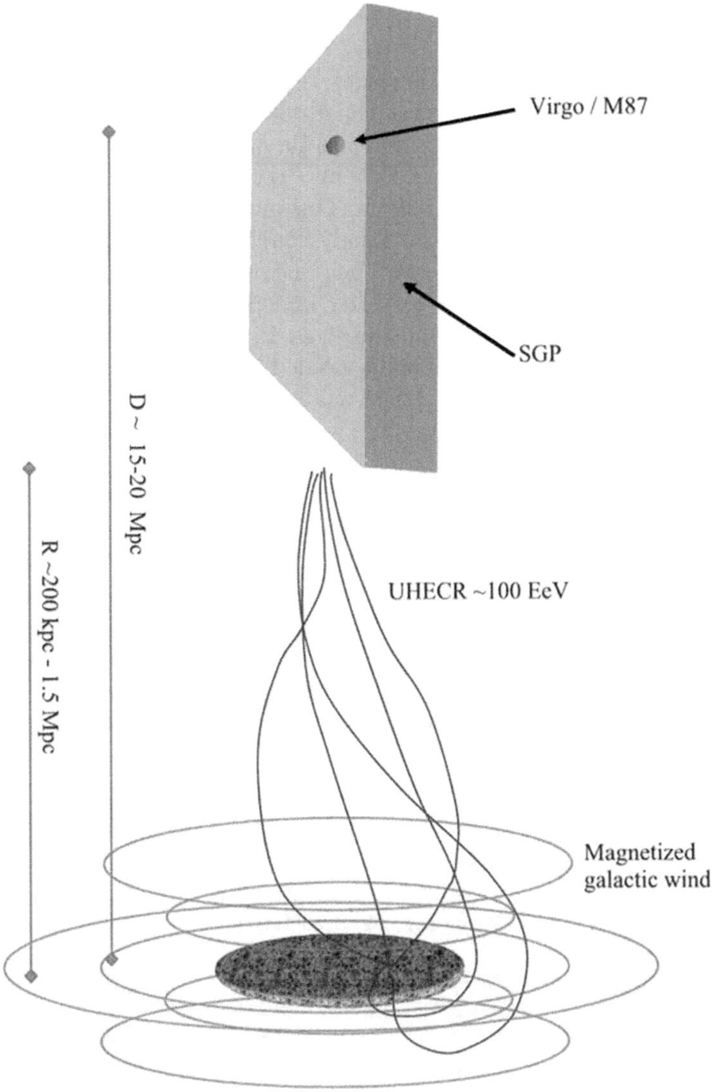

Fig. 16. Schematic view of UHECR focusing by a magnetized Galactic wind in the local universe

combined with the almost isotropic flux originated in a large galactic halo. Note, however, that the solid angle does not need to point exactly in the direction of Virgo, depending on the large-scale structure of the intervening magnetic field. Furthermore, the present data is not even enough to constraint theoretical halo models, being able to rule out only the most unlikely (very flattened or very small) halos [90].

4 Conclusions

UHECR are most likely charged particles; therefore, there is probably not such a thing as an easily identifiable optical counterpart to these extremely high energy events. Intervening galactic and intergalactic magnetic fields can affect propagation at energies of hundreds of EeV in a very significant way depending on the intensity and topology of the fields. This means that both, cosmic magnetic fields and UHECR are probably strongly coupled and will have to be tackled simultaneously as different aspects of the same problem. We will have to increase our knowledge on cosmic fields to better understand UHECR and, conversely, as soon as we go through a some threshold in our understanding of UHECR astrophysics, an invaluable and badly needed new diagnostic tool for probing magnetic fields inside the GZK-sphere will be available.

Acknowledgements

I would like to thank the local organizing committee for the financial support and several colleagues for useful discussions and suggestions: Alan Watson, Peter Biermann, Michel Hillas, Todor Stanev, Torsten Enßlin, Johannes Knapp, Carlos Escobar, John Linsley, Angela Olinto, Pasquale Blasi, Elihu Boldt, Arnold Wolfendale and Eli Waxman among others. Guenter Sigl's patience with successive delays in handing over the manuscript is much appreciated. This work is partially supported by the Brazilian agencies FAPESP and CNPq.

References

1. K. Greisen: Phys. Rev. Lett. **16**, 748 (1966)
2. G. T. Zatsepin, V. A. Kuzmin: Sov. Phys. JETP Lett. **4**, 78 (1966)
3. P. P. Kronberg: Rep. Prog. Phys. **57**, 325 (1994)
4. L. O'C. Drury: Rep. Prog. Phys. **46**, 973 (1983)
5. H. Moraal, W. I. Axford: Astron. Astrophys. **125**, 204 (1983)
6. H. J. Völk, P. L. Biermann: Astrophys. J. **333**, L65 (1988)
7. W. J. Markiewicz, L. O'C. Drury: Astron. Astrophys. **236**, 487 (1990)
8. G. A. Medina Tanco, R. Opher: Astrophys. J **411**, 690 (1993)
9. P. L. Biermann: Astron. Astrophys. **271**, 649 (1993)
10. M. Nagano, A. A. Watson: Rev. Mod. Phys. **72**, 689 (2000)
11. M. Takeda, et al.: Astrophys. J. **522**, 225 (1999)
12. N. Hayashida, et al.: Astropart. Phys. **10**, 303 (1999)
13. Y. Uchihori, et al.: Proc. *International Symposium on EHECR: Astrophysics and Future Observatories*, ed. M. Nagano (ICRR, Univ. Tokyo), p.50 (1996)
14. R. M. Baltrusaitis, G. L. Cassiday, J. W. Elbert, P. R. Gerhardy, E. C. Loh, Y. Mizumoto, P. Sokolsky, D. Steck: Phys. Rev. D **31**, 2192 (1985)
15. N. Inoue, et al.: *Proc. 26th International Cosmic Ray Conference*, Salt Lake City, vol.**1**, 361 (1999)
16. J. A. Hinton: private communication (1999)
17. F. Halzen, R. A. Vázques, T. Stanev, H. P. Vankov: Astropart. Phys. **3**, 151 (1995)
18. J. L. Puget, F. W. Stecker, J. H. Bredekamp: Astrophys. J. **205**, 638 (1976)

19. M. A. Malkan, F. W. Stecker: Astrophys. J. **496**, 13 (1998)
20. F. W. Stecker: Proc. *Workshop on observing Giant Air Showers from Space*, eds. J. F. Krizmanic, et al., AIP Conference Proceedings 433, pp.212 (1998)
21. L. N. Epele, E. Roulet: Phys. Rev. Lett. **81**, 3295 (1998)
22. V. S. Berezinsky, S. I. Grigor'eva: Astron. Astrophys. **199**, 1 (1988)
23. S. Yoshida, M. Teshima: Prog. Theor. Phys. **89**, 833 (1993)
24. R. J. Protheroe, P. Johnson: Astropart. Phys. **4**, 253 (1996)
25. G. A. Medina Tanco, E. M. de Gouveia dal Pino, J. E. Horvath: Astropart. Phys. **6**, 337 (1997)
26. M. Hillas: Nature **395**, 15 (1998)
27. G. A. Medina Tanco: Astrophys. J. **510**, L91 (1999)
28. M. Takeda, et al.: Phys. Rev. Lett. **81**, 1163 (1998)
29. M. Takeda, et al.: Proc. *26th International Cosmic Ray Conference*, Salt Lake City, vol.**3**, p.252 (1999)
30. E. Waxman, J. Miralda-Escudé, Astrophys. J. **472**, L89 (1996)
31. M. Lemoine, G. Sigl, A. V. Olinto, D. N. Schramm: Astrophys. J. **486**, L115 (1997)
32. G. Sigl, M. Lemoine, A. V. Olinto: Phys. Rev. D **56**, 4470 (1997)
33. T. Stanev, H. P. Vankov: Phys. Rev. D **55**, 1365 (1997)
34. G. A. Medina Tanco, A. A. Watson: Astropart. Phys. **10**, 157 (1999)
35. R. Beck, A. Brandenburg, D. Moss, A. Shukurov, D. Sokoloff: Ann. Rev. Astron. Astrophys. **34**, 155 (1996)
36. J. P. Vallée: Astrophys. J. **366**, 450 (1991)
37. J. L. Han, G. J. Qiao: Astron. Astrophys. **288**, 759 (1994)
38. N. Neininger, U. Klein, R. Beck, R. Wielebinski: Nature **352**, 781 (1991)
39. G. Golla, K. Hummel: Astron. Astrophys **284**, 777 (1994)
40. R. J. Rand, S. R. Kulkarni: Astrophys. J. **343**, 760 (1989)
41. R. J. Rand, A. G. Lyne: Mon. Not. R. Astron. Soc. **268**, 497 (1994)
42. Y. Sofue, M. Fujimoto: Astrophys. J. **265**, 722 (1983)
43. H. Ohno, S. Shibata: Mon. Not. R. Astron. Soc. **262**, 953 (1993)
44. A. Shukurov: Proc. *Plasma Turbulence and Energetic Particles in Astrophysics*, eds. M. Ostrowski, R. Schlickeiser, Univ. Jagiellonski, Krakow, 1999, pp. 66-73, astro-ph/0012444
45. J. L. Han: Astrophys. Sp. Sc., in press, astro-ph/0010537
46. G. A. Medina Tanco: Proc. *25th International Cosmic Ray Conference*, vol. **4**, 485 (1997)
47. G. A. Medina Tanco, E. M. de Gouveia dal Pino, J. E. Horvath: Astrophys. J **492**, 200 (1998)
48. T. Stanev: Astrophys. J **479**, 290 (1997)
49. D. Harari, S. Mollerach, E. Roulet: J. High Energy Phys. **08**, 022 (1999)
50. D. Harari, S. Mollerach, E. Roulet: J. High Energy Phys. **02**, 035 (2000)
51. D. Harari, S. Mollerach, E. Roulet: J. High Energy Phys. **10**, 047 (2000)
52. E.-J. Ahn, G. A. Medina Tanco, P. L. Biermann, T. Stanev: astro-ph/9911123
53. P. Blasi, S. Burles, A. V. Olinto: Astrophys. J **514**, L79 (1999)
54. D. Ryu, H. Kang, P. L. Biermann: Astron. Astrophys. **335**, 19 (1998)
55. H. Arp: Phys. Lett. A **129**, 135 (1988)
56. K.-T. Kim, P. P. Kronberg, G. Giovannini, T. Venturini: Nature **341**, 720 (1989)
57. G. A. Medina Tanco: Astrophys. J **505**, L79 (1998)
58. G. A. Medina Tanco: Proc. *16th ECRC*, Alcalá de Henares, Spain, pp.295-298 (1999)
59. G. Sigl, M. Lemoine, P. L. Biermann: Astropart. Phys. **10**, 141 (1999)

60. M. Lemoine, G. Sigl, P. L. Biermann: astro-ph/9903124
61. Auger Design Report (1996)
62. J. Miralda-Escudé, E. Waxman: Astrophys. J **462**, L59 (1996)
63. E. Waxman, P. Coppi: Astrophys. J. **464**, L75 (1996)
64. G. A. Medina Tanco: Proc. *25th International Cosmic Ray Conference*, vol. **4**, 481 (1997)
65. G. A. Medina Tanco: Proc. *25th International Cosmic Ray Conference*, vol. **4**, 477 (1997)
66. G. A. Medina Tanco: Astrophys. J **495**, L71 (1998)
67. A. Achterberg, Y. A. Gallant, C. A. Norman, D. B. Melrose: Mon. Not. R. Astron. Soc. submitted, astro-ph/9907060
68. T. Stanev, R. Engel, A. Muecke, R. J. Protheroe, J. P. Rachen: Phys. Rev. D **62**, 093005 (2000)
69. G. A. Medina Tanco, T. A. Enßlin: Astropart. Phys. **16**, 47 (2001)
70. J. P. Vallée: Fund. Cosm. Phys. **19**, 1 (1997)
71. G. A. Medina Tanco: Proc. *Topics in Cosmic Ray Astrophysics*, ed. M. DuVernois, 299 (Nova Science Publisher, 2000)
72. M. Blanton, P. Blasi, A. V. Olinto: Astropart. Phys. **15**, 275 (2000)
73. V. S. Berezinsky, S. I. Grigor'eva: Astron. Astrophys. **199**, 1 (1988)
74. S. Yoshida, M. Teshima: Prog. Theor. Phys. **89**, 833 (1993)
75. F. W. Stecker, M. H. Salamon: Astrophys. J. **512**, 521 (1999)
76. P. Bhattacharjee, G. Sigl: Phys. Rep. **327**, 110 (2000)
77. J. Linsley: Phys. Rev. Lett. **10**, 146 (1963)
78. J. Linsley: Scient. Amer. **239**, 60 (1978)
79. A. A. Watson: Proc. *12th European Cosmic Ray Symposium*, Nuc. Phys. B Supp. **22B**, 116 (1991)
80. M. A. Lawrence, R. J. O. Reid, A. A. Watson: J. Phys. G: Nucl. Part. Phys. **17**, 733 (1991)
81. D. J. Bird: Astrophys. J **441**, 114 (1995)
82. N. Hayashida, et al.: Phys. Rev. Lett. **63**, 3491 (1994)
83. M. Nagano: Proc. *Workshop on giant cosmic ray air showers*, eds. J. F. Krizmanic, J. F. Ormes and R. E. Streitmatter, AIP Conf. Proc. vol. **433**, 76 (1998)
84. W. Sounders, et al.: Mon. Not. R. Astron. Soc. **317**, 55 (2000)
85. R. J. O. Reid, A. A. Watson: *Catalogue of Highest Energy Cosmic Rays*, ed. M. Wada (IPCR, Tokyo), **1**, 61 (1980)
86. B. N. Afanasiev, et al.: Proc. *24th International Cosmic Ray Conference*, Roma (Italy), vol. **2**, 796 (1995)
87. J. Linsley: *Catalogue of Highest Energy Cosmic Rays*, ed. M. Wada (IPCR, Tokyo), vol. **1**, 61 (1980)
88. G. A. Medina Tanco: Astrophys. J, in press (2000)
89. J. Linsley: Phys. Rev. Lett. **34**, 1530 (1975)
90. G. A. Medina Tanco, A. A. Watson: Astropart. Phys. **12**, 25 (1999)
91. N. I. Fisher, T. Lewis, B. J. J. Embleton: *Statistical analysis of spherical data* (Cambridge University Press, 1993)
92. T. A. Enßlin: Proc. *Ringberg workshop on diffuse thermal and relativistic plasma in galaxy clusters*, eds. H. Böhringer, L. Feretti and P. Schücker, MPE Report No.271 (1999).
93. P. L. Biermann, E.-J. Ahn, G. A. Medina Tanco, T. Stanev: Nuc. Phys. B (Proc. Supp.) **87B**, 417 (2000)
94. A. Benson, A. Smialkowski, A. W. Wolfendale: Astropart. Phys. **10**, 313 (1999)
95. A. Benson, A. Smialkowski, A. W. Wolfendale: Phys. Lett. **B449**, 237 (1999)
96. S. L. Dubovsky, P. G. Tinyakov: JETP Lett. **68**, 107 (1998)

A Possible Nearby Origin for the Highest-Energy Events Observed

Peter L. Biermann[1,2], Eun-Joo Ahn[3], Philipp P. Kronberg[4], Gustavo Medina-Tanco[5], and Todor Stanev[6]

[1] Max-Planck Institute for Radioastronomy
[2] Department for Physics and Astronomy, University of Bonn, Bonn, Germany
[3] Physics Department, Seoul National University, Seoul, Korea
[4] Physics Department, University of Toronto, Toronto, Canada
[5] Sao Paolo University, Sao Paolo, Brasil
[6] Bartol Research Institute, University of Delaware, Newark, DE, USA

Abstract. The detection of ultra high energy cosmic ray (UHECR) events beyond the cutoff expected from proton interactions with the microwave background has continued to defy an explanation on simple terms. Assuming that powerful radio galaxies are the source of UHECR's at $E \geq 10^{19}$ eV, and that these are protons, susceptible to photon interactions over propagation paths longer than $\simeq 30$ Mpc, we present a possible solution that explains both the puzzling lack of a GZK cutoff due to photon interactions and the observed energy-independent isotropy of UHECR arrival directions. Our model has three essential ingredients: a) The highest energy particles come from the nearest powerful radio galaxy M87, b) they are isotropized in the magnetic field of a halo-wind of our Galaxy, and c) this isotropization can be naturally explained by a k^{-2} wavenumber spectrum of magnetic irregularities in the halo wind, which makes the bending of particle orbits independent of energy. All three concepts are well established, and in essence we have combined them. At progressively lower energies near 5×10^{19} eV, some UHECR arrivals from additional radio galaxies at moderate distances will also get mixed in. Of all proposed solutions to the challenge presented by the high energy CR events this is the simplest one, and is it based on known physical processes.

1 Introduction

Very high energy events have now been detected for four decades, and their energy determination has become increasingly certain, although of course each new experiment needs to verify once again, that our understanding of the interactions in the air shower development is sufficient. Recent discoveries are in [1,2,3,4,5,6,7], and [8,9]; the most recent review of the experimental uncertainties is in [10], see also the contributions by P. Billoir and S. Yoshida on experimental detection methods.

After the first evidence appeared of events with an energy near 10^{20} eV, the microwave background, originally predicted in the late 1940ies, was discovered, and it was immediately realized, that high energy protons would interact.This interaction would be at the pion production threshold, and so a cutoff in the observed cosmic ray spectrum should be visible near 5×10^{19} eV [11,12], commonly referred to as the GZK-cutoff.All subsequent reanalyses of this expected

Martin Lemoine, Günter Sigl (Eds.): LNP 576, pp. 181–195, 2001.

cutoff have only deepened the difficulty (e.g., [13,14]). On the one hand, we have now events with energies up to 6 times the GZK-cutoff energy, and several dozen events beyond 10^{20} eV, already a factor of two beyond the GZK-cutoff energy. So even uncertainties of a factor of 2 would not change the problem. The history especially of the Fly's Eye experiment has clearly shown [15,16,17,18,19,20,21,22], when taken together with the Haverah Park and AGASA data, that these extreme energies are unavoidable [10]; fluorescence data are a very important part of the argument in favor of very energetic protons. On the other hand, any calculation with a spatially homogeneous source population finds the same result: there should be a very clear cutoff. However, if the source distribution is not homogeneous, then the situation does improve, but cannot be solved either [23,24,25] easily.

Here we focus on the what may be the oldest suggestion, due to Ginzburg [26,27,28].

Basic arguments on cosmic rays and especially high energy cosmic rays and their physics have been reviewed in, e.g., various review articles [29,30,31,32,33], [34,35], and in books [26,36,37].

If we accept that there is a single specific source in the nearby universe responsible for the highest energy events, then immediately we are faced with the question, why the events do not come from that source in direction on the sky?

And here we explore the possibility that the paths of the particles, assumed to be protons, may be bent by the magnetic fields in our halo, in a Galactic wind, interpreted in analogy to the Parker magnetic field structure of the solar wind.

And then such a solution immediately leads to the third problem, and that is why do the particles come here at any appreciable flux, and at energies well below that energy at which the bending is important. The analogy with the solar wind would imply that at an energy somewhat below where bending is relevant, all external flux is cut off altogether, clearly in conflict with many previous arguments on extragalactic sources. Here we focus on the spectrum of the magnetic irregularities expected in the halo-wind, and show that a k^{-2} spectrum (isotropic, energy per volume, per wavenumber k) influences charged particles independently of energy; so there is no specific cutoff at all. And in fact, such a spectrum has been predicted for the halo wind (A. Bykov, 2000, priv. comm.).

Finally, we will focus on predictions in such a model, which will be testable with upcoming data, and arrays such as HiRes, Pierre Auger, EUSO and OWL.

This model has been briefly described in [38] and [39,40]. Related calculations have been done by [41,42].

2 Radiogalaxy Hot Spots and Jets

The radio galaxy M87 has long been suspected to produce very high energy events, with Ginzburg probably the first to identify it as a plausible source in the early 1960's [26,27,28].

Radio galaxy hotspots specifically have been suggested by [29], and were worked out in detail using observations of radio quasars and their jets by [43]. The key argument is that the nonthermal spectra of radio galaxy hotspots, and also sometimes jets and compact flat spectrum radio sources sometimes show a cutoff at or below 3×10^{14} Hz. Such a cutoff, seen in sources of very different properties, can readily be interpreted as due to the limiting energy of electrons going back and forth across a shock in the process of their acceleration, subject to losses from synchrotron emission and photon interactions. The electrons are scattered in the magnetic irregularities of a wavefield in the magnetic plasma which has cascaded down from large scale waves excited by similarly accelerated protons [44]. In such a case it has been shown [43] that the maximum emission frequency is approximately

$$\nu_{\max} \simeq 0.01 \, \frac{c}{r_0} \left(\frac{m_e}{mp} \right)^2 \tag{1}$$

where r_0 is the classical electron radius, c is the speed of light, m_e and m_p are of course the masses of the electron and proton, and the numerical factor in front has several powers of π and 2, so as to give a maximum frequency near 3×10^{14} Hz. This result is independent of the local parameters such as the magnetic field, and the reason is that both electrons and protons move in the same disturbed magnetic field fluctuations, but of course, in different parts of the wave-number spectrum. The basic and key idea is that the protons excite the waves in the magnetic plasma on the large wavelength scale, and the wave energy then cascades down to those wavelengths which resonate with the electron's Larmor motion. Here it can be shown from the observed spectral behaviour of the emission region behind the shock, that a Kolmogorov spectrum is appropiate [45], and this has been used. Putting in photon interaction losses modifies this result only slightly. Therefore we have here a general and rather simple explanation for the ubiquitous optical cutoff in the spectrum of Active Galactic Nuclei (AGN) and their jets, observed since the mid 1970's (for references see [43]). Using numbers for the knot A in the jet of the active galactic nucleus in the radio galaxy M87 gives then a suggested maximum proton energy of 10^{21} eV.

This is the only site, radio galaxy hotspots and knots in jets, of all proposed, in which *protons of such energy are required in the source* by other independent observations.

The magnetic field in jets scales with jet- and so disk power [46,47] as $L_{\rm jet}^{1/2}$. Since the maximum particle energy as derived from losses scales as $1/B^{1/2}$, this leads to a limit in the maximum energy of protons, which cuts off the highest energies with larger source power, due to synchrotron and photon interaction losses:

$$E_{\max} \simeq 10^{21} \, L_{\rm disk,46}^{-1/4} \, {\rm eV} \, . \tag{2}$$

The losses limit the maximum energy of a charged particle, but of course the spatial limitation can be more relevant: One can then proceed to estimate analogously the maximum energy in other powerful radio galaxy jets and hot spots, disregarding the distance to us: This then leads to a scaling of maximum energy with jet- and so disk power [47], which is analogous to that derived by Lovelace [46], but here with more certain numbers:

$$E_{\mathrm{max}} \simeq 5 \times 10^{20}\, L_{\mathrm{disk},46}^{1/2}\, \mathrm{eV} \tag{3}$$

Using the most powerful radio galaxy in the sample considered by [47], and allowing for some boosting in a weakly relativistic boundary shock we obtain

$$E_{\mathrm{max}} = 10^{22} \mathrm{eV}\,. \tag{4}$$

This is a serious upper limit; the more plausible maximum energy for protons should be below this number.

An important aspect is here, that in order to be able to produce particles anywhere near 10^{20} to 10^{21} eV at all, the source must have appreciable power. This eliminates sources such as Centaurus A; Centaurus A is too weak.

It is possible to go through this argument [43] again in order to bring out the possible relevance of heavier nuclei, following [44].

Putting acceleration [48] and loss rates due to synchrotron emission equal yields for the maximum energy

$$\gamma_{A,\mathrm{max}} A = \left[\frac{27}{80} b(\beta - 1)\right]^{1/2} \left(\frac{e}{r_0^2 B}\right)^{1/2} \left(\frac{U_{\mathrm{sh}}}{c}\right) \left(\frac{m_p}{m_e}\right) \frac{A^2}{Z^{3/2}}\,. \tag{5}$$

Here b is the energy density of the turbulence integrated over all wavenumbers, relative to the energy density in the overall magnetic field, so $b \leq 1$. U_{sh} is the upstream shock speed, and β is the spectral index of the turbulence i.e. 5/3 for a Kolmogorov spectrum. A is the number of nucleons in the nucleus considered, and Z is its charge number. For $U_{\mathrm{sh}}/c = 0.1$, $B = 10^{-4}$ Gauss, $b = 1$, and $\beta = 5/3$ this means that

$$\gamma_{A,\mathrm{max}}\, A\, m_p\, c^2 = 4 \times 10^{11} \frac{A^2}{Z^{3/2}}\, \mathrm{GeV}\,. \tag{6}$$

In this example the synchrotron time scale is

$$\tau_{A,\mathrm{syn}} = 5 \times 10^7 \frac{A^3}{Z^4}\, \mathrm{yrs} \tag{7}$$

showing, that Helium nuclei lose energy more slowly to synchrotron emission than protons. This corresponds to a mean free path of loss of about 50 Mpc. The mean free path for spallation due to photon interaction [49,50] is very much less, and so will dominate. The energy loss of Helium nuclei due to spallation in photon encounters has a mean path of interaction in the microwave background of about 1 Mpc [50]. However, since the cross-section for Helium-photon interaction has a strong threshold at 20 MeV in the nucleus frame [50], there must a strong cutoff in the observable Helium contribution somewhere near 10^{20} eV. A survival of an appreciable flux of Helium nuclei well above 10^{20} eV coming from a source such as M87 appears unlikely at present.

The general flux at about 10^{19} eV can easily be accounted for with very modest assumptions about radio galaxies, and the spectrum up to about 5×10^{19} eV can be completely understood in flux and shape [13,14].

The main difficulty is, again as noted, that there are very few radio galaxies in our cosmic neighborhood. M87 is the only one with sufficient power, Centaurus A is too weak, and NGC315 is too far. M87 is close enough and also powerful enough, and as such the only serious candidate radio galaxy. The main problem, prima facie, with M87 is the near isotropy of the arriving events.

3 A Galactic Wind Model

Here we explore the concept that our Galaxy has a wind akin to the solar wind [51,52,53]; the notion of galactic winds is quite old by now [54,55,56]. In such a wind the dominant magnetic field can be approximated by an Archimedian spiral, with

$$B_\phi \sim \sin\left(\frac{\theta}{r}\right) \tag{8}$$

in polar coordinates, while $B_r \sim 1/r^2$, and $B_\theta = 0$. We adopt the nomenclature that index 0 refers to a reference radius r_0, and the values of all parameters, such as density ρ_0, magnetic field B_0 at r_0. The Ulysses data show very clearly, e.g. [57], that the real solar wind is dominated by irregularities as one goes towards the poles, and so the θ-dependence of the Parker field is clearly a poor approximation for the overall field strength.

What drives such a wind, and what are its properties?

3.1 Power

The Galactic wind clearly needs driving, and the most powerful source available is the energy of supernovae and of normal cosmic rays L_{CR}, about $\lesssim 3 \times 10^{41}$ erg/s. Of course, it is commonly taken for granted that cosmic rays derive their energy from supernova explosions; cosmic rays have the advantage that their energy can be transported relative to the gas with velocities up to the local Alfvén velocity; that is, cosmic rays' energy deposition rate into the wind is comparable with that of the thermal Galactic wind:

$$2\pi\rho_0 r_0^2 V_W^3 \simeq L_{CR} \,. \tag{9}$$

3.2 Rotation Measure

The line of sight integral of electron density multiplied with the line-of-sight component of the magnetic field through this wind cannot exceed what the data show, about 30 rad/m^2 [58,59]. This gives the constraint

$$B_0 r_0 n_0 \lesssim 1.2 \times 10^{14} \,\mathrm{Gauss\,cm}^{-2} \tag{10}$$

(using the cgs system here as everywhere else). The density has to match one of the hotter phases of the gas, either that seen by ASCA and RXTE [60,61], or by ROSAT [62,63], as implied by the power-requirement condition above: The density and temperature found by ROSAT was $3 \times 10^{-3}\,\mathrm{cm}^{-3}$, and 4×10^6 K, respectively, while the other two missions found evidence for an even hotter phase, with a temperature around 3×10^7 K possible, suggesting at pressure equilibrium a density of around $4 \times 10^{-4}\,\mathrm{cm}^{-3}$. Any density and temperature combination with approximately the same pressure in between these two extremes may well be also possible.

3.3 Alfvén Velocity

The wind driving needs coupling from the relativistic fluid, the cosmic rays, to the gas, and this can be achieved best via magnetic fields, as proposed for WR stars [64], and this suggests that the wind may be only slightly super-Alfvénic:

$$\frac{B_0}{\sqrt{4\pi\rho_0}} \lesssim V_W \,. \tag{11}$$

3.4 Potential

Then the wind of velocity V_W has to escape the gravitational potential of the Galaxy, and so has to be faster than about 500 km/s.

$$V_W \gtrsim 5 \times 10^7 \,\mathrm{cm/s} \,. \tag{12}$$

3.5 Mass Balance

The mass loss in the wind should be less than the total gaseous mass turnover in the Galactic disk, and so less than about 10 $M_\odot$/yr [65]. Here $\dot{M}_{\mathrm{gas}}$ is the mass loss in the wind, and $\dot{M}_{\mathrm{stars}}$ is the star formation mass turnover.

$$4\pi\rho_0 r_0^2 V_W \;=\; \dot{M}_{\mathrm{gas}} \lesssim \dot{M}_{\mathrm{stars}} \,. \tag{13}$$

We assume here that the accretion time scale in the disk is the same as the star formation time scale in normal galaxies [66,67,68]. Accretion can be dominated either by angular momentum loss or by angular momentum transport, and we are using here the notion that the latter is a limit for the first.

3.6 Angular Momentum Loss

And, finally, the angular momentum loss in the wind should be less than the angular momentum transport in the Galactic disk, and so its time scale cannot be shorter than the star formation time scale. Using the Alfvén radius r_A (i.e. that radius out to which the angular velocity is constant, giving the lever arm for angular momentum loss) we then have the condition

$$\dot{M}_{\text{gas}}\, r_A^2 \; < \; \dot{M}_{\text{stars}} r^2 \; , \tag{14}$$

where now r is the typical radius for the gas in the disk. This means, if we fix our attention on the solar neighborhood in the Galaxy, that the lever arm for losing angular momentum can be larger than the local radius itself by just the factor by which the mass loss in the wind is smaller than the mass transport rate through disk viscosity, which in turn has to be equal to the local star formation rate. Or, to turn the argument around, if the mass transport in the wind becomes comparable to the star formation rate, then the lever arm for losing angular momentum cannot be larger than the local radius, and this then forces the wind to assume the asymptotic state, a tightly wound spiral pattern, very close to its starting region.

This says, if

$$\dot{M}_{\text{gas}} \; \simeq \; \dot{M}_{\text{stars}} \; . \tag{15}$$

then

$$r_A \; \simeq \; r \tag{16}$$

and then indeed the magnetic field has to approximate the true state of the wind already close to the disk, an Archimedian spiral.

3.7 Wind Zone Size

The ram pressure falls off rapidly with radius r, and is

$$P_{\text{wind}} \; = \; 4\pi\, \rho_0 \, \left(\frac{r_0}{r}\right)^2 \, V_W^2 \; . \tag{17}$$

We first need to estimate the parameters plausible for the local intergalactic medium, a density of 10^{-6} cm^{-3} and a temperature of 10^7 K. These fiducial values correspond to a transition through an accretion shock in a local cosmological filament [69], an increase from an outside temperature of about 3×10^5 K [70]. We can thus estimate the wind size to about 1 Mpc, as long as no other galactic wind is encountered. The next galaxy is M31, but its cosmic ray output can be estimated through the far-infrared emission [65,71], since cosmic ray production and thermal dust emission directly scale with each other; the next galaxy which is equivalent to ours in expected wind power is M81, at about 3 Mpc. This implies that the wind size is less than about 1.5 Mpc.

3.8 Wind Parameters

To make matters simple at first, we assume the wind to be fairly close to its asymptotic state early on, at small radii. Obviously, we take as reference radius the galactocentric radius of the Sun, of about 8 kpc. These five conditions suggest a magnetic field of 5 - 7 microGauss at the wind-base consistent with the numbers derived for the solar neighborhood [72], and a density of 5×10^{-4} cm^{-3}, as reference values at the wind-base. This rather low density is consistent with the high ASCA and RXTE temperatures [60,61], and asuming pressure equilibrium with the hot gas detected by ROSAT [62,63]. A related wind has been described in the starburst galaxy M82 by [73]. The key result is that there is a unique set of values at all, that is consistent with all the above observational constraints.

3.9 Other Models

Comparing such a wind model with those by Breitschwerdt et al. [74,75,76,77,78], we note that our wind is consistent with their approach, but just has a much higher sustained magnetic field, since we use the asymptotic regime from small radii on, following [64]. Also, they did not incorporate all of the conditions which we have used, such as the RM-limit, and the mass flow limit.

4 Transport of Charged Particles through the Wind

In the following we endeavour to describe the transport of charged particles through the wind from the outside: there are several conditions to be met, 1) the near isotropy of arriving events, 2) the smooth transition from galactic to extragalactic cosmic rays at about 3×10^{18} eV, and 3) the observed smooth continuation of the flux all the way through the events beyond 5×10^{19} eV, the expected GZK-cutoff. We start from the premise that at the highest energy we have a single source responsible, here the radio galaxy M87. This implies a specific sign of the dominant magnetic field, which is a fourth condition.

Here we refer to the basic information [59,72,79,80,81,82,83] and understanding [84,85,86,87,88] about cosmic magnetic fields. The transport not just in the halo wind, but also in the local supergalactic sheet connecting the Virgo cluster [89,90,91,92,93] and [94,95,96,97,98] and us, as well as the propagation in our own Galaxy [99] should be considered in a final analysis. Here we focus on the halo wind and its effects, since that may be the key difficulty.

4.1 Magnetic Field Symmetry

There is an important point as regards the sign of the magnetic field: In the Galactic disk the azimuthal component changes sign every now and then [100], [101,102], and so one might expect that this sign change carries over into the wind; these sign changes would then nullify to first order all systematic bending. However, the obervations of [103] of disk galaxies show, that the sign of the magnetic field and its pattern have a very clear symmetry: the magnetic field is spiral-like, and the dominant component always points along the spiral pattern inwards. This means that there is a dominant sign, and is consistent actually with the local sign of the magnetic field measured near the Sun in our Galaxy. This also means that, different from the solar wind, the azimuthal component does not change sign in mid-plane. In order to satisfy $\boldsymbol{\nabla B} = 0$ and at the same time follow a consistent spiral pattern, the radial component must change sign at higher Galactic latitude. This pattern entails that all very high energy events observed at Earth, if interpreted as protons, point ultimately upwards, when traced backwards through the halo wind.

4.2 Magnetic Irregularities

The irregularities are strong in the magnetic field of the solar wind [57], and do not decay towards the poles with the azimuthal component, therefore in the pole regions the irregularities do dominate over the regular systematic component.

The wind is ultimately driven by the energy from supernova explosions, and the cosmic rays accelerated in them; this entails that strong shocks propagating into the nearby thick disk of the hot gas and into the wind dominate the irregularities. And a multitude of shocks implies a turbulence spectrum of $I(k) \sim k^{-2}$. Such a spectrum of irregularities has the consequence that the diffusion coefficient becomes independent of energy:

$$\kappa = \frac{1}{3} r_g \, c \, \frac{B^2/8\pi}{I(k)k} \,, \tag{18}$$

where $I(k)$ is the spectrum of the magnetic irregularities, here $I(k) \sim k^{-2}$, and $k = 2\pi/r_g$, where r_g in turn is the Larmor radius of a particle in resonant Larmor motion. We take k_0 to be the smallest wave number, corresponding to the largest Larmor radius. Here of course we have that $8\pi I(k_0)k_0/B^2 \leq 1$, and is taken itself to be independent of radius. In such a case κ becomes independent of the particle's energy, and proportional to the radial distance in the halo-wind of the Galaxy.

4.3 The Bending of Particle Paths by the Magnetic Field

Such a wind bends orbits of energetic particles fairly close by, and so does not add substantially to the travel distance for the particle. The Archimedian spiral also has the advantage that the bending, which is really an integral over the Lorentz-force, gives a logarithmic divergence, and so the bending is considerably more - by the logarithm of the ratio of the outer and inner radius of the wind - than if just trivially estimated.

This wind as the key bending agent has the disadvantage that it is a priori not clear how to avoid extreme anisotropies [41], and how to let lower energy particles through from the outside, down to the disk. The analogy with solar wind modulation of cosmic rays below a few hundred MeV would suggest that, below a critical energy no particles get through down to the bottom of the wind anymore. And this critical energy cannot be very different from that where bending becomes important.

However, the solar wind also provides tentative answers to these problems: The irregularities are strong in the magnetic field of the solar wind, and they do not decay towards the poles with the azimuthal component [57], therefore in the pole regions the irregularities do dominate over the regular systematic component. This means that the transport downwards of lower energy particles is more diffusive, and not on straight line orbits, and what we need to require is that the diffusive time scale is actually shorter than the convective time scale on which particles would be carried outwards again by the wind. This condition should hold over the energy range from below 3 10^{18} eV through to above 10^{20} eV, where diffusive transport surely has changed over to a direct path-bending. This is a condition on the irregularities, the dependence on zenith distance angle, and their spectrum.

Therefore, the condition is that at any radius R we have that

$$\tau_{\rm conv} = \frac{R}{V_W} > \tau_{\rm diff} = \frac{R^2}{\kappa_W} \; . \tag{19}$$

This requires a) that the scaling of the two time scales with radius R is about the same, that b) the condition is independent of energy, c) that we can do this over a large range of 4π, and d) that we meet the condition that at the largest energy the bending is strong. Putting in numbers as given earlier we see that these conditions are all met, assuming $I(k)$ to be independent of polar angle θ as suggested by the solar wind data: Taking the inner radius and wind velocity as a beginning we see that

$$\tau_{\rm conv} = 1.5 \times 10^7 \, {\rm yrs} \; . \tag{20}$$

The turbulence in the wind can be written as

$$I(k) = I_0 \left(\frac{k}{k_0}\right)^{-\beta} \tag{21}$$

here with $\beta = 2$, and k_0 and $k_{\rm max}$ the minimum and maximum wavenumbers. The energy contained in the turbulence can be parametrized as

$$b = \int_{k_0}^{k_{\rm max}} \frac{I(k)dk}{B^2/8\pi} = \frac{8\pi I_0 k_0}{(\beta-1)B^2} \tag{22}$$

for $k_{\rm max} \ll k_0$. The diffusion coefficient is then given for parallel shocks (i.e. where the shock normal is parallel to the flow on both sides) by

$$\kappa_W \simeq \frac{c}{3}\frac{r_g}{b}\left(\frac{r_{g,\rm max}}{r_g}\right) \simeq \frac{c}{3}\frac{r_{g,\rm max}}{b} \tag{23}$$

which is independent of the energy of the particle. Obviously, b can be estimated to be near unity, but less than unity, and the maximum Larmor radius in resonance with turbulence can at most be the radius itself, and so (with $b = 1/3$)

$$\kappa_W \gtrsim 1. \times 10^{33}\,\mathrm{cm}^2\mathrm{s}^{-1}\,. \tag{24}$$

Here we use the notion that energetic particles transport the information of the turbulence around the halo in the poloidal direction, and so effectively guarantee that the relevant lengths scale with R, while shocks transport the information radially of course. This entails a diffusion time scale of

$$\tau_{\mathrm{diff}} \lesssim \frac{R^2}{\kappa_W} = 3 \times 10^4\,\mathrm{yrs} \tag{25}$$

just obeying the limit that

$$\tau_{\mathrm{diff}} > \frac{R}{c}\,. \tag{26}$$

We also need to check that the radial dependence of the two time scales is the same, so that the statement is true for all radii, once it is true for one radius, here the innermost radius.

The convective time scale scales with R. The diffusion coefficient is derived from the maximum particle energy relevant and so scales inversely with the magnetic field, which in turn scales inversely with the radius R. Therefore the diffusive time scale scales also linearly with radius R, since we are using here approximately a Parker field. And so we argue that the the two time scales indeed scale both linearly with radius R.

This means then, that indeed

$$\tau_{\mathrm{conv}} > \tau_{\mathrm{diff}} \tag{27}$$

and so the conditions are obeyed, that no particles are kept out, at any energy. Therefore, there is no systematic exclusion, the orbits are just bent in a statistical way. At all energies, particles can come in, and do so with the same time scale, transmitting the same flux down as comes in from outside. Hence we expect the flux at Earth to be derivable just from scrambling the orbits at the boundary of the halo wind, and so there should be no flux diminuition. There is no spectral distortion due to this turbulence.

Now, the estimates have been made here using the equatorial values for the magnetic field; the Ulysses data for the solar wind [57] suggest that the irregularities in the solar wind are nearly independent of zenith angle θ, and so this may be expected to be similar in the Galactic wind. Therefore, the Parker field θ-dependence cannot adequately describe the true state of such a wind, and we actually use the irregularities here.

5 Tests and Predictions

At very high statistics of arriving events, such as soon expected from Pierre Auger, the caustics of propagation may become visible, where different paths to the observer become possible: This happens when the underlying smoothed-out field causes the orbits to loop back across the observer as a function of energy. These caustics are stripes in the sky in terms of arrival directions, and of course are shifting around with energy. One critical test then is the possibility to explain the fair number of pair and triplet events with very different particle energies: These would occur where caustics at different energies overlap or join.

Also, the highest energy particles should show some preference for the northern hemisphere of the sky, while the southern sky would be expected to be depleted at the highest energies, at least in the most simple version of this model. Therefore the Southern sky promises to be the most fruitful for finding events which require less mundane physics, such as expected from the decay of supermassive relic particles.

Ultimately, we may detect neutrinos from sites such as M87, and other similar sites, with higher optical depth for hadronic interactions.

6 Conclusions - Future

We are beginning to be able to ask much more specific questions now about cosmic rays at all energies, and we may expect to obtain answers in the next 10 - 20 years.

The effects of the magnetic fields, with their pronounced asymmetries, should clearly be visible, if these particles are normal protons.

The scenario in which M87 is the source for all the most energetic particles, suggests that any extremely energetic particle population detected in the deep southern hemisphere must be a different type of population. In this case AUGER will probably reveal some new physics, or physical processes which are beyond the very simple concept discussed here.

The existing data base, such as from Haverah Park, and all the arrays, Yakutsk, AGASA, HiRes, Pierre Auger, in the mid-term future, EUSO,and in the long-term future, OWL, will clearly allow us to test all the models proposed; they may be all wrong. By Occams razor, we should eliminate the simple models first. The deep South, Auger and EUSO, will give the strongest tests yet for new physics.

Acknowledgements

We have learnt a lot from talking extensively with P. Blasi, D. Breitschwerdt, A. Bykov, J. Cronin, T. Enßlin, H. Falcke, G. Farrar, T.K. Gaisser, G. Gelmini, R. Jokipii, K.-H. Kampert, H. Kang, T. Kephardt, A. Kusenko, K. Mannheim, W. Matthaeus, U. Mebold, H. Meyer, A. Olinto, R.J. Protheroe, V. Ptuskin, J. Rachen, L. Resvanis, W. Rhode, D. Ryu, G. Schatz, E.-S. Seo, G. Sigl, S. Ter-Antonyan, Y. Wang, A.A. Watson, T. Weiler and many others.

References

1. M. Ave, et al.: Proc. *26th International Cosmic Ray Conference*, Salt Lake City (USA), eds. D. Kieda, et al., vol. **1**, p. 365 (1999)
2. M. Ave, J.A. Hinton, R.A. Vazquez, A.A. Watson, E. Zas: Phys. Rev. Lett. **85**, 2244 (2000)
3. M. Takeda, et al.: Phys. Rev. Lett. **81**, 1163 (1998)
4. M. Takeda, et al.: Astrophys. J. **522**, 225 (1999)
5. N. Hayashida, et al.: Astrophys. J. **522**, 225 (1999)
6. E.E. Antonov, et al.: JETP Lett. **69**, 650 (1999)
7. A.A. Ivanov, et al.: Proc. *26th International Cosmic Ray Conference*, Salt Lake City, eds. D. Kieda, et al., vol.**1**, 403 (1999)
8. M. Ave, et al.: Astropart. Phys. **14**, 109 (2000)
9. Y. Uchihori, et al.: Astropart. Phys. **13**, 151 (2000)
10. M. Nagano, A.A. Watson: Rev. Mod. Phys. **72**, 689 (2000)
11. K. Greisen: Phys. Rev. Lett. **16**, 748 (1966)
12. G. T. Zatsepin, V. A. Kuzmin: Sov. Phys. JETP Lett. **4**, 78 (1966)
13. V. S. Berezinsky, S. I. Grigor'eva: Astron. Astrophys. **199**, 1 (1988)
14. J.P. Rachen, T. Stanev, P. L. Biermann: Astron. Astrophys. **273**, 377 (1993)
15. D. J. Bird, et al.: Phys. Rev. Lett. **71**, 3401 (1993)
16. T. K. Gaisser, et al.: Phys. Rev. D **47**, 1919 (1993)
17. D. J. Bird, et al.: Astrophys. J. **424**, 491 (1994)
18. D. J. Bird, et al.: Astrophys. J. **441**, 144 (1995)
19. J. W. Elbert, P. Sommers: Astrophys. J. **441**, 151 (1995)
20. L. K. Ding, et al.: Astrophys. J. **474**, 490 (1997)
21. T. O'Halloran, P. Sokolsky, S. Yoshida: Physics Today **51**, 31 (1998)
22. D. J. Bird, et al.: Astrophys. J. **511**, 739 (1999)
23. G. A. Medina Tanco: Astrophys. J. Lett. **510**, L91 (1999)
24. G. A. Medina-Tanco: Astrophys. J. **549**, 711 (2001)
25. M. Blanton, P. Blasi, A. V. Olinto: Astropart. Phys. **15**, 275 (2001)
26. V. L. Ginzburg S. I. Syrovatskii: *The origin of cosmic rays*, Pergamon Press, Oxford (1964), Russian edition (1963).
27. G. Cunningham, et al.: Astrophys. J. **236**, L71 (1980)
28. A. A. Watson: Proc. *Cosmology and Particles - 16th Rencontres de Moriond*, eds. J. Audouze, et al., Editions Frontieres, p.49 (1981)
29. A. M. Hillas: Ann. Rev. Astron. Astrophys. **22**, 425 (1984)
30. P. L. Biermann: Proc. *Cosmic winds and the Heliosphere*, eds. J. R. Jokipii, et al., Univ. of Arizona Press, p. 887 (1997)
31. B. Wiebel-Sooth, P. L. Biermann: Landolt-Börnstein, vol. VI/3c, Springer Publ. Comp., p. 37 (1999)
32. T. Piran: Phys. Rep. **314**, 575 (1999)
33. P. Bhattacharjee, G. Sigl: Phys. Rep. **327**, 109 (2000)
34. A. V. Olinto: *David Schramm Memorial Volume*, Phys. Rep. **333-334**, 329 (2000)
35. J. G. Learned, K. Mannheim: Ann. Rev. Nucl. Part. Sci. **50**, 679 (2000)
36. V. S. Berezinskii, et al.: *Astrophysics of Cosmic Rays*, North-Holland, Amsterdam (especially chapter IV) (1990).
37. T. K. Gaisser: *Cosmic Rays and Particle Physics*, Cambridge Univ. Press (1990)
38. E.-J. Ahn, G. Medina-Tanco, P. L. Biermann, T. Stanev, (1999), astro-ph/9911123

39. P. L. Biermann, E.-J. Ahn, G. Medina-Tanco, T. Stanev: Proc. *TAUP99*, eds. J. Dumarchez, M. Froissart, D. Vignaud, Nucl. Phys. B (Proc. Suppl.) **87B**, p. 417 (2000)
40. P. L. Biermann, E.-J. Ahn, G. Medina-Tanco, T. Stanev: Proc. *Erice school (Dec. 1999): Current Topics in Astrofundamental Physics: The Cosmic Microwave background*, ed. N. Sánchez, p. 515 (2001)
41. P. Billoir, A. Letessier-Selvon: astro-ph/0001427
42. D. Harari, S. Mollerach, E. Roulet, E.: *Proc. International Workshop on Observing Ultra High Energy Cosmic Rays from Space and Earth*, (Puebla, Mexico, Aug. 2000), astro-ph/0010068
43. P. L. Biermann, P. A. Strittmatter: Astrophys. J. , **322**, 643 (1987)
44. P. L. Biermann: Proc. *Erice school (Nov. 2000)*, eds. T. Stanev, et al., Kluwer (2001)
45. P. L. Biermann: Proc. *Cosmic Gamma Rays, neutrinos, and related Astrophysics*, eds. M. M. Shapiro and J. P. Wefel, NATO ASI ser. C, vol. 270, Kluwer Acad. Publ., p. 21 (1988)
46. R. V. E. Lovelace: Nature **262**, 649 (1976)
47. H. Falcke, M. A. Malkan, P. L. Biermann: Astron. Astrophys. **298**, 375 (1995)
48. J. R. Jokipii: Astrophys. J. **313**, 842 (1987)
49. J.-L. Puget, F. W. Stecker, J. H. Bredekamp: Astrophys. J. **205**, 638 (1976)
50. J. P. Rachen: PhD thesis, Univ. Bonn (1996)
51. L. Biermann: Zeitschr. für Astrophys. **29**, 274 (1951)
52. E. N. Parker: Astrophys. J. **128**, 664 (1958)
53. E. J. Weber, L. Davis Jr.: Astrophys. J. **148**, 217 (1967)
54. J. A. Burke: Mon. Not. R. Astron. Soc. **140**, 241 (1968).
55. H. E. Johnson, W. I. Axford: Astrophys. J. **165**, 381 (1971).
56. W. G. Mathews, J. C. Baker: Astrophys. J. **170**, 241 (1971).
57. R. J. Forsyth, T. S. Horbury, A. Balogh, E. J. Smith: Geophys. Res. Lett. **23**, 595 (1996)
58. M. Simard-Normandin, P. P. Kronberg: Nature **279**, 115 (1979)
59. P. P. Kronberg: Rep. Prog. Phys. **57**, 325 (1994)
60. H. Kaneda, K. Makishima, S. Yamaguchi, K. Koyama, K. Matsuzaki, N. Y. Yamasaki: Astrophys. J. **491**, 638 (1997)
61. A. Valinia, F. E. Marshall: Astrophys. J. **505**, 134 (1998)
62. S. L. Snowden, et al.: Astrophys. J. **485**, 125 (1997)
63. J. Pietz, J. Kerp, P. M. W. Kalberla, W. B. Burton, D. Hartmann, U. Mebold, U.: Astron. Astrophys. **332**, 55 (1998).
64. H. Seemann, P. L. Biermann: Astron. Astrophys. **327**, 273 (1997)
65. P. G. Mezger, W. J. Duschl, R. Zylka: Astron. Astrophys. Rev. **7**, 289 (1996)
66. D. N. C. Lin, J. E. Pringle: Astrophys. J. Lett. **320**, L87 (1987)
67. Y.-P. Wang, P. L. Biermann: Astron. Astrophys. **334**, 87 (1998)
68. W. J. Duschl, P. A. Strittmatter, P. L. Biermann: Astron. Astrophys. **357**, 1123 (2000)
69. T. Enßlin, et al.: Astrophys. J. Lett. **549**, L39 (2001)
70. B. B. Nath, P. L. Biermann: Mon. Not. R. Astron. Soc. **265**, 241 (1993)
71. S. Niklas: Astron. Astrophys. **322**, 29 (1997)
72. R. Beck, A. Brandenburg, D. Moss, A. Shukurov, A., D. Sokoloff: Ann. Rev. Astron. Astrophys. **34**, 155 (1996)
73. H. Reuter, et al.: Astron. Astrophys. **256**, 10 (1992)
74. D. Breitschwerdt, J. F. McKenzie, H. J. Völk: Astron. Astrophys. **245**, 79 (1991)

75. D. Breitschwerdt, J. F. McKenzie, H. J. Völk: Astron. Astrophys. **269**, 54 (1993)
76. V. N. Zirakashvili, D. Breitschwerdt, V. S. Ptuskin, H. J. Völk: Astron. Astrophys. **311**, 113 (1996)
77. V. S. Ptuskin, H. J. Völk, H.J., D. Breitschwerdt, V. N. Zirakashvili, V.N.: Astron. Astrophys. **321**, 434 (1997)
78. D. Breitschwerdt, S. Komossa: *Proc. Astrophysical Dynamics*, eds. D. Berry, et al., Astrophys. Sp. Sc. **272**, 3 (2000)
79. K. T. Kim, et al.: Nature **341**, 720 (1989)
80. J. P. Vallée: Astron. J. **99**, 459 (1990)
81. P. P. Kronberg, H. Lesch, U. Hopp: Astrophys. J. **511**, 56 (1999)
82. T. Clarke, P. P. Kronberg, H. Böhringer, H.: Proc. *Ringberg Conference on Diffuse Thermal and Relativistic Plasma in Galaxy Clusters*, eds. H. Böhringer, L. Feretti, P. Schuecker, Max-Planck Institut für Extraterrestrische Physik, p. 82 (1999)
83. T. E. Clarke, P. P. Kronberg, H. Böhringer: Astrophys. J. Lett. **547**, L111 (2001)
84. P. L. Biermann, H. Kang, J. P. Rachen, D. Ryu: Proc. *Moriond Meeting on Very High Energy Phenomena in the Universe*, eds. Y. Giraud-Heraud, J. Tran Thanh Van, Editions Frontieres, p. 227 (1997)
85. D. Ryu, H. Kang, P. L. Biermann: Astron. Astrophys. **335**, 19 (1998)
86. R. M. Kulsrud, R. Cen, J. P. Ostriker, J.P., D. Ryu: Astrophys. J. **480**, 481 (1997)
87. R. M. Kulsrud: Ann. Rev. Astron. Astrophys. **37**, 37 (1999)
88. G. Farrar, T. Piran: Phys. Rev. Lett. **84**, 3527 (2000)
89. T. Stanev, et al.: Phys. Rev. Lett. **75**, 3056 (1995)
90. G. Sigl, M. Lemoine, A. V. Olinto: Phys. Rev. D **56**, 4470 (1997)
91. G. Sigl, M. Lemoine: Astropart. Phys. **9**, 65 (1998)
92. G. A. Medina-Tanco, E. M. De Gouveia Dal Pino, J. E. Horvath: Astrophys. J. **492**, 200 (1998)
93. G. A. Medina Tanco: Astrophys. J. Lett. **505**, L79 (1998)
94. M. Lemoine, G. Sigl, A. V. Olinto, D. N. Schramm: Astrophys. J. Lett. **486**, L115 (1997)
95. M. Lemoine, G. Sigl, P. L. Biermann, P.L., (1999), astro-ph/9903124
96. P. Blasi, S. Burles, A. V. Olinto: Astrophys. J. Lett. **514**, L79 (1999)
97. P. Blasi, A. V. Olinto: Phys. Rev. D **59**, 023001 (1999)
98. T. Stanev, R. Engel, A. Muecke, R. J. Protheroe, J. P. Rachen: Phys. Rev. D **62**, 093005 (2000)
99. T. Stanev: Astrophys. J. **479**, 290 (1997)
100. J. L. Han, G. J. Qiao: Astron. Astrophys. **288**, 759 (1994)
101. J. L. Han, R. N. Manchester, E. M. Berkhuijsen, R. Beck: Astron. Astrophys. **322**, 98 (1997)
102. J. L. Han, R. N. Manchester, G. J. Qiao: Mon. Not. R. Astron. Soc. **306**, 371 (1999)
103. F. Krause, R. Beck: Astron. Astrophys. **335**, 789 (1998)

Propagation of Ultra-High-Energy Radiation

Günter Sigl

Institut d'Astrophysique de Paris, 98bis Boulevard Arago, CNRS
75014 Paris, France

Abstract. A systematic account is given of interactions and propagation of nucleons, nuclei, $\gamma-$rays, and neutrinos of energies above a few hundred MeV between production and detection.

1 Introduction

Since implications and predictions of the spectrum especially of ultra high energy cosmic rays above 10^{18} eV (UHECRs) depend on their composition which is uncertain, we will in this chapter review the propagation of all types of particles that could play the role of UHECRs. We start with the hadronic component, and continue with a discussion on electromagnetic cascades initiated by ultra high energy (UHE) photons in extragalactic space, as well as UHE neutrinos which are searched for independent on whether they are secondaries or primaries of UHECRs. We also discuss more speculative options such as new neutral particles predicted in certain supersymmetric models of particle physics. We then discuss how propagation can be influenced by cosmic magnetic fields and what constraints on the location of UHECR sources are implied. The role played by these constraints in the search for sources of extremely high energy cosmic rays (EHECRs) beyond 10^{20} eV is discussed. Finally, the formal description of cosmic ray (CR) propagation by transport equations is briefly reviewed, with an account of the literature on analytical and numerical approaches to their solution. We close with a short discussion of more exotic effects such as anomalous kinematics and violation of Lorentz invariance. We use units in which $c = \hbar = 1$ in this contribution.

Before proceeding, we set up some general notation. The interaction length $l(E)$ of a CR of energy E and mass m propagating through a background of particles of mass m_b is given by

$$l(E)^{-1} = \int d\varepsilon n_b(\varepsilon) \int_{-1}^{+1} d\mu \frac{1 - \mu\beta\beta_b}{2} \sigma(s) , \tag{1}$$

where $n_b(\varepsilon)$ is the number density of the background particles per unit energy at energy ε, $\beta_b = (1 - m_b^2/\varepsilon^2)^{1/2}$ and $\beta = (1 - m^2/E^2)^{1/2}$ are the velocities of the background particle and the CR, respectively, μ is the cosine of the angle between the incoming momenta, and $\sigma(s)$ is the total cross section of the relevant process for the squared center of mass (CM) energy

$$s = m_b^2 + m^2 + 2\varepsilon E \left(1 - \mu\beta\beta_b\right) . \tag{2}$$

Martin Lemoine, Günter Sigl (Eds.): LNP 576, pp. 196–254, 2001.

The most important background particles turn out to be photons with energies in the infrared and optical (IR/O) range or below, so that we will usually have $m_b = 0$, $\beta_b = 1$. A review of the universal photon background has been given in Ref. [1].

It proves convenient to also introduce an energy attenuation length $l_E(E)$ that is obtained from Eq. (1) by multiplying the integrand with the inelasticity, i.e. the fraction of the energy transferred from the incoming CR to the recoiling final state particle of interest. The inelasticity $\eta(s)$ is given by

$$\eta(s) \equiv 1 - \frac{1}{\sigma(s)} \int dE' E' \frac{d\sigma}{dE'}(E', s) \,, \tag{3}$$

where E' is the energy of the recoiling particle considered in units of the incoming CR energy E. Here by recoiling particle we usually mean the "leading" particle, i.e. the one which carries most of the energy.

If one is mostly interested in this leading particle, the detailed transport equations (see Sect. 8.1) for the local density of particles per unit energy, $n(E)$, are often approximated by the simple "diffusion equation"

$$\partial_t n(E) = -\partial_E \left[b(E) n(E) \right] + \Phi(E) \tag{4}$$

in terms of the energy loss rate $b(E) = E/l_E(E)$ and the local injection spectrum $\Phi(E)$. Equation (4) applies to a particle which loses energy at a rate $dE/dt = b(E)$, and is often referred to as the continuous energy loss (CEL) approximation. The CEL approximation is in general good if the non-leading particle is of a different nature than the leading particle, and if the inelasticity is small, $\eta(s) \ll 1$. For an isotropic source distribution $\Phi(E, z)$ in the matter-dominated regime for a flat Universe ($\Omega_0 = 1$), Eq. (4) yields a differential flux today at energy E, $j(E)$, as

$$j(E) = \frac{3}{8\pi} t_0 \int_0^{z_{i,\max}} dz_i (1 + z_i)^{-11/2} \frac{dE_i(E, z_i)}{dE} \Phi[E_i(E, z_i), z_i] \,, \tag{5}$$

where t_0 is the age of the Universe, $E_i(E, z_i)$ is the energy at injection redshift z_i in the CEL approximation, i.e. the solution of $dE/dt = b(E)$ (with $b(E)$ including loss due to redshifting), $E_i(E, 0) = E$ with $t = t_0/(1 + z)^{3/2}$. The maximum redshift $z_{i,\max}$ corresponds either to an absolute cutoff of the source spectrum at $E_{\max} = E_i(E, z_{i,\max})$ or to the earliest epoch when the source became active, whichever is smaller. For a homogeneous production spectrum $\Phi(E)$, this simplifies to

$$j(E) \simeq \frac{1}{4\pi} l_E(E) \Phi(E) \,, \tag{6}$$

if $l_E(E)$ is much smaller than the horizon size such that redshift and evolution effects can be ignored. Equations (5) and (6) are often used in the literature for approximate flux calculations.

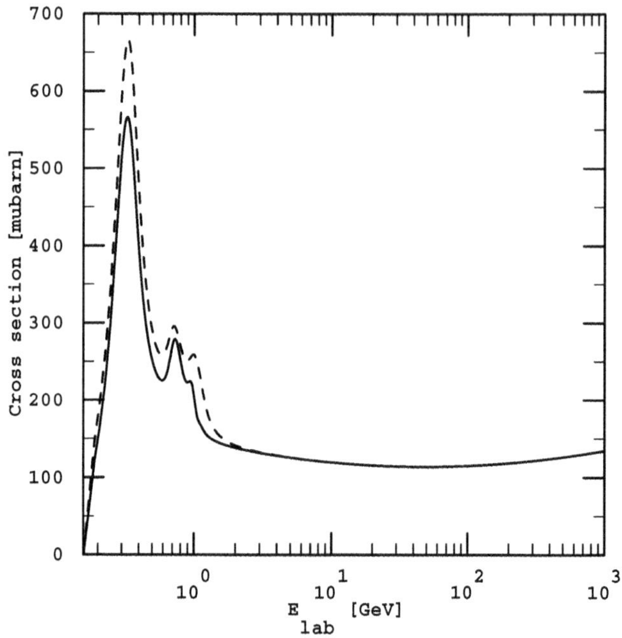

Fig. 1. The total photo-pion production cross section for protons (solid line) and neutrons (dashed line) as a function of the photon energy in the nucleon rest frame, E_{lab}

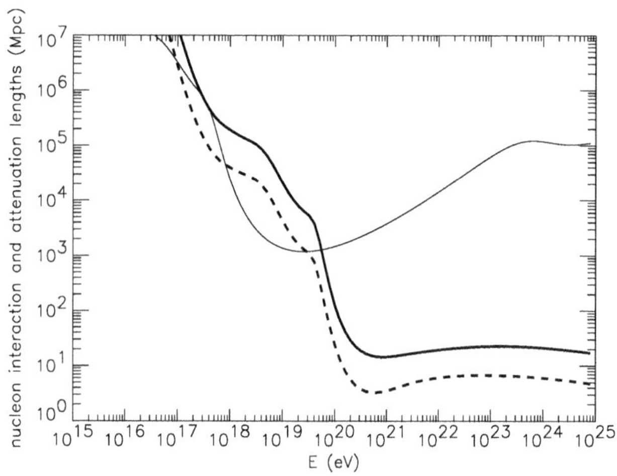

Fig. 2. The nucleon interaction length (dashed line) and attenuation length (solid line) for photo-pion production and the proton attenuation length for pair production (thin solid line) in the combined CMB and the estimated total extragalactic radio background intensity shown in Fig. 3 below

2 Nucleons, Nuclei, and the Greisen-Zatsepin-Kuzmin Cutoff

Shortly after its discovery, it was pointed out by Greisen [2] and by Zatsepin and Kuzmin [3] that the cosmic microwave background (CMB) radiation field has profound consequences for UHECRs: With respect to the rest frame of a nucleon that has a sufficiently high energy in the cosmic rest frame (CRF, defined as the frame in which the CMB is isotropic), a substantial fraction of the CMB photons will appear as γ−rays above the threshold energy for photo-pion production, $E_\gamma^{\rm lab,thr} \equiv m_\pi + m_\pi^2/(2m_N) \simeq 160\,{\rm MeV}$. The total cross section for this process as a function of the γ−ray energy in the nucleon rest frame, $E_\gamma^{\rm lab}$, is shown in Fig. 1. Near the threshold the cross section exhibits a pronounced resonance associated with single pion production, whereas in the limit of high energies it increases logarithmically with $s = m_N^2 + 2m_N E_\gamma^{\rm lab}$ [4]. The long tail beyond the first resonance is essentially dominated by multiple pion production, $N\gamma_b \to N(n\pi)$, $n > 1$ (γ_b stands for the background photon). For a background photon of energy ε in the CRF, the threshold energy $E_\gamma^{\rm lab,thr}$ translates into a corresponding threshold for the nucleon energy,

$$E_{\rm th} = \frac{m_\pi(m_N + m_\pi/2)}{\varepsilon} \simeq 6.8 \times 10^{16} \left(\frac{\varepsilon}{\rm eV}\right)^{-1} {\rm eV}\,. \qquad (7)$$

Typical CMB photon energies are $\varepsilon \sim 10^{-3}\,{\rm eV}$, leading to the so called Greisen-Zatsepin-Kuzmin (GZK) "cutoff" at a few tens of EeV where the nucleon interaction length drops to about 6 Mpc as can be seen in Fig. 2. Detailed investigations of differential cross sections, extending into the multiple pion production regime, have been performed in the literature, mainly for the purpose of calculating secondary γ−ray and neutrino production; for recent discussions and references to earlier literature see, e.g., Refs. [5,6,8].

Below this energy range, the dominant loss mechanism for protons is production of electron-positron pairs on the CMB, $p\gamma_b \to pe^+e^-$, down to the corresponding threshold

$$E_{\rm th} = \frac{m_e(m_N + m_e)}{\varepsilon} \simeq 4.8 \times 10^{14} \left(\frac{\varepsilon}{\rm eV}\right)^{-1} {\rm eV}\,. \qquad (8)$$

Therefore, pair production by protons (PPP) in the CMB ensues at a proton energy $E \sim 5 \times 10^{17}\,{\rm eV}$. The first detailed discussion of PPP in astrophysics was given by Blumenthal [9]. PPP is very similar to triplet pair production by electrons, $e\gamma_b \to ee^+e^-$ (see Sect. 3), where "electron", e, means either an electron or a positron in the following. Away from the threshold the total cross section for a nucleus of charge Z is well approximated by the one for triplet pair production, multiplied by Z^2. Parametric fits to the total cross section and the inelasticity for PPP over the whole energy range were given in Ref. [10]. The resulting proton attenuation length is shown in Fig. 2. Inverse Compton scattering (ICS) of CMB photons by protons can play a certain role for nucleon propagation in the energy range $10^{15} - 10^{17}\,{\rm eV}$ [11]. The next important loss mechanism which starts to dominate near and below the PPP threshold is redshifting due to the cosmic expansion. Indeed, all other loss processes are negligible, except possibly in very dense central regions of galaxies: The interaction length due to hadronic processes which have total cross sections of the order of 0.1 barn in the energy range of interest, for example, is $l \simeq 3 \times 10^5(\Omega_b h^2)^{-1}\,\mathrm{Mpc} \gtrsim 10^7\,\mathrm{Mpc}$, where $0.009 \lesssim \Omega_b h^2 \lesssim 0.02$ [12], with Ω_b the average cosmic baryon density in units of the critical density, and h the Hubble constant H_0 in units of $100\,\mathrm{km\,sec^{-1}Mpc^{-1}}$.

For neutrons, β–decay ($n \to pe^-\bar{\nu}_e$) is the dominant loss process for $E \lesssim 10^{20}\,\mathrm{eV}$. The neutron decay rate $\Gamma_n = m_N/(\tau_n E)$, with $\tau_n \simeq 888.6 \pm 3.5\,\mathrm{sec}$ the laboratory lifetime, implies a neutron range of propagation

$$R_n = \tau_n \frac{E}{m_N} \simeq 0.9 \left(\frac{E}{10^{20}\,\mathrm{eV}}\right) \mathrm{Mpc}\,. \tag{9}$$

The dominant loss process for nuclei of energy $E \gtrsim 10^{19}\,\mathrm{eV}$ is photodisintegration [13,14,15,16] in the CMB and the IR background due to the giant dipole resonance. Early calculations [14] suggested a loss length of a few Mpc. Recent observations of multi-TeV γ–rays from the BL Lac objects Mrk 421 and Mrk 501 suggest [17,18], however, an IR background roughly a factor 10 lower than previously assumed, which is also consistent with recent independent calculation [19] of the intensity and spectral energy distribution of the IR background based on empirical data primarily from IRAS galaxies. This tends to increase the loss length for nuclei [20]. Recent detailed Monte Carlo simulations [21,22,23] indicate that, with the reduced IR background, the CMB becomes the dominant photon background responsible for photodisintegration and, for example, leads to a loss length of $\simeq 10\,\mathrm{Mpc}$ at $2 \times 10^{20}\,\mathrm{eV}$. This loss length plays an important role for scenarios in which the highest energy events observed are heavy nuclei that have been accelerated to UHE (see, e.g., Ref. [24]): The accelerators can not be much further away than a few tens of Mpc. Specific flux calculations for the source NGC 253 have been performed in Ref. [25]. Apart from photodisintegration, nuclei are subject to the same loss processes as nucleons, where the respective thresholds are given by substituting m_N by the mass of the nucleus in Eqs. (7) and (8).

3 UHE Photons and Electromagnetic Cascades

As in the case of UHE nucleons and nuclei, the propagation of UHE photons (and electrons/positrons) is also governed by their interaction with the cosmic photon background. The dominant interaction processes are the attenuation (absorption) of UHE photons due to pair production (PP) on the background photons γ_b: $\gamma\gamma_b \to e^+e^-$ [26], and ICS of the electrons (positrons) on the background photons. Early studies of the effect of PP attenuation on the cosmological UHE γ-ray flux can be found, e.g., in Refs. [27].

The $\gamma-$ray threshold energy for PP on a background photon of energy ε is

$$E_{\rm th} = \frac{m_e^2}{\varepsilon} \simeq 2.6 \times 10^{11} \left(\frac{\varepsilon}{\rm eV}\right)^{-1} \,{\rm eV}\,, \tag{10}$$

whereas ICS has no threshold. In the high energy limit, the total cross sections for PP and ICS are

$$\sigma_{\rm PP} \simeq 2\sigma_{\rm ICS} \simeq \frac{3}{2}\sigma_T \frac{m_e^2}{s} \ln \frac{s}{2m_e^2} \quad (s \gg m_e^2)\,. \tag{11}$$

For $s \ll m_e^2$, $\sigma_{\rm ICS}$ approaches the Thomson cross section $\sigma_T \equiv 8\pi\alpha^2/3m_e^2$ (α is the fine structure constant), whereas $\sigma_{\rm PP}$ peaks near the threshold Eq. (10). Therefore, the most efficient targets for electrons and $\gamma-$rays of energy E are background photons of energy $\varepsilon \simeq m_e^2/E$. For UHE this corresponds to $\varepsilon \lesssim 10^{-6}\,{\rm eV} \simeq 100\,{\rm MHz}$. Thus, radio background photons play an important role in UHE γ-ray propagation through extragalactic space.

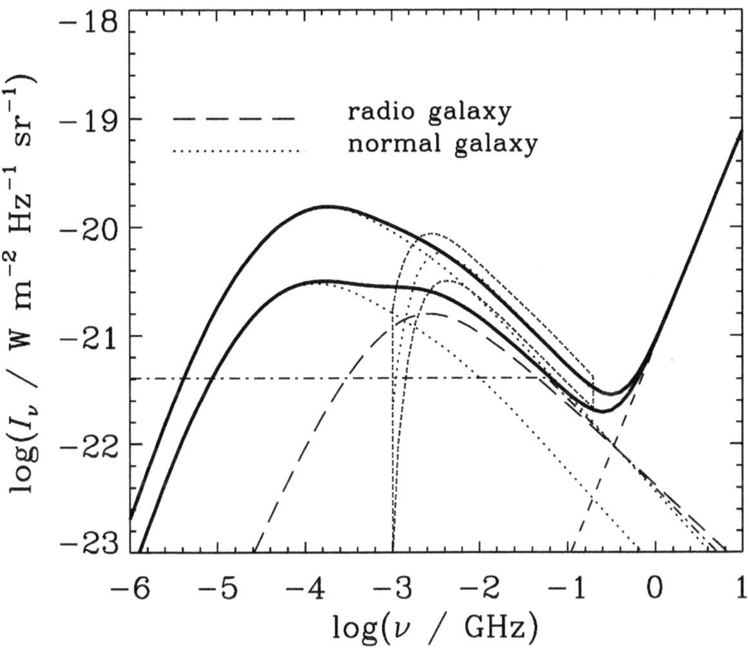

Fig. 3. Contributions of normal galaxies (dotted curves), radio galaxies (long dashed curve), and the cosmic microwave background (short dashed curve) to the extragalactic radio background intensity (thick solid curves) with pure luminosity evolution for all sources (upper curves), and for radio galaxies only (lower curves), from Ref. [28]. Dotted band gives an observational estimate of the total extragalactic radio background intensity [29] and the dot-dash curve gives an earlier theoretical estimate [30] (From Protheroe and Biermann [28])

Unfortunately, the universal radio background (URB) is not very well known mostly because it is difficult to disentangle the Galactic and extragalactic components. Observational estimates have been given in Ref. [29], and an early theoretical estimate was given in Ref. [30]. Recently, an attempt has been made to calculate the contribution to the URB from radio galaxies and Active Galactic Nuclei (AGNs) [28], and also from clusters of galaxies which tends to give higher estimates. The issue does not seem to be settled, however. At frequencies somewhere below 1 MHz the URB is expected to cut off exponentially due to free-free absorption. The exact location of the cut-off depends on the abundance and clustering of electrons in the intergalactic medium and/or the radio source and is uncertain between about 0.1 – 2 MHz. Figure 3 compares results from Ref. [28] with Ref. [30] and the observational estimate from Ref. [29].

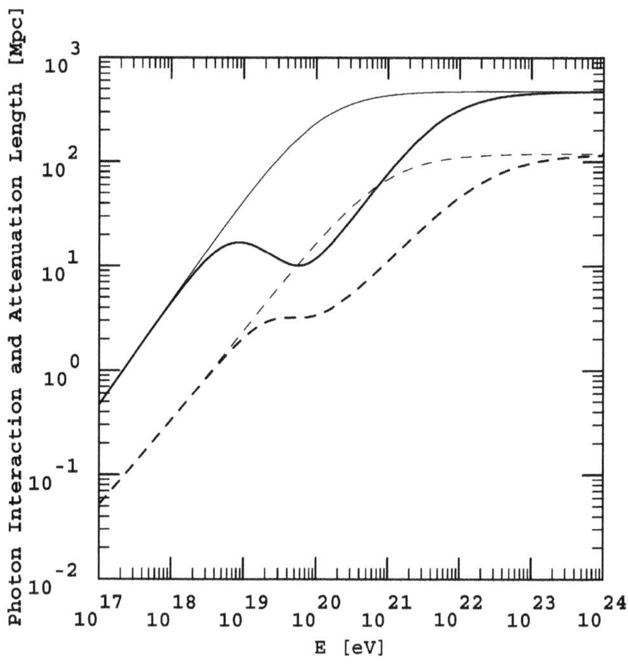

Fig. 4. Interaction lengths (dashed lines) and energy attenuation lengths (solid lines) of γ−rays in the CMB (thin lines) and in the total low energy photon background spectrum shown in Fig. 3 with the observational URB estimate from Ref. [29] (thick lines), respectively. The interactions taken into account are single and double pair production

In the extreme Klein-Nishina limit, $s \gg m_e^2$, either the electron or the positron produced in the process $\gamma\gamma_b \to e^+e^-$ carries most of the energy of the initial UHE photon. This leading electron can then undergo ICS whose inelasticity (relative to the electron) is close to 1 in the Klein-Nishina limit. As a

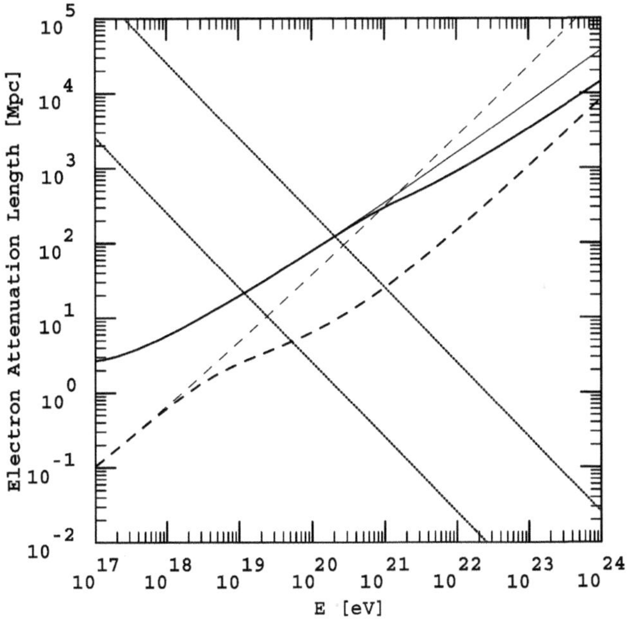

Fig. 5. Energy attenuation lengths of electrons for various processes: Solid lines are for triplet pair production, and dashed lines for inverse Compton scattering in the CMB (thin lines) and in the total low energy photon background spectrum shown in Fig. 3 with the observational URB estimate from Ref. [29] (thick lines). The dotted lines are for synchrotron emission losses in a large-scale extragalactic magnetic field of r.m.s. strength of 10^{-11} G (upper curve) and 10^{-10} G (lower curve), respectively

consequence, the upscattered photon which is now the leading particle after this two-step cycle still carries most of the energy of the original $\gamma-$ray, and can initiate a fresh cycle of PP and ICS interactions. This leads to the development of an *electromagnetic (EM) cascade* which plays an important role in the resulting observable $\gamma-$ray spectra. An important consequence of the EM cascade development is that the effective penetration depth of the EM cascade, which can be characterized by the energy attenuation length of the leading particle (photon or electron/positron), is considerably greater than just the interaction lengths [31]; see Figs. 4 and 5). As a result, the predicted flux of UHE photons can be considerably larger than that calculated by considering only the absorption of UHE photons due to PP.

EM cascades play an important role particularly in some exotic models of UHECR origin such as collapse or annihilation of topological defects (see contribution by P. Bhattacharjee and G. Sigl in this volume) in which the EHECR injection spectrum is predicted to be dominated by γ-rays [35]. Even if only UHE nucleons and nuclei are produced in the first place, for example, via conventional shock acceleration (see contribution by G. Pelletier in this volume), EM cascades can be produced by the secondaries coming from the decay of pions which are created in interactions of UHE nucleons with the low energy photon background.

The EM cascading process and the resulting diffuse γ-ray fluxes in the conventional acceleration scenarios of UHECR origin were calculated in the 1970s; see, e.g., Refs. [32,33,34]. The EM cascades initiated by "primary" γ-rays and their effects on the diffuse UHE γ-ray flux in the topological defect scenario of UHECRs were first considered in Ref. [35]. All these calculations were performed within the CEL approximation which, as described above, deals with only the leading particle. However, the contribution of non-leading particles to the flux can be substantial for cascades that are not fully developed. A reliable calculation of the flux at energies much smaller than the maximal injection energy should therefore go beyond the CEL approximation, i.e., one should solve the relevant Boltzmann equations for propagation; this is discussed in Sect. 8.

Cascade development accelerates at lower energies due to the decreasing interaction lengths (see Figs. 4 and 5) until most of the $\gamma-$rays fall below the PP threshold on the low energy photon background at which point they pile up with a characteristic $E^{-1.5}$ spectrum below this threshold [36,37,38,39]. The source of these $\gamma-$rays are predominantly the ICS photons of average energy $\langle E_\gamma\rangle = E_e(1-4\langle s\rangle/3m_e^2)$ arising from interactions of electrons of energy E_e with the background at average squared CM energy $\langle s\rangle$ in the Thomson regime. The relevant background for cosmological propagation is constituted by the universal IR/O background, corresponding to $\varepsilon \lesssim 1\,\mathrm{eV}$ in Eq. (10), or $E_{\mathrm{th}} \simeq 10^{11}\,\mathrm{eV}$. Therefore, most of the energy of fully developed EM cascades ends up below $\simeq 100\,\mathrm{GeV}$ where it is constrained by measurements of the diffuse $\gamma-$ray flux by EGRET on board the CGRO [40] and other effects. For an injection rate $\propto t^{-p}$ in cosmic age t, the EGRET measurement implies the limit $Q^0_{\mathrm{EM}} \lesssim 2.2\times10^{-23}\,h(3p-1)\,\mathrm{eV\,cm^{-3}\,sec^{-1}}$ on the electromagnetic injection rate at zero redshift. Constraints from limits on CMB distortions and light element abundances from ^{4}He-photodisintegration are comparable to the bound from the directly observed diffuse GeV γ-rays [41].

Flux predictions involving EM cascades are therefore an important source of constraints of UHE energy injection on cosmological scales.

It should be mentioned here that the development of EM cascades depends sensitively on the strength of the extragalactic magnetic fields (EGMFs) which is rather uncertain. The EGMF typically inhibits cascade development because of the synchrotron cooling of the e^+e^- pairs produced in the PP process. For a sufficiently strong EGMF the synchrotron cooling time scale of the leading electron (positron) may be small compared to the time scale of ICS interaction, in which case, the electron (positron) synchrotron cools before it can undergo ICS, and thus cascade development stops. In this case, the UHE γ-ray flux is determined mainly by the "direct" γ-rays, i.e., the ones that originate at distances less than the absorption length due to PP process. The energy lost through synchrotron cooling does not, however, disappear; rather, it reappears at lower energies and can even initiate fresh EM cascades there depending on the remaining path length and the strength of the relevant background photons. Thus, the overall effect of a relatively strong EGMF is to deplete the UHE γ-ray flux above some energy and increase the flux below a corresponding energy in the "low" (typically few tens to hundreds of GeV) energy region. These issues are further discussed in Sect. 5.1.

The lowest order cross sections, Eq. (11), fall off as $\ln s/s$ for $s \gg m_e^2$. Therefore, at EHE, higher order processes with more than two final state particles start to become important because the mass scales of these particles can enter into the corresponding cross section which typically is asymptotically constant or proportional to powers of $\ln s$.

Double pair production (DPP), $\gamma\gamma_b \to e^+e^-e^+e^-$, is a higher order QED process that affects UHE photons. The DPP total cross section is a sharply rising function of s near the threshold that is given by Eq. (10) with $m_e \to 2m_e$, and quickly approaches its asymptotic value [42]

$$\sigma_{\rm DPP} \simeq \frac{172\alpha^4}{36\pi m_e^2} \simeq 6.45\,\mu\text{barn} \quad (s \gg m_e^2)\,. \tag{12}$$

DPP begins to dominate over PP above $\sim 10^{21} - 10^{23}$ eV, where the higher values apply for stronger URB (see Fig. 4).

For electrons, the relevant higher order process is triplet pair production (TPP), $e\gamma_b \to ee^+e^-$. This process has been discussed in some detail in Refs. [43] and its asymptotic high energy cross section is

$$\sigma_{\rm TPP} \simeq \frac{3\alpha}{8\pi}\sigma_T \left(\frac{28}{9}\ln\frac{s}{m_e^2} - \frac{218}{27}\right) \quad (s \gg m_e^2)\,, \tag{13}$$

with an inelasticity of

$$\eta \simeq 1.768 \left(\frac{s}{m_e^2}\right)^{-3/4} \quad (s \gg m_e^2)\,. \tag{14}$$

Thus, although the total cross section for TPP on CMB photons becomes comparable to the ICS cross section already around 10^{17} eV, the energy attenuation is not important up to $\sim 10^{22}$ eV because $\eta \lesssim 10^{-3}$ (see Fig. 5). The main effect of TPP between these energies is to create a considerable number of electrons and channel them to energies below the UHE range. However, TPP is dominated over by synchrotron cooling (see Sect. 5.1), and therefore negligible, if the electrons propagate in a magnetic field of r.m.s. strength $\gtrsim 10^{-12}$ G, as can be seen from Fig. 5.

Various possible processes other than those discussed above — e.g., those involving the production of one or more muon, tau, or pion pairs, double Compton scattering ($e\gamma_b \to e\gamma\gamma$), $\gamma-\gamma$ scattering ($\gamma\gamma_b \to \gamma\gamma$), Bethe-Heitler pair production ($\gamma X \to Xe^+e^-$, where X stands for an atom, an ion, or a free electron), the process $\gamma\gamma_b \to e^+e^-\gamma$, and photon interactions with magnetic fields such as magnetic pair production ($\gamma B \to e^+e^-$) — are in general negligible in EM cascade development. The total cross section for the production of a single muon pair

$(\gamma\gamma_b \to \mu^+\mu^-)$, for example, is smaller than that for electron pair production by about a factor 10 and thus does not play a significant role in the resulting EM fluxes. Muon and pion pair production by photons, however, could lead to an observable "bump" in the neutrino spectrum around 10^{17} eV [44]. A rough estimate implies that the process $\gamma\gamma_b \to \pi_0 \to \gamma\gamma$ is at most as imoprtant as DPP, but the author is not aware of a detailed discussion of this process in the literature. Energy loss rate contributions for TPP involving pairs of heavier particles of mass m are suppressed by a factor $\simeq (m_e/m)^{1/2}$ for $s \gg m^2$. Similarly, DPP involving heavier pairs is also negligible [42]. The cross section for double Compton scattering is of order α^3 and must be treated together with the radiative corrections to ordinary Compton scattering of the same order. Corrections to the lowest order ICS cross section from processes involving m_γ additional photons in the final state, $e\gamma_b \to e+(m_\gamma+1)\gamma$, $m_\gamma \geq 1$, turn out to be smaller than 10% in the UHE range [45]. A similar remark applies to corrections to the lowest order PP cross section from the processes $\gamma\gamma_b \to e^+e^- + m_\gamma\gamma$, $m_\gamma \geq 1$. Photon−photon scattering can only play a role at redshifts beyond $\simeq 100$ and at energies below the redshift-dependent pair production threshold given by Eq. (10) [46,47,48]. A similar remark applies to Bethe-Heitler pair production [47]. Photon interactions with magnetic fields of typical galactic strength, $\sim 10^{-6}$ G, are only relevant for $E \gtrsim 10^{24}$ eV [49]. For EGMFs the critical energy for such interactions is even higher.

4 Propagation and Interactions of Neutrinos and "Exotic" Particles

4.1 Neutrinos

Neutrino Propagation

The propagation of UHE neutrinos is governed mainly by their interaction with the relic neutrino background (RNB). In this section we give a short overview over the relevant interactions within the general framework for particle propagation used in the present contribution. A more detailed presentation of the resulting neutrino fluxes can be found in the contribution by S. Yoshida on neutrino cascades in this volume.

The average squared CM energy for interaction of an UHE neutrino of energy E with a relic neutrino of energy ε is given by

$$\langle s\rangle \simeq (45\,\mathrm{GeV})^2 \left(\frac{\varepsilon}{10^{-3}\,\mathrm{eV}}\right)\left(\frac{E}{10^{15}\,\mathrm{GeV}}\right) . \tag{15}$$

If the relic neutrino is relativistic, then $\varepsilon \simeq 3T_\nu(1+\eta_b/4)$ in Eq. (15), where $T_\nu \simeq 1.9(1+z)\,\mathrm{K} = 1.6\times 10^{-4}(1+z)\,\mathrm{eV}$ is the temperature at redshift z and $\eta_b \lesssim 50$ is the dimensionless chemical potential of relativistic relic neutrinos. For nonrelativistic relic neutrinos of mass $m_\nu \lesssim 20\,\mathrm{eV}$, $\varepsilon \simeq \max[3T_\nu, m_\nu]$. Note that Eq. (15) implies interaction energies that are typically smaller than electroweak energies even for UHE neutrinos, except for energies near the Grand Unification scale, $E \gtrsim 10^{15}\,\mathrm{GeV}$, or if $m_\nu \gtrsim 1\,\mathrm{eV}$. In this energy range, the cross sections are given by the Standard Model of electroweak interactions which are well confirmed experimentally. Physics beyond the Standard Model is, therefore, not expected to play a significant role in UHE neutrino interactions with the low energy relic backgrounds.

The dominant interaction mode of UHE neutrinos with the RNB is the exchange of a $W^\pm$ boson in the t-channel ($\nu_i+\bar{\nu}_j \to l_i+\bar{l}_j$), or of a Z^0 boson in either the s-channel ($\nu_i + \bar{\nu}_i \to f\bar{f}$) or the t-channel ($\nu_i + \bar{\nu}_j \to \nu_i + \bar{\nu}_j$) [50,51,52,53]. Here, i, j stands for either the electron, muon, or tau flavor, where $i \neq j$ for the first reaction, l denotes a charged lepton, and f any charged fermion. If the latter is a quark, it will, of course, subsequently fragment into hadrons. As an example, the differential cross section for s-channel production of Z^0 is given by

$$\frac{d\sigma_{\nu_i+\bar{\nu}_j\to Z^0\to f\bar{f}}}{d\mu^*} = \frac{G_\mathrm{F}^2 s}{4\pi}\,\frac{M_Z^4}{(s-M_Z^2)^2+M_Z^2\Gamma_Z^2}\left[g_L^2(1+\mu^*)^2+g_R^2(1-\mu^*)^2\right] , \tag{16}$$

where G_F is the Fermi constant, M_Z and Γ_Z are mass and lifetime of the Z^0, g_L and g_R are the usual dimensionless left- and right-handed coupling constants for f, and μ^* is the cosine of the scattering angle in the CM system.

The t-channel processes have cross sections that rise linearly with s up to $s \simeq M_W^2$, with M_W the $W^\pm$ mass, above which they are roughly constant with a value $\sigma_t(s \gtrsim M_W) \sim G_\mathrm{F}^2 M_W^2 \sim 10^{-34}\,\mathrm{cm}^2$. Using Eq. (15) this yields the rough estimate

$$\begin{aligned}\sigma_t(E,\varepsilon) &\sim \min\left[10^{-34}, 10^{-44}\left(\frac{s}{\mathrm{MeV}^2}\right)\right]\,\mathrm{cm}^2 \\ &\sim \min\left[10^{-34}, 3\times10^{-39}\left(\frac{\varepsilon}{10^{-3}\,\mathrm{eV}}\right)\left(\frac{E}{10^{20}\,\mathrm{eV}}\right)\right]\,\mathrm{cm}^2 .\end{aligned} \tag{17}$$

In contrast, within the Standard Model neutrino-nucleon cross sections roughly behave as

$$\sigma_{\nu N}(E) \sim 10^{-31}\left(\frac{E}{10^{20}\,\mathrm{eV}}\right)^{0.4}\,\mathrm{cm}^2 \tag{18}$$

for $E \gtrsim 10^{15}\,\mathrm{eV}$, see Eq. (19) below. Interactions of UHE neutrinos with nucleons are, however, still negligible compared to interactions with the RNB because the RNB particle density is about ten orders of magnitude larger than the baryon density. The only exception could occur near Grand Unification scale energies and at high redshifts and/or if contributions to the neutrino-nucleon cross section from physics beyond the Standard Model dominate at these energies (see at end of Sect. 4.1 below).

It has recently been pointed out [54] that above the threshold for $W^\pm$ production the process $\nu+\gamma \to lW^+$ becomes comparable to the $\nu\nu$ processes discussed above. Figure 6 compares the cross sections relevant for neutrino propagation at CM energies around the electroweak scale. Again, for UHE neutrino interactions with the RNB the relevant CM energies can only be reached if (a) the

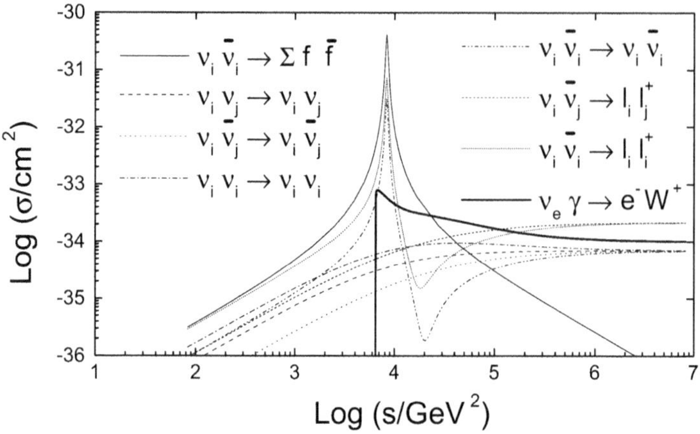

Fig. 6. Various cross sections relevant for neutrino propagation as a function of s [51,54]. The sum $\sum_j f_j \bar{f}_j$ does not include $f_j = \nu_i, l_i, t, W$, or Z. (From Ref. [54])

UHE neutrino energy is close to the Grand Unification scale, or (b) the RNB neutrinos have masses in the eV regime, or (c) at redshifts $z \gtrsim 10^3$. Even then the $\nu\gamma$ process never dominates over the $\nu\nu$ process.

At lower energies there is an additional $\nu\gamma$ interaction that was recently discussed as potentially important besides the $\nu\nu$ processes: Using an effective Lagrangian derived from the Standard Model, Ref. [55] obtained the result $\sigma_{\gamma+\nu\to\gamma+\gamma+\nu}(s) \simeq 9 \times 10^{-56}\,(s/\mathrm{MeV}^2)^5\,\mathrm{cm}^2$, supposed to be valid at least up to $s \lesssim 10\,\mathrm{MeV}^2$. Above the electron pair production threshold the cross section has not been calculated because of its complexity but is likely to level off and eventually decrease. Nevertheless, if the s^5 behavior holds up to $s \simeq$ a few hundred MeV^2, comparison with Eq. (17) shows that the process $\gamma + \nu \to \gamma + \gamma + \nu$ would start to dominate and influence neutrino propagation around $E \sim 3 \times 10^{17}\,(\varepsilon/10^{-3}\,\mathrm{eV})\,\mathrm{eV}$, as was pointed out in Ref. [56].

For a given source distribution, the contribution of the "direct" neutrinos to the flux can be computed by integrating Eq. (5) up to the interaction redshift $z(E)$, i.e. the average redshift from which a neutrino of present day energy E could have propagated without interacting. This approximation neglects the secondary neutrinos and the decay products of the leptons created in the neutral current and charged current reactions of UHE neutrinos with the RNB discussed above. Similarly to the EM case, these secondary particles can lead to neutrino cascades developing over cosmological redshifts [52] (see contribution by S. Yoshida on neutrino cascades in this volume for more details).

Approximate expressions for the interaction redshift for the processes discussed above have been given in Refs. [36,57] for CM energies below the electroweak scale, assuming relativistic, nondegenerate relic neutrinos, $m_\nu \lesssim T_\nu$, and $\eta_b \ll 1$. Approaching the electroweak scale, a resonance occurs in the interaction cross section for s-channel Z^0 exchange at the Z^0 mass, $s = M_Z^2 \simeq (91\,\mathrm{GeV})^2$, see Eq. (16). The absorption redshift for the corresponding neutrino energy, $E \simeq 10^{15}\,\text{GeV}(\varepsilon/10^{-3}\,\text{eV})^{-1}$ drops to a few (or less for a degenerate, relativistic RNB) and asymptotically approaches constant values of a few tens at higher energies.

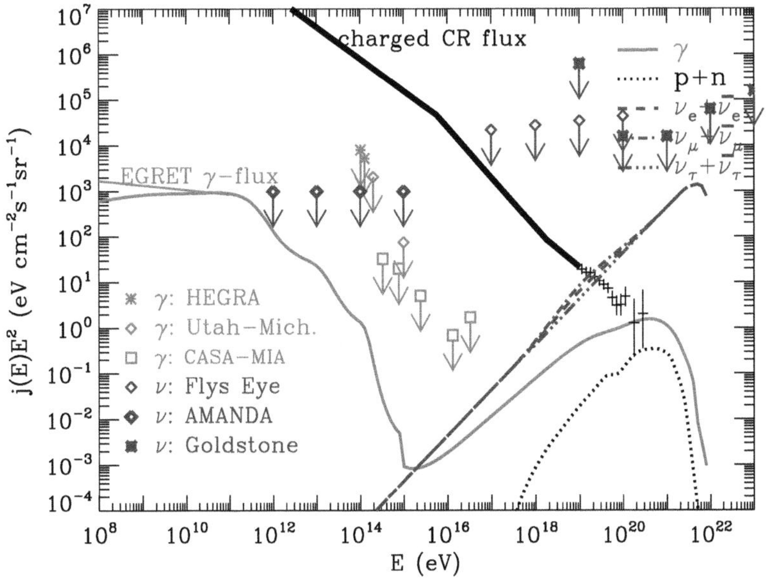

Fig. 7. Fluxes of neutrinos (dashed and dashed-dotted, as indicated), γ−rays (solid), and nucleons (dotted) predicted by the Z-burst mechanism for $m_{\nu_e} = 0.1\,\text{eV}$, $m_{\nu_\mu} = m_{\nu_\tau} = 1\,\text{eV}$, for homogeneously distributed sources emitting neutrinos with an E^{-1} spectrum (equal for all flavors) up to $10^{22}\,\text{eV}$ and an E^{-2} γ−ray spectrum of equal power up to 100 TeV. Injection rates were assumed comovingly constant up to $z = 2$. The relic neutrino overdensity was assumed to be 200 over 5 Mpc. The calculation used the code described in Ref. [58] and assumed the lower limit of the URB [28] (see Fig. 3) and a vanishing extragalactic magnetic field. 1 sigma error bars are the combined data from the Haverah Park [59], the Fly's Eye [60], and the AGASA [61] experiments above $10^{19}\,\text{eV}$. Also shown are piecewise power law fits to the observed charged CR flux (thick solid line) and the EGRET measurement of the diffuse γ−ray flux between 30 MeV and 100 GeV [40] (solid line on left margin). Upper limits on the γ−ray flux below $10^{17}\,\text{eV}$ and on neutrino fluxes above $10^{17}\,\text{eV}$ (except for AMANDA) from various experiments are as indicated (see text and Ref. [62] for more details)

An interesting situation arises if the RNB consists of massive neutrinos with $m_\nu \sim 1\,\text{eV}$: Such neutrinos would constitute hot dark matter which is expected to cluster [63], for example, in galaxy clusters. This would potentially increase the interaction probability for any neutrino of energy within the width of the Z^0 resonance at $E = M_Z^2/2m_\nu = 4 \times 10^{21}(\text{eV}/m_\nu)\,\text{eV}$. Recently it has been suggested that the stable end products of the "Z-bursts" that would thus be induced at close-by distances ($\lesssim 50\,\mathrm{Mpc}$) from Earth may explain the highest energy cosmic rays [64,65] and may also provide indirect evidence for neutrino hot dark matter. These end products would be mostly nucleons and $\gamma-$rays with average energies a factor of $\simeq 5$ and $\simeq 40$ lower, respectively, than the original UHE neutrino. As a consequence, if the UHE neutrino was produced as a secondary of an accelerated proton, the energy of the latter would have to be at least a few $10^{22}\,\mathrm{eV}$ [64], making Z-bursts above GZK energies more likely to play a role in the context of non-acceleration scenarios (see contribution by P. Bhattacharjee and G. Sigl in this volume). Moreover, it has subsequently been pointed out [66] that Z production is dominated by annihilation on the non-clustered massive RNB compared to annihilation with neutrinos clustering in the Galactic halo or in nearby galaxy clusters. As a consequence, for a significant contribution of neutrino annihilation to the observed EHECR flux, a new class of neutrino sources, unrelated to UHECR sources, seems necessary. This has been confirmed by more detailed numerical simulations [67] where it has, however, also been demonstrated that the most significant contribution could come from annihilation on neutrino dark matter clustering in the Local Supercluster by amounts consistent with expectations. In the absence of any assumptions on the neutrino sources, the minimal constraint comes from the unavoidable production of secondary $\gamma-$rays contributing to the diffuse flux around 10 GeV measured by EGRET : If the Z-burst decay products are to explain EHECR, the massive neutrino overdensity f_ν over a length scale l_ν has to satisfy $f_\nu \gtrsim 20\,(l_\nu/5\,\mathrm{Mpc})^{-1}$, provided that only neutrinos leave the source, a situation that may arise in top-down models if the X particles decay exclusively into neutrinos (see Ref. [58] for a model involving topological defects and Ref. [68] for a scenario involving decaying superheavy relic particles). If, instead, the total photon source luminosity is comparable to the total neutrino luminosity, as in most models, the EGRET constraint translates into the more stringent requirement $f_\nu \gtrsim 10^3(l_\nu/5\,\mathrm{Mpc})^{-1}$. This bound can only be relaxed if most of the EM energy is radiated in the TeV range where the Universe is more transparent [67]. Figure 7 shows an example of this situation.

Furthermore, the Z-burst scenario requires sources that are optically thick for accelerated protons with respect to photo-pion production because otherwise the observable proton flux below the GZK cutoff would be comparable to the neutrino flux [70,66]. This argument can be generalized: If neutrinos are produced as secondaries of protons accelerated in astrophysical sources and if these protons are not absorbed in the sources, but rather contribute to the UHECR flux observed, then the energy content in the neutrino flux can not be higher than the one in UHECRs, leading to the so called Waxman-Bahcall bound [71,72,73] (shown in Fig. 8 below). If one of these assumptions does not apply, such as for acceleration sources that are opaque to nucleons or in the top-down scenarios where X particle decays produce much fewer nucleons than $\gamma-$rays and neutrinos, the Waxman-Bahcall bound does not apply, but the neutrino flux is still constrained by the observed diffuse $\gamma-$ray flux in the GeV range. This is true as long as the energy fluences produced in $\gamma-$rays and neutrinos are comparable, which follows from isospin symmetry if neutrinos are produced by pion production, because $\gamma-$rays injected above the pair production threshold on the CMB will cascade down to the GeV regime (see Sect. 3 above).

A systematic parameter study of RNB overdensities required in the Z-burst scenario, based on analytical flux estimates, has been performed in Ref. [74]. Recently it has been noted that a degenerate relic neutrino background would increase the interaction probability and thereby make the Z-burst scenario more promising [75]. A neutrino asymmetry of order unity is not excluded phenomenologically [76] and can be created in the early Universe, for example, through the Affleck-Dine baryogenesis mechanism [77] or due to neutrino oscillations. The authors of Ref. [75] pointed out that for a neutrino mass $m_\nu \simeq 0.07\,\text{eV}$, a value suggested by the Super-Kamiokande experiment [78], and for sources at redshifts of a few, the flux of secondary Z-decay products is maximal for a RNB density parameter $\Omega_\nu \simeq 0.01$. Such neutrino masses, however, require the sources to produce neutrinos at least up to $10^{22}\,\text{eV}$.

UHE neutrinos from the decay of pions, that are produced by interactions of accelerated protons in astrophysical sources, must have originated within redshifts of a few. Moreover, in most conventional models their flux is expected to fall off rapidly above $10^{20}\,\text{eV}$. Examples are production in AGNs within hadronic models [79,80,81,82,83,71], and "cosmogenic" neutrinos from interactions of UHECR nucleons (near or above the GZK cutoff) with the CMB (see, e.g.,Refs. [84,85]). The latter source is the only one that is guaranteed to exist due to existence of UHECRs near the GZK cutoff, but the fluxes are generally quite small. Therefore, interaction of these UHE neutrinos with the RNB, that could reveal the latter's existence, can, if at all, be important only if the relic neutrinos have a mass $m_\nu \gtrsim 1\,\text{eV}$ [50]. Due to the continuous release of UHE neutrinos up to much higher redshifts, most top-down scenarios would imply substantially higher fluxes that also extend to much higher energies [57]. Certain features in the UHE neutrino spectrum predicted within such top-down scenarios, such as a change of slope for massless neutrinos [52] or a dip structure for relic neutrino masses of order $1\,\text{eV}$ [53,65], have therefore been proposed as possibly the only way to detect the RNB. However, some of the scenarios at the high end of neutrino flux predictions have recently been ruled out based on constraints on the accompanying energy release into the EM channel (see contribution by P. Bhattacharjee and G. Sigl in this volume).

The recent claim that the Z-burst mechanism can explain the EHECR flux even without overdensity, $f_\nu = 1$, and even allows to determine the neutrino mass $m_\nu \sim 1\,\text{eV}$ [86] was based on assumptions that we characterized here as highly unrealistic: Sources accelerating nuclei to $\gtrsim 10^{23}\,\text{eV}$ while being completely opaque to both the primaries and the secondary photons. The energy fluence of the latter, however, after possibly being reprocessed to lower energies by EM cascading, must be comparable to the neutrino energy fluence by simple isospin symmetry in the production of charged and neutral pions.

Since in virtually all models UHE neutrinos are created as secondaries from pion decay, i.e. as electron or muon neutrinos, $\tau-$neutrinos can only be produced by a flavor changing $W^\pm$ t-channel interaction with the RNB. The flux of UHE $\tau-$neutrinos is therefore usually expected to be substantially smaller than the one of electron and muon neutrinos, if no neutrino oscillations take place at these energies. However, the recent evidence from the Superkamiokande experiment for nearly maximal mixing between muon and $\tau-$neutrinos with $|\Delta m^2| = |m^2_{\nu_\mu} - m^2_{\nu_\tau}| \simeq 5 \times 10^{-3}\,\mathrm{eV}^2$ [78] would imply an oscillation length of $L_{osc} = 2E/|\Delta m^2| = 2.6 \times 10^{-6}(E/\mathrm{PeV})(|\Delta m^2|/5 \times 10^{-3}\,\mathrm{eV}^2)^{-1}\,\mathrm{pc}$ and, therefore, a rough equilibration between muon and $\tau-$neutrino fluxes from any source at a distance larger than L_{osc} [88]. Turning this around, one sees that a source at distance d emitting neutrinos of energy E is sensitive to neutrino mixing with $|\Delta m^2| = 2E/d \simeq 1.3 \times 10^{-16}\,(E/\mathrm{PeV})(d/100\,\mathrm{Mpc})^{-1}\,\mathrm{eV}^2$ [89,90]. Under certain circumstances, resonant conversion in the potential provided by the RNB clustering in galactic halos may also influence the flavor composition of UHE neutrinos from extraterrestrial sources [91]. In addition, such huge cosmological baselines can be sensitive probes of neutrino decay [92].

Neutrino Detection

We now turn to a discussion of UHE neutrino interactions with matter relevant for neutrino detection. UHE neutrinos can be detected by detecting the muons produced in ordinary matter via charged current reactions with nucleons; see, e.g., Refs. [93,94,95] for recent discussions. Corresponding cross sections are calculated by folding the fundamental Standard Model quark-neutrino cross section with the distribution function of the partons in the nucleon. These cross sections are most sensitive to the abundance of partons of fractional momentum $x \simeq M_W^2/2m_N E$, where E is the neutrino energy. For the relevant squared momentum transfer, $Q^2 \sim M_W^2$, these parton distribution functions have been measured down to $x \simeq 0.02$ [96]. (It has been suggested that observation of the atmospheric neutrino flux with future neutrino telescopes may probe parton distribution functions at much smaller x currently inaccessible to colliders [97]). Currently, therefore, neutrino-nucleon cross sections for $E \gtrsim 10^{14}\,\mathrm{eV}$ can be obtained only by extrapolating the parton distribution functions to lower x. Above $10^{19}\,\mathrm{eV}$, the resulting uncertainty has been estimated to be a factor 2 [94], whereas within the dynamical radiative parton model it has been claimed to be at most 20 % [95]. An intermediate estimate of the charged current neutrino-nucleon cross section using the CTEQ4-DIS distributions can roughly be parameterized by [94]

$$\sigma_{\nu N}(E) \simeq 2.36 \times 10^{-32} \left(\frac{E}{10^{19}\,\mathrm{eV}}\right)^{0.363} \mathrm{cm}^2 \quad (10^{16}\,\mathrm{eV} \lesssim E \lesssim 10^{21}\,\mathrm{eV})\,. \quad (19)$$

The neutral current cross sections are a factor 2–3 smaller than this. Improved calculations including non-leading logarithmic contributions in $1/x$ have recently been performed in Ref. [98]. The results for the neutrino-nucleon cross section differ by less than a factor 1.5 with Refs. [94,95] even at $10^{21}\,\mathrm{eV}$.

However, more recently it has been argued that the neutrino-nucleon cross-section calculated within the Standard Model becomes unreliable for $E \gtrsim 2 \times 10^{17}$ eV: the authors of [99] used the $O(g^2)$ expression for the elastic forward scattering amplitude to derive via the optical theorem the bound $\sigma_{\nu N} \leq 9.3 \times 10^{-33}\,\mathrm{cm}^2$ for the total cross-section. Current parton distribution functions (pdf's) predict a violation of this unitarity bound above $E \gtrsim 2 \times 10^{17}$ eV. The authors of [99] argue that the large $O(g^4)$ corrections to the forward scattering amplitude necessary to restore unitarity signal a breakdown of electroweak perturbation theory. Alternatively, large changes in the evolution of the parton distribution functions have to set in soon after the kinematical range probed by HERA. A large $O(g^4)$ correction to the forward scattering amplitude and cross section is, however, not surprising because the $O(g^2)$ amplitude contains no resonant contribution and is real. In particular, the total cross-section is therefore not only bounded by a constant but zero at $O(g^2)$, and the imaginary part of the box diagram of $O(g^4)$ is the *first* contribution to the total neutrino-nucleon cross-section [100].

Interestingly, it has been shown that the increasing target mass provided by the Earth for increasing zenith angles below the horizontal implies that the rate of up-going air showers in UHECR detectors does not decrease with decreasing neutrino-nucleon cross section but may even increase [101]. Thus, cross sections smaller than Eq. (19) do not lead to reduced event rates in UHECR detectors and can be measured from the angular distribution of events. UHECR and neutrino experiments can thus contribute to measure cross sections at energies inaccessible in accelerator experiments!

Neutral current neutrino-nucleon cross sections are expected to be a factor 2-3 smaller than charged current cross sections at UHE and interactions with electrons only play a role at the Glashow resonance, $\bar{\nu}_e e \to W$, at $E = 6.3 \times 10^{15}$ eV. Furthermore, cross sections of neutrinos and anti-neutrinos are basically identical at UHE. Radiative corrections influence the total cross section negligibly compared to the parton distribution uncertainties, but may lead to an increase of the average inelasticity in the outgoing lepton from $\simeq 0.19$ to $\simeq 0.24$ at $E \sim 10^{20}$ eV [102], although this would probably hardly influence the shower character.

Neutrinos propagating through the Earth start to be attenuated above $\simeq$ 100 TeV due to the increasing Standard Model cross section as indicated by Eq. (19). Detailed integrations of the relevant transport equations for muon neutrinos above a TeV have been presented in Ref. [98], and, for a general cold medium, in Ref. [103]. In contrast, $\tau-$neutrinos with energy up to $\simeq$ 100 PeV can penetrate the Earth due to their regeneration from τ decays [90]. As a result, a primary UHE $\tau-$neutrino beam propagating through the Earth would cascade down below $\simeq$ 100 TeV and in a neutrino telescope could give rise to a higher total rate of upgoing events as compared to downgoing events for the same beam arriving from above the horizon. As mentioned above, a primary $\tau-$neutrino beam could arise even in scenarios based on pion decay, if $\nu_\mu - \nu_\tau$ mixing occurs with the parameters suggested by the Super-Kamiokande results [88]. In the PeV range, $\tau-$neutrinos can produce characteristec "double-bang" events where the first bang would be due to the charged current production by the $\tau-$neutrino of a τ whose decay at a typical distance $\simeq 100$ m would produce the second bang [89]. These effects have also been suggested as an independent astrophysical test of the neutrino oscillation hypothesis. In addition, isotropic neutrino fluxes in the energy range between 10 TeV and 10 PeV have been suggested as probes of the Earth's density profile, whereby neutrino telescopes could be used for neutrino absorption tomography [104].

New Interactions

It has been suggested that the neutrino-nucleon cross section, $\sigma_{\nu N}$, can be enhanced by new physics beyond the electroweak scale in the center of mass (CM) frame, or above about a PeV in the nucleon rest frame [105,106,107]. Neutrino induced air showers may therefore rather directly probe new physics beyond the electroweak scale.

The lowest partial wave contribution to the cross section of a point-like particle is constrained by unitarity to be not much larger than a typical electroweak cross section [108]. However, at least two major possibilities allowing considerably larger cross sections have been discussed in the literature for which unitarity bounds need not be violated. In the first, a broken SU(3) gauge symmetry dual to the unbroken SU(3) color gauge group of strong interaction is introduced as the "generation symmetry" such that the three generations of leptons and quarks represent the quantum numbers of this generation symmetry. In this scheme, neutrinos can have close to strong interaction cross sections with quarks. In addition, neutrinos can interact coherently with all partons in the nucleon, resulting in an effective cross section comparable to the geometrical nucleon cross section. This model lends itself to experimental verification through shower development altitude statistics [105].

The second possibility consists of a large increase in the number of degrees of freedom above the electroweak scale [109]. A specific implementation of this idea is given in theories with n additional large compact dimensions and a quantum gravity scale $M_{4+n} \sim$TeV that has recently received much attention in the literature [110] because it provides an alternative solution (i.e., without supersymmetry) to the hierarchy problem in grand unifications of gauge interactions. The cross sections within such scenarios have not been calculated from first principles yet. Within the field theory approximation which should hold for squared CM energies $s \lesssim M_{4+n}^2$, the spin 2 character of the graviton predicts $\sigma_g \sim s^2/M_{4+n}^6$ [111]. For $s \gg M_{4+n}^2$, several arguments based on unitarity within field theory have been put forward. The emission of massive Kaluza-Klein (KK) graviton modes associated with the increased phase space due to the extra dimensions leads to the rough estimate (derived for $n = 2$) [111]

$$\sigma_g \simeq \frac{4\pi s}{M_{4+n}^4} \simeq 10^{-27} \left(\frac{M_{4+n}}{\text{TeV}}\right)^{-4} \left(\frac{E}{10^{20}\,\text{eV}}\right) \text{cm}^2 \,, \tag{20}$$

where in the last expression we specified to a neutrino of energy E hitting a nucleon at rest. A more detailed calculation taking into account scattering on individual partons leads to similar orders of magnitude [107]. Note that a neutrino would typically start to interact in the atmosphere for $\sigma_{\nu N} \gtrsim 10^{-27}\,\mathrm{cm}^2$, i.e. in the case of Eq. (20) for $E \gtrsim 10^{20}\,\mathrm{eV}$, assuming $M_{4+n} \simeq 1\,\mathrm{TeV}$. For cross sections such large the neutrino therefore becomes a primary candidate for the observed EHECR events. However, since in a neutral current interaction the neutrino transfers only about 10% of its energy to the shower, the cross section probably has to be at least a few $10^{-26}\,\mathrm{cm}^2$ to be consistent with observed showers which start within the first $50\,\mathrm{g\,cm}^{-2}$ of the atmosphere [112,113]. A specific signature of this scenario would be the absence of any events above the energy where σ_g grows beyond $\simeq 10^{-27}\,\mathrm{cm}^2$ in neutrino telescopes based on ice or water as detector medium [114], and a hardening of the spectrum above this energy in atmospheric detectors such as the Pierre Auger Project [115] and the proposed space based AirWatch type detectors [118,119,120]. Furthermore, according to Eq. (20), the average atmospheric column depth of the first interaction point of neutrino induced air showers in this scenario is predicted to depend linearly on energy. This should be easy to distinguish from the logarithmic scaling of the elongation rate expected for nucleons, nuclei, and $\gamma-$rays. To test such scalings one can, for example, take advantage of the fact that the atmosphere provides a detector medium whose column depth increases from $\sim 1000\,\mathrm{g/cm}^2$ towards the zenith to $\sim 36000\,\mathrm{g/cm}^2$ towards horizontal arrival directions. This probes cross sections in the range $\sim 10^{-29} - 10^{-27}\,\mathrm{cm}^2$. Due to the increased column depth, water/ice detectors would probe cross sections in the range $\sim 10^{-31} - 10^{-29}\,\mathrm{cm}^2$ [121] which could be relevant for TeV scale gravity models [122].

From a perturbative point of view within string theory, $\sigma_{\nu N}$ can be estimated as follows: Individual amplitudes are expected to be suppressed exponentially above the string scale M_s which again for simplicity we assume here to be comparable to M_{4+n}. This can be interpreted as a result of the finite spatial extension of the string states. In this case, the neutrino nucleon cross section would be dominated by interactions with the partons carrying a momentum fraction $x \sim M_s^2/s$, leading to [112]

$$\sigma_{\nu N} \simeq \frac{4\pi}{M_s^2}\ln\left(\frac{s}{M_s^2}\right)\left(\frac{s}{M_s^2}\right)^{0.363} \simeq 6\times 10^{-29}\left(\frac{M_s}{\mathrm{TeV}}\right)^{-4.726}\left(\frac{E}{10^{20}\,\mathrm{eV}}\right)^{0.363}$$
$$\times\left[1+0.08\ln\left(\frac{E}{10^{20}\,\mathrm{eV}}\right)-0.16\ln\left(\frac{M_s}{\mathrm{TeV}}\right)\right]^2\,\mathrm{cm}^2 \qquad (21)$$

This is probably too small to make neutrinos primary candidates for the highest energy showers observed, given the fact that complementary constraints from accelerator experiments result in $M_s \gtrsim 1\,\mathrm{TeV}$ [133]. On the other hand, general arguments on the production of "string balls" or small black holes from two point particles represented by light strings [134] leads to the asymptotic scaling $\sigma \simeq (s/M_s^2)^{1/(n+1)}$ for $s \gtrsim M_s^2$ for the fundamental neutrino-parton cross section. This could lead to values for $\sigma_{\nu N}$ larger than Eq. (21) [135]. Some other recent

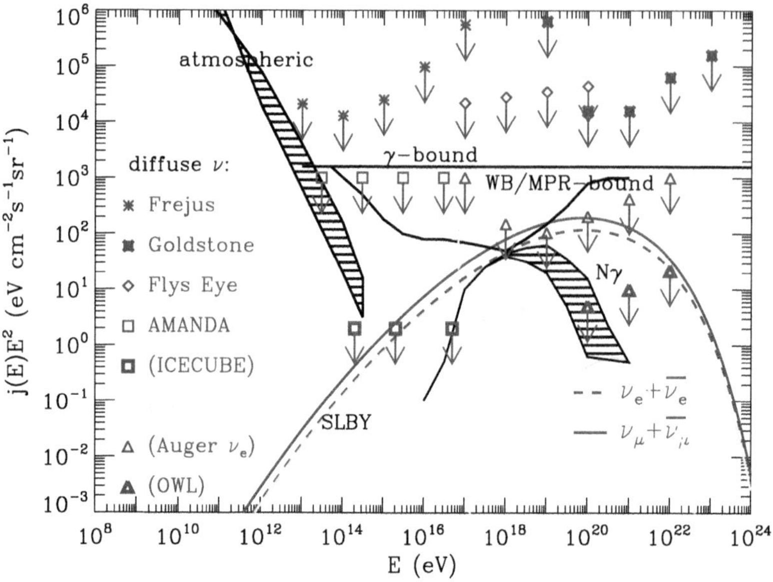

Fig. 8. Various neutrino flux predictions and experimental upper limits or projected sensitivities. Shown are upper limits from the Frejus underground detector [123], the Fly's Eye experiment [124], the Goldstone radio telescope [127], and the Antarctic Muon and Neutrino Detector Array (AMANDA) neutrino telescope [128], as well as projected neutrino flux sensitivities of ICECUBE, the planned kilometer scale extension of AMANDA [129], the Pierre Auger Project [130] (for electron and tau neutrinos separately) and the proposed space based OWL [118] concept. Neutrino fluxes are shown for the atmospheric neutrino background [131] (hatched region marked "atmospheric"), for EHECR interactions with the CMB [132] ("$N\gamma$", dashed range indicating typical uncertainties for moderate source evolution), and for the "top-down" model (marked "SLBY"), where EHECR and neutrinos are produced by decay of superheavy relics (see Ref. [58] and contribution by P. Bhattacharjee and G. Sigl in this volume for more details). The top-down fluxes are shown for electron-, muon, and tau-neutrinos separately, assuming no (lower ν_τ-curve) and maximal $\nu_\mu - \nu_\tau$ mixing (upper ν_τ-curve, which would then equal the ν_μ-flux), respectively. The Waxman-Bahcall bound in the version of Mannheim, Protheroe, and Rachen [72] ("WB/MPR-bound") for sources optically thin for the proton primaries, and the γ−ray bound ("γ−bound") are also shown

work seems to imply that in TeV string models cross sections, if not sufficient to make neutrinos UHECR primary candidates, could at least be significantly larger than Standard Model cross sections [136,137]. Thus, an experimental detection of the signatures discussed in this section [138] could lead to constraints on some string-inspired models of extra dimensions.

There are, however, severe astrophysical and cosmological constraints on M_{4+n} which result from limiting the emission of bulk gravitons into the extra dimensions. The strongest constraints in this regard come from the production due to nucleon-nucleon bremsstrahlung in type II supernovae [139] and their subsequent decay into a diffuse background of $\gamma-$rays in the MeV range [140]. The latter read $M_6 \gtrsim 84\,\mathrm{TeV}$, $M_7 \gtrsim 7\,\mathrm{TeV}$, for $n = 2, 3$, respectively, and, therefore, $n \geq 5$ is required if neutrino primaries are to serve as a primary candidate for the EHECR events observed above $10^{20}\,\mathrm{eV}$. This assumes a toroidal geometry of the extra dimensions with equal radii given by

$$r_n \simeq M_{4+n}^{-1} \left(\frac{M_{\mathrm{Pl}}}{M_{4+n}}\right)^{2/n} \simeq 2 \times 10^{-17} \left(\frac{\mathrm{TeV}}{M_{4+n}}\right) \left(\frac{M_{\mathrm{Pl}}}{M_{4+n}}\right)^{2/n} \mathrm{cm}\,, \qquad (22)$$

where M_{Pl} denotes the Planck mass. The above lower bounds on M_{4+n} thus translate into the corresponding upper bounds $r_2 \lesssim 0.9 \times 10^{-4}\,\mathrm{mm}$, $r_3 \lesssim 0.19 \times 10^{-6}\,\mathrm{mm}$, respectively. Still stronger but somewhat more model dependent bounds result from the production of KK modes during the reheating phase after inflation. For example, a bound $M_6 \gtrsim 500\,\mathrm{TeV}$ has been reported [141] based on the contribution to the diffuse $\gamma-$ray background in the 100 MeV region. We note, however, that all astrophysical and cosmological bounds are changed in more complicated geometries of extra dimensions [142].

The neutrino primary hypothesis of EHECR together with other astrophysical and cosmological constraints thus provides an interesting testing ground for theories involving large compact extra dimensions representing one possible kind of physics beyond the Standard Model. In this context, we mention that in theories with large compact extra dimensions mentioned above, Newton's law of gravity is expected to be modified at distances smaller than the length scale given by Eq. (22). Indeed, there are laboratory experiments measuring gravitational interaction at small distances (for a recent review of such experiments see Ref. [143]), which also probe these theories. Thus, future EHECR experiments and gravitational experiments in the laboratory together have the potential of providing rather strong tests of these theories. These tests would be complementary to constraints from collider experiments [133].

Independent of theoretical arguments, the EHECR data can be used to put constraints on cross sections satisfying $\sigma_{\nu N}(E \gtrsim 10^{19}\,\mathrm{eV}) \lesssim 10^{-27}\,\mathrm{cm}^2$. Particles with such cross sections would give rise to horizontal air showers. The Fly's Eye experiment established an upper limit on horizontal air showers [124]. The non-observation of the neutrino flux expected from pions produced by EHECRs interacting with the CMB the results in the limit [144,121]

$$\begin{aligned}
&\sigma_{\nu N}(10^{17}\,\mathrm{eV}) \lesssim 1 \times 10^{-29}/\bar{y}^{1/2}\,\mathrm{cm}^2\\
&\sigma_{\nu N}(10^{18}\,\mathrm{eV}) \lesssim 8 \times 10^{-30}/\bar{y}^{1/2}\,\mathrm{cm}^2\\
&\sigma_{\nu N}(10^{19}\,\mathrm{eV}) \lesssim 5 \times 10^{-29}/\bar{y}^{1/2}\,\mathrm{cm}^2\,, \qquad (23)
\end{aligned}$$

where $\bar{y}$ is the average energy fraction of the neutrino deposited into the shower ($\bar{y} = 1$ for charged current reactions and $\bar{y} \simeq 0.1$ for neutral current reactions). Neutrino fluxes predicted in various scenarios are shown in Fig. 8. The projected sensitivity of future experiments such as the Pierre Auger Observatories and the AirWatch type satellite projects indicate that the cross section limits Eq. (23)

could be improved by up to four orders of magnitude, corresponding to one order of magnitude in M_s or M_{4+n}. This would close the window between cross sections allowing horizontal air showers, $\sigma_{\nu N}(E \gtrsim 10^{19}\,\mathrm{eV}) \lesssim 10^{-27}\,\mathrm{cm}^2$, and the Standard Model value Eq. (19).

We note in this context that, only assuming 3 + 1 dimensional field theory, consistency of the UHE νN cross section with data at electroweak energies does not lead to very stringent constraints: Relating the cross section to the νN elastic amplitude in a model independent way yields [145]

$$\sigma(E) \lesssim 3 \times 10^{-24} \left(\frac{E}{10^{19}\,\mathrm{eV}}\right) \mathrm{cm}^2 . \tag{24}$$

However, it has been argued [99] (see discussion above), that cross sections $\sigma_{\nu N} \gtrsim 9.3 \times 10^{-33}\,\mathrm{cm}^2$ could signal a breakdown of perturbation theory.

In the context of conventional astrophysical sources, the relevant UHE neutrino primaries could, of course, only be produced as secondaries in interactions with matter or with low energy photons of protons or nuclei accelerated to energies of at least 10^{21} eV. This implies strong requirements on the possible sources. In addition, neutrino primaries with new interactions would predict a significant correlation of UHECR arrival directions with high redshift objects. Indeed, possible correlations of that type have recently been discussed (see Sect. 7).

4.2 Supersymmetric Particles

Certain supersymmetric particles have been suggested as candidates for the EHECR events. For example, if the gluino is light and has a lifetime long compared to the strong interaction time scale, because it carries color charge, it will bind with quarks, anti-quarks and/or gluons to form color-singlet hadrons, so-called R-hadrons. This can occur in supersymmetric theories involving gauge-mediated supersymmetry (SUSY) breaking [146] where the resulting gluino mass arises dominantly from radiative corrections and can vary between ~ 1 GeV and ~ 100 GeV. In these scenarios, the gluino can be the lightest supersymmetric particle (LSP). There are also arguments against a light quasi-stable gluino [147], mainly based on constraints on the abundance of anomalous heavy isotopes of hydrogen and oxygen which could be formed as bound states of these nuclei and the gluino. However, the case of a light quasi-stable gluino does not seem to be settled.

In the context of such scenarios a specific case has been suggested in which the gluino mass lies between 0.1 and 1 GeV [148]. The lightest gluino-containing baryon, $uds\tilde{g}$, denoted S^0, could then be long-lived or stable, and the kinematical threshold for $\gamma_b - S^0$ "GZK" interaction would be higher than for nucleons, at an energy given by substituting the S^0 mass M_{S^0} for the nucleon mass in Eq. (7) [150]. Furthermore, the cross section for $\gamma_b - S^0$ interaction peaks at an energy higher by a facor $(m_{S^0}/m_N)(m_* - m_{S^0})/(m_\Delta - m_N)$ where the ratio of the mass splittings between the primary and the lowest lying resonance of the S^0 (of mass m_*) and the nucleon satisfies $(m_* - m_{S^0})/(m_\Delta - m_N) \gtrsim 2$. As a result of this and a somewhat smaller interaction cross section of S^0 with photons, the effective GZK threshold is higher by factors of a few and sources of events above $10^{19.5}$ eV could be 15-30 times further away than for the case of nucleons. The existence of such events, whose arrival directions should be correlated with their sources, as for the case of neutrino primaries discussed in the previous section, was therefore proposed as a signal of supersymmetry [150] (see Sect. 7 for a discussion of angular correlation studies).

Meanwhile, however, accelerator constraints have become more stringent (see Refs. [151,152]) and seem to be inconsistent with the scenario from Ref. [148]. However, the scenario with a "tunable" gluino mass [146] still seems possible and suggests either the gluino–gluon bound state $g\tilde{g}$, called glueballino R_0, or the isotriplet $\tilde{g} - (u\bar{u} - d\bar{d})_8$, called $\tilde{\rho}$, as the lightest quasi-stable R-hadron. For a summary of scenarios with light gluinos consistent with accelerator constraints see Ref. [153].

Similar to the neutrino primary hypothesis in the context of acceleration sources (see Sect. 4.1), a specific difficulty of this scenario is the fact that, of course, the neutral R-hadron can not be accelerated, but rather has to be produced as a secondary of an accelerated proton interacting with the ambient matter. As a consequence, protons must be accelerated to at least 10^{21} eV at the source in order for the secondary S^0 particles to explain the EHECR events. Furthermore, secondary production would also include neutrinos and especially $\gamma-$rays, leading to fluxes from powerful discrete acceleration sources that may be detectable in the GeV range by space-borne $\gamma-$ray instruments such as EGRET and GLAST, and in the TeV range by ground based $\gamma-$ray detectors such as HEGRA and WHIPPLE and the planned VERITAS, HESS, and MAGIC projects. At least the latter three ground based instruments should have energy thresholds low enough to detect $\gamma-$rays from the postulated sources at redshift $z \sim 1$. Such observations in turn imply constraints on the required branching ratio of proton interactions into the R-hadron which, very roughly, should be larger than ~ 0.01. These constraints, however, will have to be investigated in more detail for specific sources. It was also suggested to search for heavy neutral baryons in the data from Čerenkov instruments in the TeV range in this context [154].

A further constraint on new, massive strongly interacting particles in general comes from the character of the air showers created by them: The observed EHECR air showers are consistent with nucleon primaries and limits the possible primary rest mass to less than $\simeq 50$ GeV [155]. With the statistics expected from upcoming experiments such as the Pierre Auger Project, this upper limit is likely to be lowered down to $\simeq 10$ GeV.

It is interesting to note in this context that in case of a confirmation of the existence of new neutral particles in UHECRs, a combination of accelerator, air shower, and astrophysics data would be highly restrictive in terms of the underlying physics: In the above scenario, for example, the gluino would have to be in a narrow mass range, 1–10 GeV, and the newest accelerator constraints on the Higgs mass, $m_h \gtrsim 90$ GeV, would require the presence of a D term of an anomalous $U(1)_X$ gauge symmetry, in addition to a gauge-mediated contribution to SUSY breaking at the messenger scale [146].

The possibility of new axion-like and supersymmetric primaries has been recently discussed from a more general point of view, with the sgoldstino, the superpartner of the fermion associated with supersymmetry breaking, as a specific example [156]: The requirement of avoiding strong interactions with the CMB and thus the GZK effect and at the same time assuring sufficiently large, almost hadronic strength interactions with air nuclei to explain observed air showers causes a tension that rules out basically all of the available parameter space.

Finally, SUSY could also play a role in top-down scenarios where it would modify the spectra of particles resulting from the decay of the X particles (see contribution by P. Bhattacharjee and G. Sigl in this volume).

4.3 Other Particles

Recently it was suggested that QCD instanton induced interactions between quarks can lead to a stable, strong bound state of two $\Lambda = uds$ particles, a so called $uuddss$ H-dibaryon state with a mass $M_H \simeq 1700\,\mathrm{MeV}$ [157]. This particle would have properties similar to the sypersymmetric S^0 particle discussed in the previous section, i.e. it is neutral and its spin is zero. Its effective GZK cutoff would, therefore, also be considerably higher than for nucleons, at approximately $7.3 \times 10^{20}\,\mathrm{eV}$, according to Ref. [157]. It would thus also be a primary candidate for the observed EHECR events that could be produced at high redshift sources.

Finally, extended field configurations, called topological defects, that are classical solutions of the equations describing the fundamental forces in Nature, could in some cases also propagate unattenuated and thus be UHECR primary candidates or produce "ordinary" particles by decay. The latter possibility is usually subsummed under "top-down" scenarios and we refer the reader to the contribution by P. Bhattacharjee and G. Sigl in this volume.

5 Signatures of Galactic and Extragalactic Magnetic Fields in UHECR Spectra and Images

Cosmic magnetic fields can have several implications for UHECR propagation that may leave signatures in the observable spectra which could in turn be used to constrain or even measure the magnetic fields in the halo of our Galaxy and/or the EGMF.

5.1 Synchrotron Radiation and Electromagnetic Cascades

As already mentioned in Sect. 3, the development of EM cascades strongly depends on presence and strength of magnetic fields via the synchrotron loss of its electronic component: For a particle of mass m and charge Ze (e is the electron charge) the energy loss rate in a field of squared r.m.s. strength B^2 is

$$\frac{dE}{dt} = -\frac{4}{3}\sigma_T \frac{B^2}{8\pi} \left(\frac{Zm_e}{m}\right)^4 \left(\frac{E}{m_e}\right)^2 . \tag{25}$$

For UHE protons this is negligible, whereas for UHE electrons the synchrotron losses eventually dominate over their attenuation (due to interaction with the background photons) above some critical energy $E_{\rm tr} \sim 10^{20}(B/10^{-10}\,{\rm G})^{-1}\,{\rm eV}$ that depends somewhat on the URB (see Fig. 5). Cascade development above that energy is essentially blocked because the electrons lose their energy through synchrotron radiation almost instantaneously once they are produced. In this energy range, $\gamma-$ray propagation is therefore governed basically by absorption due to PP or DPP, and the observable flux is dominated by the "direct" or "first generation" $\gamma-$rays, and their flux can be calculated by integrating Eq. (5) up to the absorption length (or redshift). Since this length is much smaller than the Hubble radius, for a homogeneous source distribution this reduces to Eq. (6), with $l_E(E)$ replaced by the interaction length $l(E)$.

Thus, for a given injection spectrum of $\gamma-$rays and electrons for a source beyond a few Mpc, the observable cascade spectrum depends on the EGMF. As mentioned in Sect. 3, the hadronic part of UHECRs is a continuous source of secondary photons whose spectrum may therefore contain information on the large scale magnetic fields [158]. This spectrum should be measurable down to $\simeq 10^{19}\,{\rm eV}$ if $\gamma-$rays can be discriminated from nucleons at the $\sim 1\%$ level. In more speculative models of UHECR origin such as the topological defect scenario that predict domination of $\gamma-$rays above $\sim 10^{20}\,{\rm eV}$, EGMFs can have even more direct consequences for UHECR fluxes and constraints on such scenarios (see contribution by P. Bhattacharjee and G. Sigl in this volume).

The photons coming from the synchrotron radiation of electrons of energy E have a typical energy given by

$$E_{\rm syn} \simeq 6.8 \times 10^{13} \left(\frac{E}{10^{21}\,{\rm eV}}\right)^2 \left(\frac{B}{10^{-9}\,{\rm G}}\right) {\rm eV} , \tag{26}$$

which is valid in the classical limit, $E_{\rm syn} \ll E$. Constraints can arise when this energy falls in a range where there exist measurements of the diffuse $\gamma-$ray flux, such as from EGRET around 1 GeV [40], or upper limits on it, such as at $50 - 100\,{\rm TeV}$ from HEGRA [159], and between $\simeq 6 \times 10^{14}\,{\rm eV}$ and $\simeq 6 \times 10^{16}\,{\rm eV}$ from CASA-MIA [160]. For example, certain strong discrete sources of UHE $\gamma-$rays such as massive topological defects with an almost monoenergetic injection spectrum in a $10^{-9}\,{\rm G}$ EGMF would predict $\gamma-$ray fluxes that are larger than the charged cosmic ray flux for some energies above $\simeq 10^{16}\,{\rm eV}$ and can therefore be ruled out [132].

5.2 Deflection and Delay of Charged Hadrons

Whereas for electrons synchrotron loss is more important than deflection in the EGMF, for charged hadrons the opposite is the case. A relativistic particle of charge Ze and energy E has a Larmor radius $r_L \simeq E/(ZeB_\perp)$ where $B_\perp$ is the field component perpendicular to the particle momentum. If this field is constant over a distance d, this leads to a deflection angle

$$\theta(E,d) \simeq \frac{d}{r_L} \simeq 0.52^\circ Z \left(\frac{E}{10^{20}\,\mathrm{eV}}\right)^{-1} \left(\frac{d}{1\,\mathrm{Mpc}}\right) \left(\frac{B_\perp}{10^{-9}\,\mathrm{G}}\right) . \tag{27}$$

Magnetic fields beyond the Galactic disk are poorly known and include a possible extended field in the halo of our Galaxy and a large scale EGMF. In both cases, the magnetic field is often characterized by an r.m.s. strength B and a correlation length λ, i.e. it is assumed that its power spectrum has a cut-off in wavenumber space at $k_c = 2\pi/\lambda$ and in real space it is smooth on scales below λ which is often also called coherence length. If we neglect energy loss processes for the moment, then the r.m.s. deflection angle over a distance d in such a field is $\theta(E,d) \simeq (2d\lambda/9)^{1/2}/r_L$, or

$$\theta(E,d) \simeq 0.8^\circ\, Z \left(\frac{E}{10^{20}\,\mathrm{eV}}\right)^{-1} \left(\frac{d}{10\,\mathrm{Mpc}}\right)^{1/2} \left(\frac{\lambda}{1\,\mathrm{Mpc}}\right)^{1/2} \left(\frac{B}{10^{-9}\,\mathrm{G}}\right) , \tag{28}$$

for $d \gtrsim \lambda$, where the numerical prefactors were calculated from the analytical treatment in Ref. [161]. There it was also pointed out that there are two different limits to distinguish: For $d\theta(E,d) \ll \lambda$, particles of all energies "see" the same magnetic field realization during their propagation from a discrete source to the observer. In this case, Eq. (28) gives the typical coherent deflection from the line-of-sight source direction, and the spread in arrival directions of particles of different energies is much smaller. In contrast, for $d\theta(E,d) \gg \lambda$, the image of the source is washed out over a typical angular extent again given by Eq. (28), but in this case it is centered on the true source direction. If $d\theta(E,d) \simeq \lambda$, the source may even have several images, similar to the case of gravitational lensing. Therefore, observing images of UHECR sources and identifying counterparts in other wavelengths would allow one to distinguish these limits and thus obtain information on cosmic magnetic fields. If d is comparable to or larger than the interaction length for stochastic energy loss due to photo-pion production or photodisintegration, the spread in deflection angles is always comparable to the average deflection angle.

Deflection also implies an average time delay of $\tau(E,d) \simeq d\theta(E,d)^2/4$, or

$$\tau(E,d) \simeq 1.5 \times 10^3\, Z^2 \left(\frac{E}{10^{20}\,\mathrm{eV}}\right)^{-2} \left(\frac{d}{10\,\mathrm{Mpc}}\right)^{2} \left(\frac{\lambda}{1\,\mathrm{Mpc}}\right) \left(\frac{B}{10^{-9}\,\mathrm{G}}\right)^{2} \mathrm{yr} \tag{29}$$

relative to rectilinear propagation with the speed of light. It was pointed out in Ref. [162] that, as a consequence, the observed UHECR spectrum of a bursting source at a given time can be different from its long-time average and would typically peak around an energy E_0, given by equating $\tau(E,d)$ with the time of observation relative to the time of arrival for vanishing time delay. Higher energy particles would have passed the observer already, whereas lower energy particles would not have arrived yet. Similarly to the behavior of deflection angles, the width of the spectrum around E_0 would be much smaller than E_0 if both d is smaller than the interaction length for stochastic energy loss and $d\theta(E,d) \ll \lambda$. In all other cases the width would be comparable to E_0.

Constraints on magnetic fields from deflection and time delay cannot be studied separately from the characteristics of the "probes", namely the UHECR sources, at least as long as their nature is unknown. An approach to the general case is discussed in Sect. 8.2.

6 Constraints on EHECR Source Locations

Nucleons, nuclei, and γ–rays above a few 10^{19} eV cannot have originated much further away than $\simeq 50$ Mpc. For nucleons this follows from the GZK effect (see Fig. 2, the range of nuclei is limited mainly by photodisintegration on the CMB (see Sect. 2), whereas photons are restricted by PP and DPP on the CMB and URB (see Fig. 4). Together with Eq. (28) this implies that above a few 10^{19} eV the arrival direction of such particles should in general point back to their source within a few degrees [163]. This argument is often made in the literature and follows from the Faraday rotation bound on the EGMF and a possible extended field in the halo of our Galaxy, which in its original form reads $B\lambda^{1/2} \lesssim 10^{-9}\,\mathrm{G\,Mpc}^{1/2}$ [164,165], as well as from the known strength and scale height of the field in the disk of our Galaxy, $B_g \simeq 3 \times 10^{-6}$ G, $l_g \lesssim 1$ kpc. Furthermore, the deflection in the disk of our Galaxy can be corrected for in order to reconstruct the extragalactic arrival direction: Maps of such corrections as a function of arrival direction have been calculated in Refs. [166,167] for plausible models of the Galactic magnetic field. The deflection of UHECR trajectories in the Galactic magnetic field may, however, also give rise to several other important effects [168] such as (de)magnification of the UHECR fluxes due to the magnetic lensing effect mentioned in the previous section (which can modify the UHECR spectrum from individual sources), formation of multiple images of a source, and apparent "blindness" of the Earth towards certain regions of the sky with regard to UHECRs. These effects may in turn have important implications for UHECR source locations. In fact, it was recently claimed [169] that, assuming a certain model of the magnetic fields in the Galactic winds, the highest energy cosmic ray events could all have originated in the Virgo cluster or specifically in the radio galaxy M87, However, as was subsequently pointed out in Ref. [170], this Galactic wind model leads to focusing of all positively charged highest energy particles to the North Galactic pole and, consequently, this can not be interpreted as evidence for a point source situated close to the North Galactic pole.

However, important modifications of the Faraday rotation bound on the EGMF have recently been discussed in the literature: The average electron density which enters estimates of the EGMF from rotation measures, can now be more reliably estimated from the baryon density $\Omega_b h^2 \simeq 0.02$, whereas in the original bound the closure density was used. Assuming an unstructured Universe and $\Omega_0 = 1$ results in the much weaker bound [171,172]

$$B \lesssim 7 \times 10^{-7} \left(\frac{\Omega_b h^2}{0.02}\right)^{-1} \left(\frac{h}{0.65}\right) \left(\frac{\lambda}{\text{Mpc}}\right)^{-1/2} \text{G}\,, \tag{30}$$

which suggests much stronger deflection. However, taking into account the large scale structure of the Universe in the form of voids, sheets, filaments etc., and assuming flux freezing of the magnetic fields whose strength then approximately scales with the 2/3 power of the local density, leads to more stringent bounds: Using the Lyman α forest to model the density distribution yields [171]

$$B \lesssim 10^{-9} - 10^{-8}\,\text{G} \tag{31}$$

for the large scale EGMF for coherence scales between the Hubble scale and 1 Mpc. This estimate is closer to the original Faraday rotation limit. However, in this scenario the maximal fields in the sheets and voids can be as high as a μG [173,171,172].

Therefore, according to Eq. (28) and (31), deflection of UHECR nucleons is still expected to be on the degree scale if the local large scale structure around the Earth is not strongly magnetized. However, rather strong deflection can occur if the Supergalactic plane is strongly magnetized, for particles originating in nearby galaxy clusters where magnetic fields can be as high as 10^{-6} G [164,165,174] (see Sect. 8.2) and/or for heavy nuclei such as iron [24]. In this case, magnetic lensing in the EGMF can also play an important role in determining UHECR source locations [175,176].

7 Source Search for EHECR Events

The identification of sources of EHECR has been attempted in it least two different ways: First, it has been tried to associate some of the EHE events with discrete sources. For the 300 EeV Fly's Eye event, potential extragalactic sources have been discussed in Ref. [24]. Prominent objects that are within the range of nuclei and nucleons typically require strong magnetic bending, such as Centaurus A at $\simeq 3$ Mpc and $\simeq 136°$ from the arrival direction (see Fig. 17 below for an explicit simulation), Virgo A (13 – 26 Mpc, $\simeq 87°$), and M82 (3.5 Mpc, $\simeq 37°$). The Seyfert galaxy MCG 8-11-11 at 62 – 124 Mpc and the radio galaxy 3C134 of Fanaroff-Riley (FR) class II are within about 10° of the arrival direction. Due to Galactic obscuration, the redshift (and thus the distance) of the latter is, however, not known with certainty, and estimates range between 30 and 500 Mpc [177]. A powerful quasar, 3C147, within the Fly's Eye event error box at redshift $z \simeq 0.5$ has been suggested as a neutrino source [178]. Recall, however, the problems associated with explaining EHECRs by powerful neutrino sources as discussed in Sect. 4.1. For the highest energy AGASA event [61], a potential source for the neutrino option is the FR-II galaxy 3C33 at $\simeq 300$ Mpc distance, whereas the FR-I galaxy NGC 315 at $\simeq 100$ Mpc is a candidate in case of a nucleon primary. A Galactic origin for both the highest energy Fly's Eye and AGASA event seems only possible in case of iron primaries and an extended Galactic halo magnetic field [179].

Second, identification of UHECR sources with classes of astrophysical objects has been attempted by testing statistical correlations between arrival directions and the locations of such objects. The Haverah park data set and some data from the AGASA, the Volcano Ranch, and the Yakutsk experiments were tested for correlation with the Galactic and Supergalactic plane, and positive result at a level of almost 3σ was found for the latter case for events above 4×10^{19} eV [180]. An analysis of the SUGAR data from the southern hemisphere, however, did not give significant correlations [181]. More recently, a possible correlation of a subset of about 20% of the events above 4×10^{19} eV among each other and with the Supergalactic plane was reported by the AGASA experiment, whereas the rest of the events seemed consistent with an isotropic distribution [182,183]. Results from a similar analysis combining data from the Volcano Ranch, the Haverah Park, the Yakutsk, and the Akeno surface arrays in the northern hemisphere [184], as well as from these and the Fly's Eye experiment [185] were found consistent with that, although no final conclusions can be drawn presently yet. These findings give support to the hypothesis that at least part of the EHECRs are accelerated in objects associated with the Supergalactic plane. However, it was subsequently pointed out [186] that the Supergalactic plane correlation at least of the Haverah Park data seems to be too strong for an origin of these particles in objects associated with the large-scale galaxy structure because, within the range of the corresponding nucleon primaries, galaxies beyond the Local Supercluster become relevant as well. As a possible resolution it was suggested [187,177] that the possible existence of strong magnetic fields with strengths up to μ G and coherence lengths in the Mpc range, aligned along the large-scale structure [173], could produce a focusing effect of UHECRs along the sheets and filaments of galaxies. A recent study claims, however, consistency of the arrival directions of UHECRs with the distribution of galaxies within 50 Mpc from the Cfa Redshift Catalog [188]. The case of UHECR correlations with the large scale structure of galaxies, therefore, does not seem to be settled yet.

Correlations between arrival directions of UHECRs above 4×10^{19} eV and $\gamma-$ray burst (GRB) locations have also been investigated. Although the arrival directions of the two highest energy events are within the error boxes of two strong GRBs detected by BATSE [189], no significant positive result was found for the larger UHECR sample [190]. This may be evidence against an association of UHECRs with GRBs if their distance scale is Galactic, but not if they have an extragalactic origin because of the implied large time delays of UHECRs relative to GRB photons (see contribution by E. Waxman in this volume). Furthermore, whereas no enhancement of the TeV $\gamma-$ray flux has been found in the direction of the Fly's Eye event in Ref. [191], a weak excess was recently reported in Ref. [192].

Recently various claims occured in the literature for significant angular correlations of UHECRs with certain astrophysical objects at distances too large for the primaries to be nucleons, nuclei, or $\gamma-$rays. Farrar and Biermann reported a possible correlation between the arrival direction of the five highest energy CR events and compact radio quasars at redshifts between 0.3 and 2.2 [193]

Undoubtedly, with the present amount of data the interpretation of such evidence for a correlation remains somewhat subjective, as is demonstrated by the criticism of the statistical analysis in Ref. [193] by Hoffman [194] and the reply by Farrar and Biermann [195]). Also, a new analysis with the somewhat larger data set now available did not support such correlations [196]. This is currently disputed since another group claims to have found a correlation on the 99.9% confidence level [197]. Most recently, a correlation between UHECRs of energy $E \gtrsim 4 \times 10^{19}$ eV and BL Lacertae objects at redshifts $z > 0.1$ was claimed [198]. None of these claims are convincing yet but confirmation or refutation should be possible within the next few years by the new experiments. Clearly, a confirmation of one of these correlations would be exciting as it would probably imply new physics such as neutrinos with new interactions or new neutral particles discussed in Sect. 4.

Finally, a statistically significant correlation between the arrival directions of UHECR events in the energy range $(0.8—4)\times10^{19}$ eV and directions of pulsars along the Galactic magnetic field lines has been claimed for the Yakutsk air shower data in Ref. [199]. It would be interesting to look for similar correlations for the data sets from other UHECR experiments.

8 Detailed Calculations of Ultra-High-Energy Cosmic Ray Propagation

In order to obtain accurate predictions of observable CR spectra for given production scenarios, one has to solve the equations of motion for the total and differential cross sections for the loss processes discussed in Sects. 2–5. If deviations from rectilinear propagation are unimportant, for example, if one is only interested in time averaged fluxes, one typically solves the coupled Boltzmann equations for CR transport in one spatial dimension either directly or by Monte Carlo simulation. In contrast, if it is important to follow 3-dimensional trajectories, for example, to compute images of discrete UHECR sources in terms of energy and time and direction of arrival in the presence of magnetic fields, the only feasible approach for most purposes is a Monte Carlo simulation. We describe both cases briefly in the following.

8.1 Average Fluxes and Transport Equations in One Dimension

Computation of time averaged fluxes from transport equations or one-dimensional Monte Carlo simulation is most relevant for diffuse fluxes from many sources and for spectra from discrete sources that emit constantly over long time periods. This is applicable at sufficiently high energies such that deflection angles in potential magnetic fields are much smaller than unity. Formally, the Boltzmann equations for the evolution of a set of species with local densities per energy $n_i(E)$ are given by

$$\frac{\partial n_i(E)}{\partial t} = \Phi_i(E) - n_i(E) \int d\varepsilon n_b(\varepsilon) \int_{-1}^{+1} d\mu \frac{1-\mu\beta_b\beta_i}{2} \sum_j \sigma_{i\to j}|_{s=\varepsilon E(1-\mu\beta_b\beta_i)}$$

$$+\int dE' \int d\varepsilon n_b(\varepsilon) \int_{-1}^{+1} d\mu \\ \times \sum_j \frac{1-\mu\beta_b\beta_j'}{2} n_j(E') \left. \frac{d\sigma_{j\to i}(s,E)}{dE} \right|_{s=\varepsilon E'(1-\mu\beta_b\beta_j)} \quad (32)$$

for an isotropic background distribution (here assumed to be only one species) with our notation, see Eqs. (7 and (8), extended to several species. We briefly summarize work on solving these equations for the propagation of nucleons, nuclei, $\gamma-$rays, electrons, and neutrinos in turn.

Nucleons and Nuclei

Motivated by conventional acceleration models (see contribution by G. Pelletier in this volume), many studies on propagation of nucleons and nuclei have been published in the literature. Approximate analytical solutions of the transport equations can only be found for very specific situations, for example, for the propagation of nucleons near the GZK cutoff (e.g., [200,201,202,203]) and/or under certain simplifying assumptions such as the CEL approximation Eqs. (4)−(6) for nucleons (e.g., [204,205]) and $\gamma-$rays (e.g., [35]). The CEL approximation is excellent for PPP because of its small inelasticity. For pion production, due to its stochastic nature implied by its large inelasticity, the CEL approximation tends to produce a sharper pile-up right below the GZK cutoff compared to exact solutions [206]. It still works reasonably well as long as many pion production events take place on average, i.e. for continuous source distributions and for distant discrete sources. Numerical solutions for nucleons solve the transport equations either directly [207,206,6] or through Monte Carlo simulation [208,24,209,210]. Monte Carlo studies of the photodisintegration histories of nuclei have first been performed in Ref. [14] and subsequently in Refs. [24,21,23].

Electromagnetic Cascades

Numerical calculations of average $\gamma-$ray fluxes from EM cascades beyond the analytical CEL approximation are more demanding due to the exponential growth of the number of electrons and photons and are usually not feasible within a pure Monte Carlo approach. Such simulations have been performed mainly in the context of topological defect models of UHECR origin (see contribution by P. Bhattacharjee and G. Sigl in this volume). Calculations of the photon flux between $\simeq 100\,\text{MeV}$ and $\simeq 10^{16}\,\text{GeV}$ (the Grand Unification Scale) have been presented in Ref. [132,211] where a hybrid Monte Carlo matrix doubling method [212] was used, and in Ref. [6,67,58] where the transport equations are solved by an implicit numerical method. Such calculations play an important role in deriving constraints on top-down models from a comparison of the predicted and observed photon flux down to energies of $\simeq 100\,\text{MeV}$ (see contribution by P. Bhattacharjee and G. Sigl in this volume). EM cascade simulations are also relevant for the secondary $\gamma-$ray flux produced from interactions of primary hadrons [206] and its dependence on cosmic magnetic fields [158]. Under certain circumstances, this secondary flux can become comparable to the primary flux [39].

Analytical calculations have been performed for saturated EM cascades [37]. These calculations show that the cascade spectrum below the pair production threshold has a generic shape. As mentioned above in Sect. 3 this has also been used to derive constraints on energy injection based on direct observation of this cascade flux or on a comparison of its side effects, for example, on light element abundances, with observations.

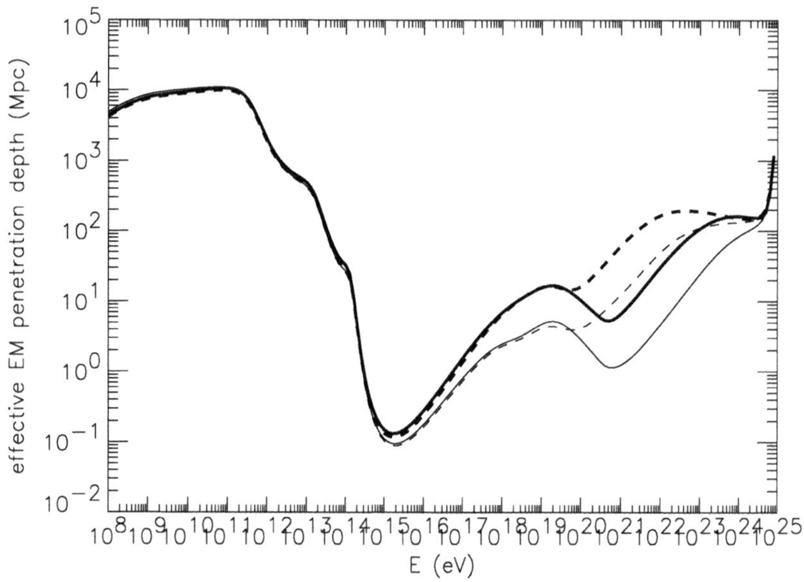

Fig. 9. Effective penetration depth of EM cascades, as defined in the text, for the strongest theoretical URB estimate (solid lines), and the observational URB estimate from Ref. [29] (dashed lines), as shown in Fig. 3, and for an EGMF $\ll 10^{-11}$ G (thick lines), and 10^{-9} G (thin lines), respectively

As a first application of numerical transport calculations we present the effective penetration depth of EM cascades, which we define as the coefficient $l_E(E)$ in Eq. (6), where $j(E)$ is the $\gamma-$ray flux resulting after propagating a homogeneous injection flux $\Phi(E)$. Figure 9 shows results computed for the new estimates of the IR background from Ref. [17], and for some combinations of the URB and the EGMF.

Neutrino Fluxes

Accurate predictions for the UHE neutrino flux have become more relevant recently due to studies and proposals for the detection of UHE neutrinos [213,126].

Fluxes of secondary neutrinos from photo-pion production by UHECRs have been calculated numerically, e.g., in Refs. [206,53,6], by solving the full transport equations for nucleons. Because of the small redshifts involved, the neutrinos can be treated as interaction-free, and the main uncertainties come from the poorly known injection history of the primary nucleons. In top-down scenarios, neutrinos are continuously produced up to very high redshifts and secondaries produced in neutrino interactions can enhance the UHE neutrino fluxes compared to the simple absorption approximation used in Refs. [57,214]. By solving the full Boltzmann equations for the neutrino cascade, unnormalized spectral shapes of neutrino fluxes from topological defects have been calculated in Ref. [52], and absolute fluxes in Ref. [53]. Semianalytical calculations of $\gamma-$ray, nucleon, and neutrino fluxes for a specific class of cosmic string models predicting an absolute normalization of the UHECR injection rate (see contribution by P. Bhattacharjee and G. Sigl in this volume) have been performed in Ref. [215].

Recently an integrated code has been developed which solves the coupled full transport equations for all species, i.e, nucleons, $\gamma-$rays, electrons, and neutrinos concurrently [67,58]. This allows, for example, to make detailed predictions for the spectra of the nucleons and $\gamma-$rays produced by resonant Z^0 production of UHE neutrinos on a massive RNB which could serve as a signature of hot dark matter [64,65] (see Sect. 4.1).

8.2 Angle-Time-Energy Images of Ultra-High-Energy Cosmic Ray Sources

In Sect. 5 we gave simple analytical estimates for the average deflection and time delay of nucleons propagating in a cosmic magnetic field. Here we review approaches that have been taken in the literature to compute effects of magnetic fields on both spectra and angular images (and their time dependence) of sources of UHE nucleons.

Strong Deflection

An exact analytical expression for the distribution of time delays that applies in the limit $d\theta(E,d) \gg \lambda$ for $E \lesssim 4 \times 10^{19}\,\text{eV}$ where photo-pion production is negligible has been given in Ref. [161]. The consequences for the spectra in this energy range and their temporal behavior, especially for the possibility of bursting sources such as cosmological GRBs (see contribution by E. Waxman in this volume), have been discussed there.

Indications for rather strong magnetic fields in the range between $10^{-8}\,$G up to $10^{-6}\,$G have been observed near large mass agglomerations such as clusters of galaxies or even the filaments and sheets connecting them [164,165]. UHECR deflection in such regions could be strong enough for the diffusion approximation to become applicable. The uncertainties in strength and spectrum of the magnetic fields translate directly into a corresponding uncertainty in the energy dependent diffusion coefficient $D(E)$ which is often obtained by simply fitting calculated fluxes to the data. At $\sim 10^{19}\,$eV estimated values of $D(E)$ range between $\simeq 5 \times 10^{33}\,\mathrm{cm}^2\,\mathrm{sec}^{-1}$ and $\simeq 3 \times 10^{35}\,\mathrm{cm}^2\,\mathrm{sec}^{-1}$, and energy dependence like $D(E) \propto E^\alpha$ with α in the range 1/3 and 2 have been suggested [36,216]. This is confirmed by numerical simulations (see below).

Several approximate treatments for calculations of fluxes in the diffusion approximation have been pursued in the literature: If pion production is treated in the CEL approximation, the problem reduces to solving Eq. (4) with an additional, in general location and energy dependent diffusion coefficient $D(\mathbf{r}, E)$:

$$\partial_t n(\mathbf{r}, E) = -\partial_E \left[b(E) n(\mathbf{r}, E)\right] + \boldsymbol{\nabla}\left[D(\mathbf{r}, E)\boldsymbol{\nabla} n(\mathbf{r}, E)\right] + \Phi(E)\,. \tag{33}$$

If $D(\mathbf{r}, E)$ is independent of $\mathbf{r}$, an analytical solution of this "energy loss-diffusion equation" given by Syrovatskii [217] can be employed. This solution or approximations to it have been used in Refs. [216,36,218] to compute the expected spectra from discrete sources in an EGMF of a few $10^{-8}\,$G for energies up to $\simeq 10^{20}\,$eV (the typical range of validity of the diffusion approximation). In some sense a complementary approach has been taken in Ref. [208] where the effects of diffusion were taken into account by using an average propagation time roughly given by $\tau(E) \simeq d^2/D(E)$ and treating pion production exactly by Monte Carlo. Ref. [219] improved on that by representing the EGMF (assumed to be homogeneous) by a finite number of modes and following trajectories explicitly. This paper did, however, not present any spectra directly.

Once Eq. (33) has been solved, the anisotropy defined by

$$|\delta| = \frac{I_{\mathrm{max}} - I_{\mathrm{min}}}{I_{\mathrm{max}} + I_{\mathrm{min}}}\,, \tag{34}$$

where I_{min} and I_{max} are the minimum and maximum CR intensity as a function of arrival direction, can be calculated from the relation [36]

$$\delta(E) = 3\,\frac{D(\mathbf{r}, E)}{n(\mathbf{r}, E)}\,\left|\boldsymbol{\nabla} n(\mathbf{r}, E)\right|\,. \tag{35}$$

Equation (33) and its generalization to an anisotropic diffusion tensor plays a prominent role also in models of Galactic CR propagation. We stress here that while this equation provides a good description of the propagation of Galactic CR for energies up to the knee, it has rather limited applicability in studying UHECR propagation which often takes place in the transition regime between diffusion and rectilinear propagation (see below).

Small Deflection

For small deflection angles and if photo-pion production is important, one has to resort to numerical Monte Carlo simulations in 3 dimensions. Such simulations have been performed in Ref. [220] for the case $d\theta(E, d) \gg \lambda$ and in Refs. [221,209,210] for the general case.

In Refs. [221,209,210] the Monte Carlo simulations were performed in the following way: The magnetic field was represented as Gaussian random field

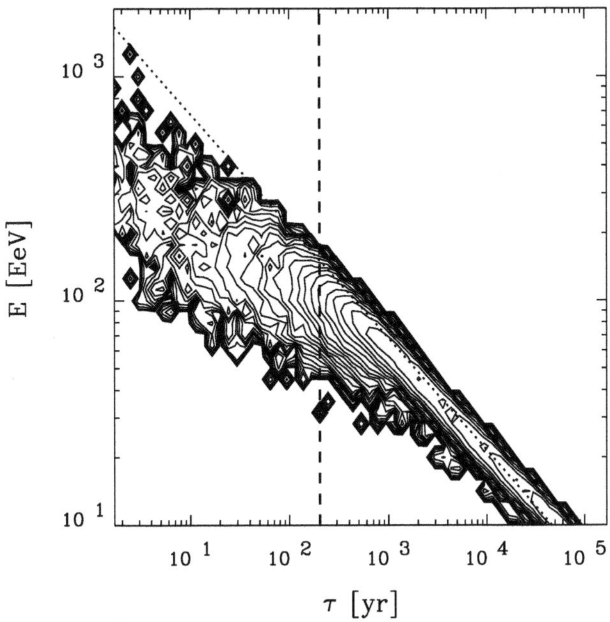

Fig. 10. Contour plot of the UHECR image of a bursting source at $d = 30\,\mathrm{Mpc}$, projected onto the time-energy plane, with $B = 2\times 10^{-10}\,\mathrm{G}$, $\lambda = 1\,\mathrm{Mpc}$, from Ref. [221]. The contours decrease in steps of 0.2 in the logarithm to base 10. The dotted line indicates the energy-time delay correlation $\tau(E,d) \propto E^{-2}$ as would be obtained in the absence of pion production losses. Clearly, $d\theta(E,d) \ll \lambda$ in this example, since for $E < 4\times 10^{19}\,\mathrm{eV}$, the width of the energy distribution at any given time is much smaller than the average (see Sect. 5). The dashed lines, which are not resolved here, indicate the location (arbitrarily chosen) of the observational window, of length $T_{obs} = 5\,\mathrm{yr}$

with zero mean and a power spectrum with $\left\langle B^2(k)\right\rangle \propto k^{n_H}$ for $k < k_c$ and $\left\langle B^2(k)\right\rangle = 0$ otherwise, where $k_c = 2\pi/\lambda$ characterizes the numerical cut-off scale and the r.m.s. strength is $B^2 = \int_0^\infty dk\, k^2 \left\langle B^2(k)\right\rangle$. The field is then calculated on a grid in real space via Fourier transformation. For a given magnetic field realization and source, nucleons with a uniform logarithmic distribution of injection energies are propagated between two given points (source and observer) on the grid. This is done by solving the equations of motion in the magnetic field interpolated between the grid points, and subjecting nucleons to stochastic production of pions and (in case of protons) continuous loss of energy due to PP. Upon arrival, injection and detection energy, and time and direction of arrival are recorded. From many (typically 40000) propagated particles, a histogram of average number of particles detected as a function of time and energy of arrival is constructed for any given injection spectrum by weighting the injection energies correspondingly. This histogram can be scaled to any desired total fluence at the detector and, by convolution in time, can be constructed for arbitrary emission

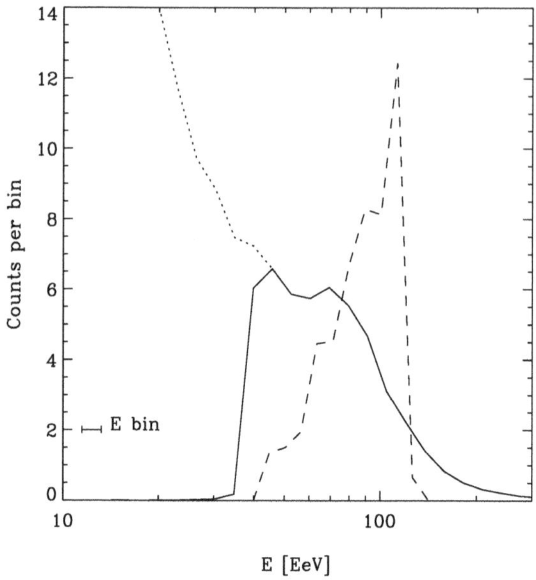

Fig. 11. Energy spectra for a continuous source (solid line), and for a burst (dashed line), from Ref. [221]. Both spectra are normalized to a total of 50 particles detected. The parameters corresponding to the continuous source case are: $T_S = 10^4$ yr, $\tau_{100} = 1.3 \times 10^3$ yr, and the time of observation is $t = 9 \times 10^3$ yr, relative to rectilinear propagation with the speed of light. A low energy cutoff results at the energy $E_S = 4 \times 10^{19}$ eV where $\tau_{E_S} = t$. The dotted line shows how the spectrum would continue if $T_S \ll 10^4$ yr. The case of a bursting source corresponds to a slice of the image in the $\tau(E) - E$ plane, as indicated in Fig. 10 by dashed lines. For both spectra, $d = 30$ Mpc, and $\gamma = 2$

time scales of the source. An example for the distribution of arrival times and energies of UHECRs from a bursting source is given in Fig. 10.

We adopt the following notation for the parameters: τ_{100} denotes the time delay due to magnetic deflection at $E = 100$ EeV and is given by Eq. (29) in terms of the magnetic field parameters; T_S denotes the emission time scale of the source; $T_S \ll 1$yr correspond to a burst, and $T_S \gg 1$yr (roughly speaking) to a continuous source; γ is the differential index of the injection energy spectrum; N_0 denotes the fluence of the source with respect to the detector, *i.e.*, the total number of particles that the detector would detect from the source on an infinite time scale; finally, $\mathcal{L}$ is the likelihood, function of the above parameters.

By putting windows of width equal to the time scale of observation over these histograms one obtains expected distributions of events in energy and time and direction of arrival for a given magnetic field realization, source distance and position, emission time scale, total fluence, and injection spectrum. Examples of the resulting energy spectrum are shown in Fig. 11. By dialing Poisson statistics on such distributions, one can simulate corresponding observable event clusters.

Conversely, for any given real or simulated event cluster, one can construct a likelihood of the observation as a function of the time delay, the emission time scale, the differential injection spectrum index, the fluence, and the distance. In order to do so, and to obtain the maximum of the likelihood, one constructs histograms for many different parameter combinations as described above, randomly puts observing time windows over the histograms, calculates the likelihood function from the part of the histogram within the window and the cluster events, and averages over different window locations and magnetic field realizations.

In Ref. [209] this approach has been applied to and discussed in detail for the three pairs observed by the AGASA experiment [182], under the assumption that all events within a pair were produced by the same discrete source. Although the inferred angle between the momenta of the paired events acquired in the EGMF is several degrees [222], this is not necessarily evidence against a common source, given the uncertainties in the Galactic field and the angular resolution of AGASA which is $\simeq 2.5°$. As a result of the likelihood analysis, these pairs do not seem to follow a common characteristic; one of them seems to favor a burst, another one seems to be more consistent with a continuously emitting source. The current data, therefore, does not allow one to rule out any of the models of UHECR sources. Furthermore, two of the three pairs are insensitive to the time delay. However, the pair which contains the 200 EeV event seems to significantly favor a comparatively small average time delay, $\tau_{100} \lesssim 10\,\mathrm{yr}$, as can be seen from the likelihood function marginalized over T_S and N_0 (see Fig. 12). According to Eq. (29) this translates into a tentative bound for the r.m.s. magnetic field, namely,

$$B \lesssim 2 \times 10^{-11} \left(\frac{\lambda}{1\,\mathrm{Mpc}}\right)^{-1/2} \left(\frac{d}{30\,\mathrm{Mpc}}\right)^{-1} \mathrm{G}\,, \tag{36}$$

which also applies to magnetic fields in the halo of our Galaxy if d is replaced by the lesser of the source distance and the linear halo extent. If confirmed by future data, this bound would be at least two orders of magnitude more restrictive than the best existing bounds which come from Faraday rotation measurements, see Eq. (31), and, for a homogeneous EGMF, from CMB anisotropies [223]. UHECRs are therefore at least as sensitive a probe of cosmic magnetic fields as other measures in the range near existing limits such as the polarization [224] and the small scale anisotropy [225] of the CMB.

More generally, confirmation of a clustering of EHECR would provide significant information on both the nature of the sources and on large-scale magnetic fields [226]. This has been shown quantitatively [210] by applying the hybrid Monte Carlo likelihood analysis discussed above to simulated clusters of a few tens of events as they would be expected from next generation experiments [117] such as the High Resolution Fly's Eye [227], the Telescope Array [228], and most notably, the Pierre Auger Project [115], provided the clustering recently suggested by the AGASA experiment [182,183] is real. The proposed AirWatch type satellite observatory concepts [118,119,120] might even allow one to detect clusters of hundreds of such events.

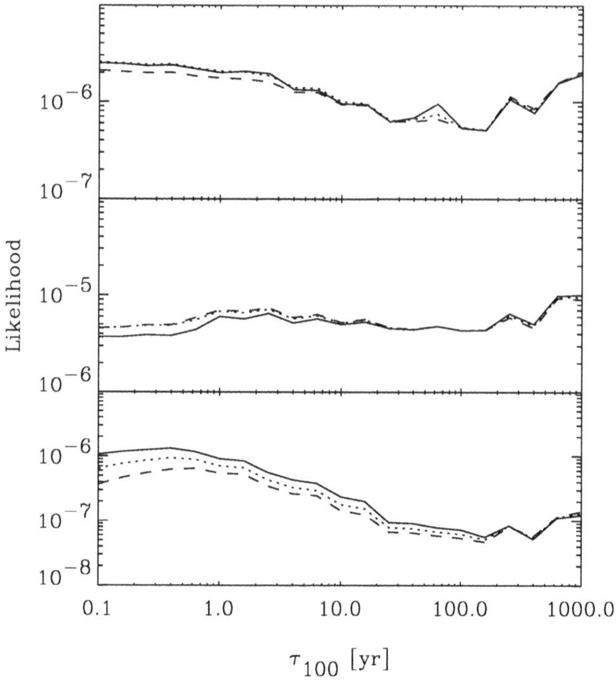

Fig. 12. The logarithm of the likelihood, $\log_{10}\mathcal{L}$, marginalized over T_S and N_0 as a function of the average time delay at 10^{20} eV, τ_{100}, assuming a source distance $d = 30\,\mathrm{Mpc}$. The panels are for pair # 3 through # 1, from top to bottom, of the AGASA pairs [182] (see Sect. 7). Solid lines are for $\gamma = 1.5$, dotted lines for $\gamma = 2.0$, and dashed lines for $\gamma = 2.5$

Five generic situations of the time-energy images of UHECRs were discussed in Ref. [210], classified according to the values of the time delay $\tau(E)$ induced by the magnetic field, the emission timescale of the source T_{S}, as compared to the lifetime of the experiment. The likelihood calculated for the simulated clusters in these cases presents different degeneracies between different parameters, which complicates the analysis. As an example, the likelihood is degenerate in the ratios N_0/T_{S}, or $N_0/\Delta\tau_{100}$, where N_0 is the total fluence, and $\Delta\tau_{100}$ is the spread in arrival time; these ratios represent rates of detection. Another example is given by the degeneracy between the distance d and the injection energy spectrum index γ. Yet another is the ratio $[d\tau(E)]^{1/2}/\lambda$, that controls the size of the scatter around the mean of the $\tau(E) - E$ correlation. Therefore, in most general cases, values for the different parameters cannot be pinned down, and generally, only domains of validity are found. In the following the reconstruction quality of the main parameters considered is summarized.

The distance to the source can be obtained from the pion production signature, above the GZK cut-off, when the emission time scale of the source dominates over the time delay. Since the time delay decreases with increasing energy, the lower the energy $E_{\rm C}$, defined by $\tau(E_{\rm C}) \simeq T_{\rm S}$, the higher the accuracy on the distance d. The error on d is, in the best case, typically a factor 2, for one cluster of $\simeq 40$ events. In this case, where the emission time scale dominates over the time delay at all observable energies, information on the magnetic field is only contained in the angular image, which was not systematically included in the likelihood analysis of Ref. [210] due to computational limits. Qualitatively, the size of the angular image is proportional to $B(d\lambda)^{1/2}/E$, whereas the structure of the image, *i.e.*, the number of separate images, is controlled by the ratio $d^{3/2}B/\lambda^{1/2}/E$. Finally, in the case when the time delay dominates over the emission time scale, with a time delay shorter than the lifetime of the experiment, one can also estimate the distance with reasonable accuracy.

Some sensitivity to the injection spectrum index γ exists whenever events are recorded over a sufficiently broad energy range. At least if the distance d is known, it is in general comparatively easy to rule out a hard injection spectrum if the actual $\gamma \gtrsim 2.0$, but much harder to distinguish between $\gamma = 2.0$ and 2.5.

If the lifetime of the experiment is the largest time scale involved, the strength of the magnetic field can only be obtained from the time-energy image because the angular image will not be resolvable. When the time delay dominates over the emission time scale, and is, at the same time, larger than the lifetime of the experiment, only a lower limit corresponding to this latter time scale, can be placed on the time delay and hence on the strength of the magnetic field. When combined with the Faraday rotation upper limit Eq. (31), this would nonetheless allow one to bracket the r.m.s. magnetic field strength within a few orders of magnitude. In this case also, significant information is contained in the angular image. If the emission time scale is larger then the delay time, the angular image is obviously the only source of information on the magnetic field strength.

The coherence length λ enters in the ratio $[d\tau(E)]^{1/2}/\lambda$ that controls the scatter around the mean of the $\tau(E) - E$ correlation in the time-energy image. It can therefore be estimated from the width of this image, provided the emission time scale is much less than $\tau(E)$ (otherwise the correlation would not be seen), and some prior information on d and $\tau(E)$ is available.

An emission time scale much larger than the experimental lifetime may be estimated if a lower cut-off in the spectrum is observable at an energy $E_{\rm C}$, indicating that $T_{\rm S} \simeq \tau_{E_{\rm C}}$. The latter may, in turn, be estimated from the angular image size via Eq. (29), where the distance can be estimated from the spectrum visible above the GZK cut-off, as discussed above. An example of this scenario is shown in Fig. 13. For angular resolutions $\Delta\theta$, time scales in the range

$$3 \times 10^3 \left(\frac{\Delta\theta}{1^\circ}\right)^2 \left(\frac{d}{10\,{\rm Mpc}}\right) {\rm yr} \lesssim T_{\rm S} \simeq \tau(E) \lesssim 10^4 \cdots 10^7 \left(\frac{E}{100\,{\rm EeV}}\right)^{-2} {\rm yr} \qquad (37)$$

could be probed. The lower limit follows from the requirement that it should be possible to estimate $\tau(E)$ from θ_E, using Eq. (29), otherwise only an upper limit on $T_{\rm S}$, corresponding to this same number, would apply. The upper bound in Eq. (37) comes from constraints on maximal time delays in cosmic

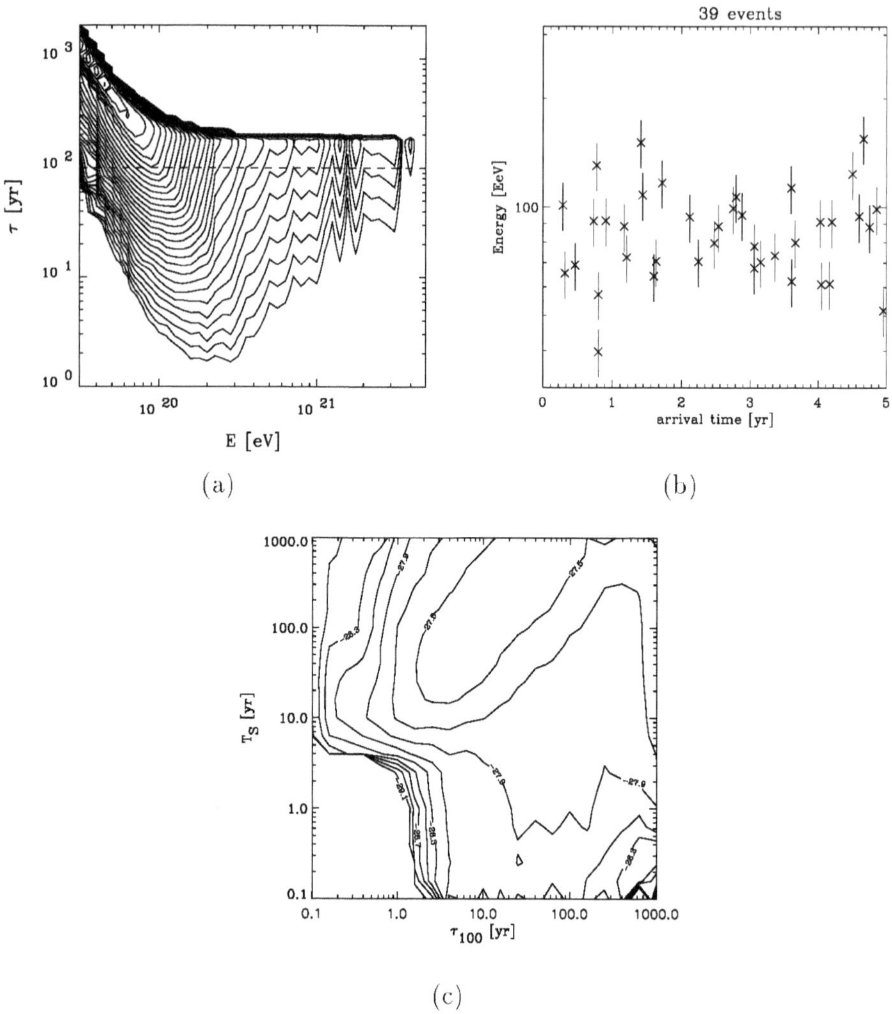

Fig. 13. (a) Arrival time-energy histogram for $\gamma = 2.0$, $\tau_{100} = 50\,\mathrm{yr}$, $T_S = 200\,\mathrm{yr}$, $\lambda \simeq 1\,\mathrm{Mpc}$, $d = 50\,\mathrm{Mpc}$, corresponding to $B \simeq 3 \times 10^{-11}\,\mathrm{G}$. Contours are in steps of a factor $10^{0.4} = 2.51$; (b) Example of a cluster in the arrival time-energy plane resulting from the cut indicated in (a) by the dashed line at $\tau \simeq 100\,\mathrm{yr}$; (c) The likelihood function, marginalized over N_0 and γ, for $d = 50\,\mathrm{Mpc}$, $\lambda \simeq \mathrm{Mpc}$, for the cluster shown in (b), in the $T_S - \tau_{100}$ plane. The contours shown go from the maximum down to about 0.01 of the maximum in steps of a factor $10^{0.2} = 1.58$. Note that the likelihood clearly favors $T_S \simeq 4\tau_{100}$. For τ_{100} large enough to be estimated from the angular image size, $T_S \gg T_{\mathrm{obs}}$ can, therefore, be estimated as well

magnetic fields, such as the Faraday rotation limit in the case of cosmological large-scale field (smaller number) and knowledge on stronger fields associated with the large-scale galaxy structure (larger number). Equation (37) constitutes an interesting range of emission time scales for many conceivable scenarios of UHECRs. For example, the hot spots in certain powerful radio galaxies that have been suggested as UHECR sources [202], have a size of only several kpc and could have an episodic activity on time scales of $\sim 10^6$ yr.

A detailed comparison of analytical estimates for the distributions of time delays, energies, and deflection angles of nucleons in weak random magnetic fields with the results of Monte Carlo simulations has been presented in Ref. [229]. In this work, deflection was simulated by solving a stochastic differential equation and observational consequences for the two major classes of source scenarios, namely continuous and impulsive UHECR production, were discussed. In agreement with earlier work [162] it was pointed out that at least in the impulsive production scenario and for an EGMF in the range $0.1 - 1 \times 10^{-9}$ G, as required for cosmological GRB sources, there is a typical energy scale $E_b \sim 10^{20.5} - 10^{21.5}$ eV below which the flux is quasi-steady due to the spread in arrival times, whereas above which the flux is intermittent with only a few sources contributing. A similar code including secondary production has been developed in Ref. [230] and has subsequently been applied to propagation of UHE protons in regular EGMFs associated with the Supergalactic plane [231].

General Case

Unfortunately, neither the diffusive limit nor the limit of nearly rectilinear propagation is likely to be applicable to the propagation of UHECRs around 10^{20} eV in general. This is because in magnetic fields in the range of a few 10^{-8} G, values that are realistic for the Supergalactic plane [187,177], the Larmor radii of charged particles is of the order of a few Mpc which is comparable to the distance to the sources. An accurate, reliable treatment in this regime can only be achieved by numerical simulation.

To this end, the Monte Carlo simulation approach of individual trajectories developed in Refs. [209,210] has recently been generalized to arbitrary deflections [175]. The Supergalactic plane was modeled as a sheet with a thickness of a few Mpc and a Gaussian density profile. The same statistical description for the magnetic field was adopted as in Refs. [209,210], but with a field power law index $n_H = -11/3$, representing a turbulent Kolmogorov type spectrum, and weighted with the sheet density profile. It should be mentioned, however, that other spectra, such as the Kraichnan spectrum [232], corresponding to $n_H = -7/2$, are also possible. The largest mode with non-zero power was taken to be the largest turbulent eddy whose size L is roughly the sheet thickness. For a Kolmogorov spectrum, it also roughly corresponds to the coherence length. In addition, a coherent field component B_c is allowed that is parallel to the sheet and varies proportional to the density profile.

When CR backreaction on the magnetic field is neglected, the diffusion coefficient of CR of energy E is dominated by the magnetic field power on wavelenghts

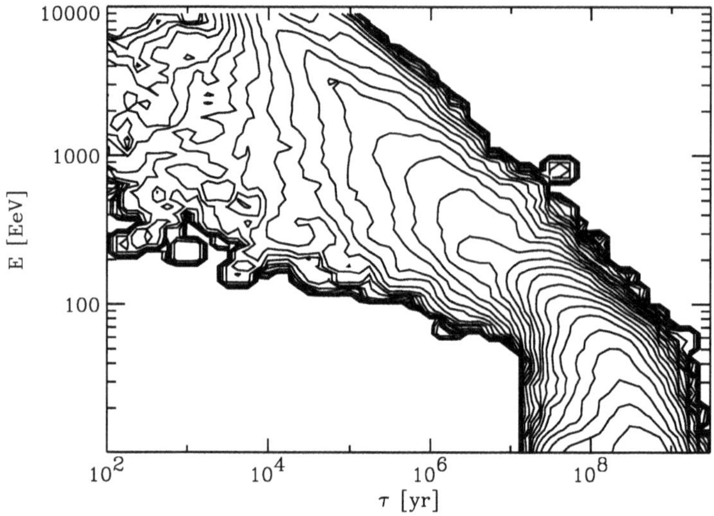

Fig. 14. The distribution of time delays $\tau(E)$ and energies E for a burst with spectral index $\gamma = 2.4$ at a distance $d = 10\,\text{Mpc}$, similar to Fig. 10, but for the Supergalactic plane scenario discussed in the text. The turbulent magnetic field component in the sheet center is $B = 3 \times 10^{-7}\,\text{G}$. Furthermore, a vanishing coherent field component is assumed. The inter-contour interval is 0.25 in the logarithm to base 10 of the distribution per logarithmic energy and time interval. The regimes discussed in the text, $\tau(E) \propto E^{-2}$ in the rectilinear regime $E \gtrsim 200\,\text{EeV}$, $\tau(E) \propto E^{-1}$ in the Bohm diffusion regime $60\,\text{EeV} \lesssim E \lesssim 200\,\text{EeV}$, and $\tau(E) \propto E^{-1/3}$ for $E \lesssim 60\,\text{EeV}$ are clearly visible

comparable to the particle Larmor radius, and in the literature is often approximated by [233,234]

$$D(E) \simeq \frac{1}{3} r_L(E) \frac{B^2}{\int_{1/r_L(E)}^{\infty} dk\, k^2 \langle B^2(k) \rangle} \,. \tag{38}$$

However, more detailed studies combining analytical and numerical techniques [235] (see also contribution by G. Pelletier in this volume) demonstrate that this approximation is in general not correct. For example, in the absence of a coherent field component, $B_c = 0$, the energy dependent diffusion coefficient D can be parametrized for Kolmogorov turbulence by (in units of Mpc^2/Myr)

$$\begin{aligned} D(E) &\simeq 0.02 \left(\frac{E}{10^{20}\,\text{eV}}\right)^{7/3} \left(\frac{B}{\mu\text{G}}\right)^{-7/3} \left(\frac{L}{\text{Mpc}}\right)^{-4/3} , \quad \left[E_{\text{cr}} \lesssim E\right] , \\ &\simeq 0.03 \left(\frac{E}{10^{20}\,\text{eV}}\right) \left(\frac{B}{\mu\text{G}}\right) , \quad \left[0.1 E_{\text{cr}} \lesssim E \lesssim E_{\text{cr}}\right] , \\ &\simeq 0.004 \left(\frac{E}{10^{20}\,\text{eV}}\right)^{1/3} \left(\frac{B}{\mu\text{G}}\right)^{-1/3} \left(\frac{L}{\text{Mpc}}\right)^{-2/3} , \quad \left[E \lesssim 0.1 E_{\text{cr}}\right] . \end{aligned} \tag{39}$$

In this expression, $E_{\rm cr} = 1.45 \times 10^{20}(B/\mu\,{\rm G})(L/{\rm Mpc})\,{\rm eV}$ corresponds to the condition $r_L = L/2\pi$, where the Larmor radius $r_L = 0.11(E/10^{20}\,{\rm eV})(B/\mu{\rm G})^{-1}\,{\rm Mpc}$. Note the difference of the above result with the formula given in Ref. [218], for which $D(E) \propto E^{1/3}$ for $E \lesssim E_{\rm c}$, and $D(E) \propto E$ for $E > E_{\rm cr}$. The dependence of $D(E)$ for $0.1E_{\rm cr} \lesssim E \lesssim E_{\rm cr}$ in Eq. (39) above agrees very well with the phenomenological Bohm diffusion coefficient $D_{\rm B} \simeq r_L$. This is reflected in the dependence of the time delay $\tau(E)$ on energy E: From the rectilinear regime, $\tau(E) \lesssim d$, hence at the largest energies, where $\tau(E) \propto E^{-2}$, Eq. (29), it changes rather smoothly to $\tau(E) \propto E^{-7/3}$ for $E \gtrsim E_{\rm cr}$, then switches to $\tau \propto E^{-1}$ in the regime which is often called Bohm diffusion, and, for $E \lesssim 0.1E_{\rm cr}$, eventually to $\tau(E) \propto E^{-1/3}$ at the smallest energies, or largest time delays. Indeed, all these regimes can be seen in Fig. 14 which shows an example of the distribution of arrival times and energies of UHECRs from a bursting source.

In a steady state situation, diffusion leads to a modification of the injection spectrum by roughly a factor $\tau(E)$, at least in the absence of significant energy loss and for a homogeneous, infinitely extended medium that can be described by a spatially constant diffusion coefficient. Since in the non-diffusive regime the observed spectrum repeats the shape of the injection spectrum, a change to a flatter observed spectrum at high energies is expected in the transition region [216]. Ignoring for the moment the question about the resulting angular distribution, this suggests the possibility of explaining the observed UHECR flux above $\simeq 10\,{\rm EeV}$ including the highest energy events with only one discrete source [218].

The more detailed Monte Carlo simulations reveal the following refinements of this qualitative picture: The presence of a non-trivial geometry where the magnetic field falls off at large distances, such as with a sheet, tends to deplete the flux in the diffusive regime as compared to the case of a homogeneous medium. This is the dominant effect as long as particles above the GZK cutoff do not diffuse, this being the case, for example, for an r.m.s. field strength of $B \lesssim 5 \times 10^{-8}\,{\rm G}$, $d \simeq 10\,{\rm Mpc}$. The simple explanation is that the fixed total amount of particles injected over a certain time scale is distributed over a larger volume in case of a non-trivial geometry due to faster diffusion near the boundary of the strong field region. With increasing field strengths the diffusive regime will extend to energies beyond the GZK cutoff and the increased pion production losses start to compensate for the low energy suppression from the non-trivial geometry. For very strong fields, for example, for $B \gtrsim 10^{-7}\,{\rm G}$, $d \simeq 10\,{\rm Mpc}$, the pion production effect will overcompensate the geometry effect and reverse the situation: In this case, the flux above the GZK cutoff is strongly suppressed due to the diffusively enhanced pion production losses and the flux at lower energies is enhanced. Therefore, there turns out to be an optimal field strength that depends on the source distance and provides an optimal fit to the data above 10 EeV. The optimal case for $d = 10\,{\rm Mpc}$, with a maximal r.m.s. field strength of $B_{\rm max} = 10^{-7}\,{\rm G}$ in the plane center is shown in Fig. 15.

Furthermore, the numerical results indicate an effective Larmor radius that is roughly a factor 10 higher than the analytical estimate, with a correspondingly

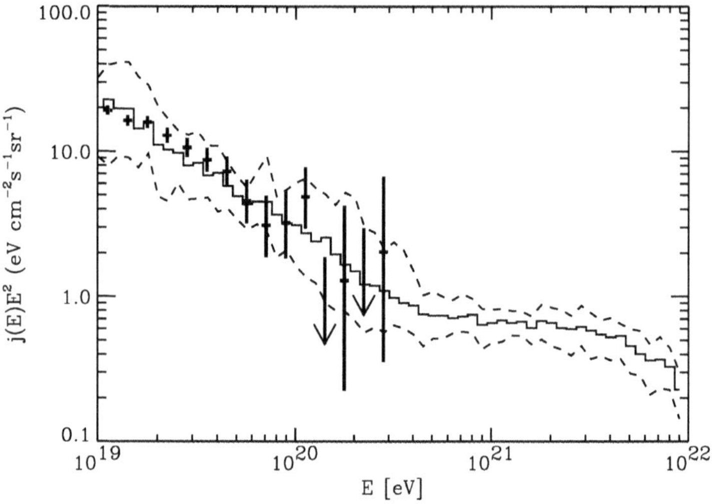

Fig. 15. The average (solid histogram) and standard deviation (dashed lines) with respect to 15 simulated magnetic field realizations of the best fit spectrum to the data above 10^{19} eV for the scenario of a single source in a magnetized Supergalactic plane. This best fit corresponds to a maximal magnetic field in the plane center, $B_{\rm max} = 10^{-7}$ G, with all other parameters as in Fig. 14. 1 sigma error bars are the combined data from the Haverah Park [59], the Fly's Eye [60], and the AGASA [61] experiments above 10^{19} eV

larger diffusion coefficient compared to Eq. (38). In addition, the fluctuations of the resulting spectra between different magnetic field realizations can be substantial, as can be seen in Fig. 15. This is a result of the fact that most of the magnetic field power is on the largest scales where there are the fewest modes. These considerations mean that the applicability of analytical flux estimates of discrete sources in specific magnetic field configurations is rather limited.

Angular images of discrete sources in a magnetized Supercluster in principle contain information on the magnetic field structure. For the recently suggested field strengths between $\sim 10^{-8}$ G and $\simeq 1\mu$G the angular images are large enough to exploit that information with instruments of angular resolution in the degree range. An example where a transition from several images at low energies to one image at high energies allows one to estimate the magnetic field coherence scale is shown in Fig. 16.

The newest AGASA data [183], however, indicate an isotropic distribution of EHECR. To explain this with only one discrete source would require the magnetic fields to be so strong that the flux beyond 10^{20} eV would most likely be too strongly suppressed by pion production, as discussed above. The recent claim that the powerful radio galaxy Centaurus A at a distance of 3.4 Mpc from Earth can explain both observed flux and angular distributions of UHECRs above $10^{18.7}$ eV for an extragalactic magnetic field of strength $\simeq 0.3\mu$G [236] was based on the diffusive approximation which is clearly shown by numerical simulations

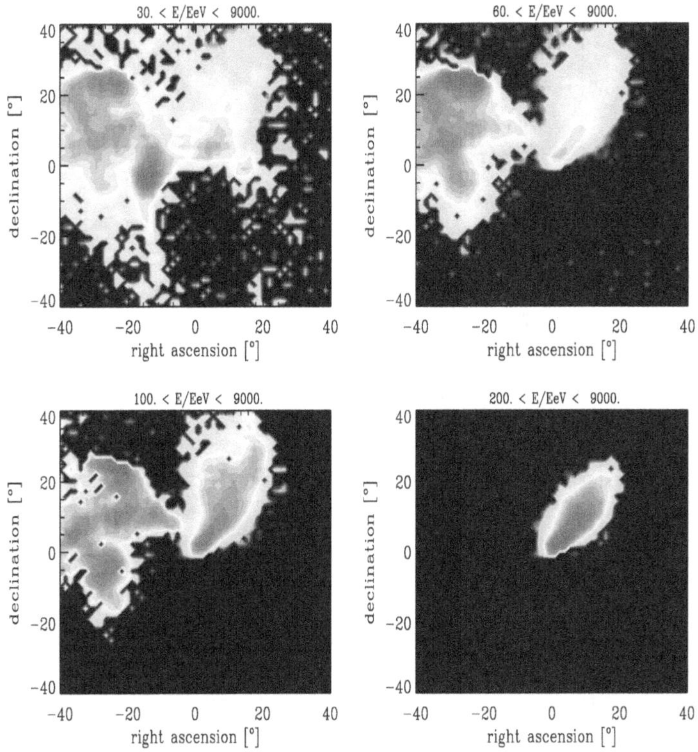

Fig. 16. Angular image of a point-like source in a magnetized Supergalactic plane, corresponding to one particular magnetic field realization with a maximal magnetic field in the plane center, $B_{\max} = 5 \times 10^{-8}$ G, all other parameters being the same as in Fig. 15. The image is shown in different energy ranges, as indicated, as seen by a detector of $\simeq 1°$ angular resolution. A transition from several images at lower energies to only one image at the highest energies occurs where the linear deflection becomes comparable to the effective field coherence length. The difference between neighboring shade levels is 0.1 in the logarithm to base 10 of the integral flux per solid angle

not to apply in this situation. For nucleons, as assumed there, the predicted distributions are too anisotropic even at 10^{19} eV. We demonstrate this in Fig. 17 from Ref. [237] which shows predictions for the declination distributions and spectra predicted by this scenario.

This suggests a more continuous source distribution which may also still reproduce the observed UHECR flux above $\simeq 10^{19}$ eV with only one spectral component [238]. Statistical studies neglecting magnetic deflection can be performed analytically and also suggest that many sources should contribute to the UHECR spectrum [239]. A more systematic parameter study of sky maps and spectra in UHECRs in different scenarios with magnetic fields and varying number of sources is now being pursued [240,176]. Such studies will also be needed to

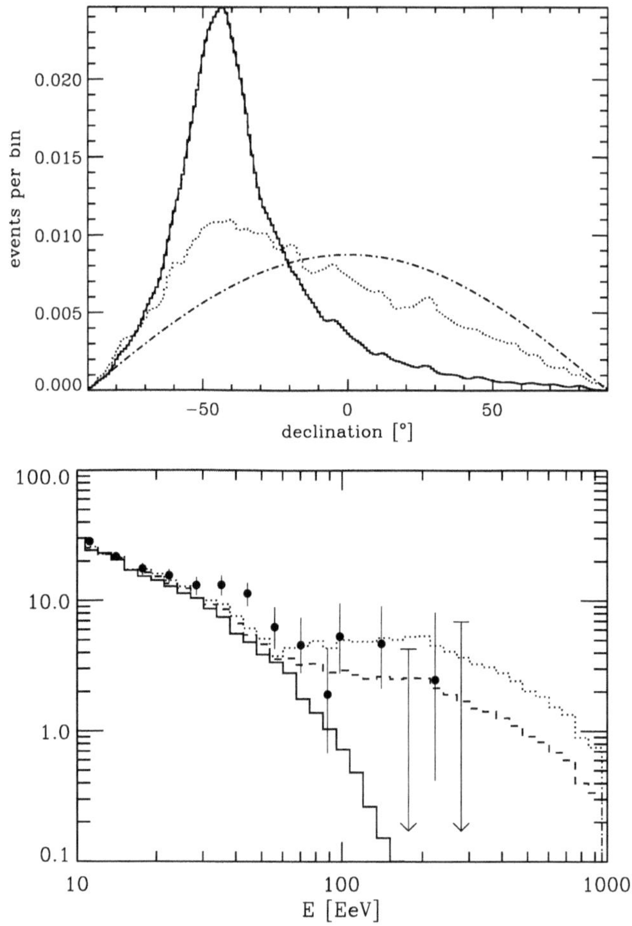

Fig. 17. Predictions for the scenario where Centaurus A at RA = 201.3^o, $\delta = -43.0^o$ in equatorial coordinates at a distance of 3.4 Mpc is the only source of UHECRs injecting a E^{-2} spectrum up to 10^{21} eV. Top panel: The distribution of arrival declination on Earth, averaged over many realizations, for $E \geq 40$ EeV (dotted line) and $E \geq 100$ EeV (solid line). The dash-dotted line represents an isotropic distribution. The pixel size is 1° and the image has again been convolved with an angular resolution of 2.4°. Bottom panel: The realization averaged differential energy spectrum multiplied by E^2 in eV/cm^{2}s sr. The solid line represents the spectrum that would have been detected by AGASA, and has been obtained by folding the simulated distributions with the AGASA exposure function. The dashed line indicates the spectrum uniformly averaged over the whole sky. The dotted line is the spectrum predicted by an AGASA type experiment in the Southern hemisphere. The one sigma error bars indicate the AGASA data [183]. The various spectra have been normalized to optimally fit the AGASA flux

generalize analytical considerations on correlations of UHECR arrival directions with the large scale structure of galaxies [241] to scenarios involving significant magnetic deflection. Furthermore, such studies should be generalized to include nuclei of arbitrary mass and their interactions discussed in Sect. 2 (the latter have been studied numerically in general in absence of deflection).

Intriguingly, scenarios in which a diffuse source distribution follows the density in the Supergalactic plane within a certain radius, can accomodate both the large scale isotropy and the small scale clustering revealed by AGASA if a magnetic field of strength $B \gtrsim 0.05\mu\,\mathrm{G}$ permeates the Supercluster [176].

Figure 18 shows the distribution of arrival times and energies, the solid angle integrated spectrum, and the angular distribution of arrival directions in Galactic coordinates in such a scenario where the UHECR sources with spectral index $\gamma = 2.4$ are distributed according to the matter density in the Local Supercluster, following a pancake profile with scale height of 5 Mpc and scale length 20 Mpc. The r.m.s. magnetic field has a Kolmogorov spectrum with a maximal field strength $B_{\mathrm{max}} = 5 \times 10^{-7}\,\mathrm{G}$ in the plane center, and also follows the matter density. The observer is within 2 Mpc of the Supergalactic plane whose location is indicated by the solid line in the lower panel and at a distance $d = 20\,\mathrm{Mpc}$ from the plane center. The absence of sources within 2 Mpc from the observer was assumed. The transition discussed above from the diffusive regime below $\simeq 2 \times 10^{20}\,\mathrm{eV}$ to the regime of almost rectilinear propagation above that energy is clearly visible.

Detailed Monte Carlo simulations performed on these distributions reveal that the anisotropy decreases with increasing magnetic field strength due to diffusion and that small scale clustering increases with coherence and strength of the magnetic field due to magnetic lensing. Both anisotropy and clustering also increase with the (unknown) source distribution radius. Furthermore, the discriminatory power between models with respect to anisotropy and clustering strongly increases with exposure [176].

As a result, a diffuse source distribution associated with the Supergalactic plane can explain most of the currently observed features of UHECRs at least for field strengths close to $0.5\,\mu\,\mathrm{G}$. The large-scale anisotropy and the clustering predicted by this scenario will allow strong discrimination against other models with next generation experiments such as the Pierre Auger Project.

9 Anomalous Kinematics, Quantum Gravity Effects, Lorentz Symmetry Violations

The existence of UHECRs beyond the GZK cutoff has prompted several suggestions of possible new physics beyond the Standard Model. We have already discussed some of these suggestions in Sect. 4 in the context of propagation of UHECRs in the extragalactic space. Further, the contribution by P. Bhattacharjee and G. Sigl in this volume will discuss suggestions regarding possible new

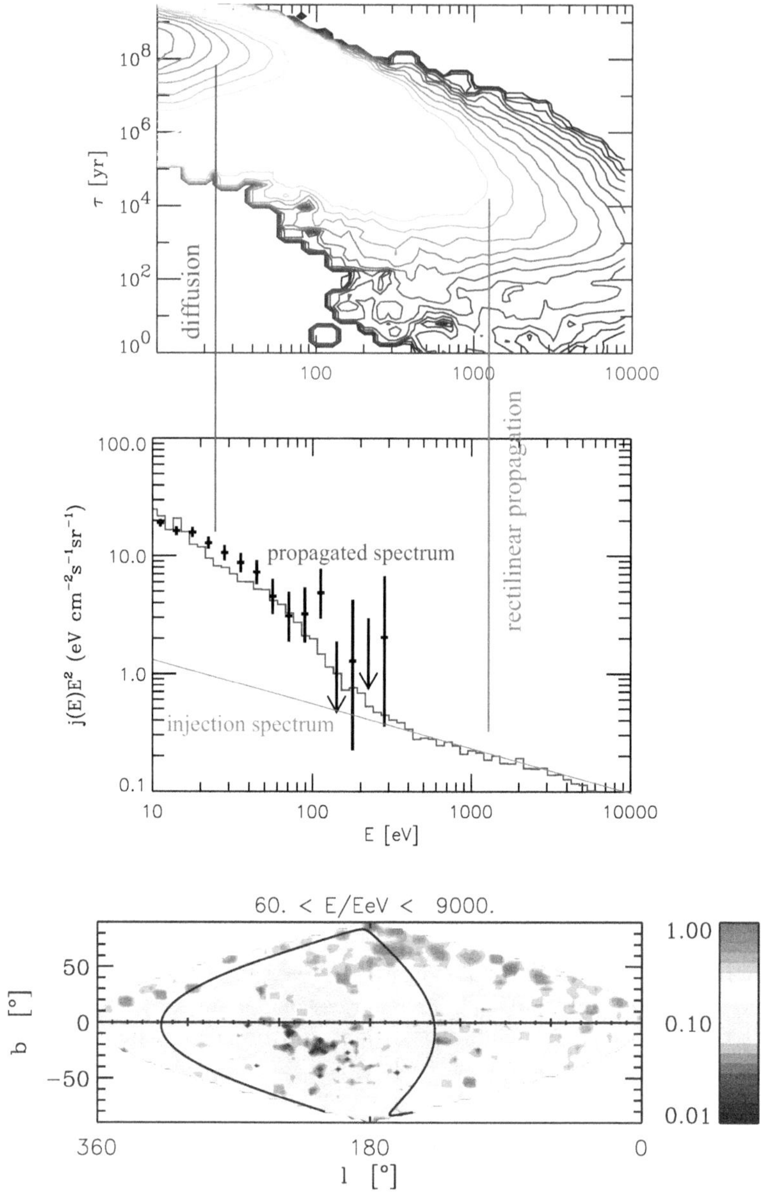

Fig. 18. The distribution of arrival times and energies (top), the solid angle integrated spectrum (middle, with 1 sigma error bars showing combined data from the Haverah Park [59], the Fly's Eye [60], and the AGASA [61] experiments above 10^{19} eV), and the angular distribution of arrival directions in Galactic coordinates (bottom, with scale showing the intensity per solid angle: higher fluxes near north pole, lower fluxes below Galactic aequator) in the Supercluster scenario with continuous source distribution explained in the text, averaged over 4 magnetic field realizations with 20000 particles each

sources of EHECR that also involve postulating new physics beyond the Standard Model. In the present section, to end our discussions on the propagation and interactions of UHE radiation, we briefly discuss some examples of possible small violations or modifications of certain fundamental tenets of physics (and constraints on the magnitude of those violations/modifications) that have also been discussed in the literature in the context of propagation of UHECRs.

For example, as an interesting consequence of the very existence of UHECRs, constraints on possible violations of Lorentz invariance (VLI) have been pointed out [242]. These constraints rival precision measurements in the laboratory: If events observed around 10^{20} eV are indeed protons, then the difference between the maximum attainable proton velocity and the speed of light has to be less than about 1×10^{-23}, otherwise the proton would lose its energy by Čerenkov radiation within a few hundred centimeters. Possible tests of other modes of VLI with UHECRs have been discussed in Ref. [243], and in Ref. [244] in the context of horizontal air showers generated by cosmic rays in general. Gonzalez-Mestres [243], Coleman and Glashow [245], and earlier, Sato and Tati [246] and Kirzhnits and Chechin [247] have also suggested that due to modified kinematical constraints the GZK cutoff could even be evaded by allowing a tiny VLI too small to have been detected otherwise. Aloisio et al. [248] have shown that a reliable experimental determination of source distances and primary composition could determine the mass scale associated with such symmetry violations or constrain it possibly more strongly than accelerator experiments [248]. Similar consequences apply to other energy loss processes such as pair production by photons above a TeV with the low energy photon background [249]. It seems to be possible to accomodate such effects within theories involving generalized Lorentz transformations [250] or deformed relativistic kinematics [251]. Furthermore, it has been pointed out [252] that violations of the principle of equivalence (VPE), while not dynamically equivalent, also produce the same kinematical effects as VLI for particle processes in a constant gravitational potential, and so the constraints on VLI from UHECR physics can be translated into constraints on VPE such that the difference between the couplings of protons and photons to gravity must be less than about 1×10^{-19}. Again, this constraint is more stringent by several orders of magnitude than the currently available laboratory constraint from Eötvös experiments.

As a specific example of VLI, we consider an energy dependent photon group velocity $\partial E/\partial k = c[1 - \chi E/E_0 + \mathcal{O}(E^2/E_0^2)]$ where c is the speed of light in the low energy limit, $\chi = \pm 1$, and E_0 denotes the energy scale where this modification becomes of order unity. This corresponds to a dispersion relation

$$c^2k^2 \simeq E^2 + \chi \frac{E^3}{E_0}\,, \tag{40}$$

which, for example, can occur in quantum gravity and string theory [253]. The kinematics of electron-positron pair production in a head-on collision of a high energy photon of energy E with a low energy background photon of energy ε then leads to the constraint

$$\varepsilon \simeq \frac{E}{4}\left(\frac{m_e^2}{E_1 E_2} + \theta_1\theta_2\right) + \chi\frac{E^2}{4E_0}\,, \tag{41}$$

where E_i and $\theta_i \sim \mathcal{O}(m/E_i)$ are respectively the energy and outgoing momentum angle (with respect to the original photon momentum) of the electron and positron ($i = 1, 2$). For the case considered by Coleman and Glashow [242] in which the maximum attainable speed c_i of the matter particle is different from the photon speed c, the kinematics can be obtained by substituting $c_i^2 - c^2$ for $\chi E/E_0$ in Eq. (41).

Let us define a critical energy $E_c = (m_e^2 E_0)^{1/3} \simeq 15(E_0/m_{\rm Pl})^{1/3}\,$TeV in the case of the energy dependent photon group velocity, and $E_c = m_e/|c_i^2 - c^2|^{1/2}$ in the case considered by Coleman and Glashow. If $\chi < 0$, or $c_i < c$, then ε becomes negative for $E \gtrsim E_c$. This signals that the photon can spontaneously decay into an electron-positron pair and propagation of photons across extragalactic distances will in general be inhibited. The observation of extragalactic photons up to $\simeq 20\,$TeV [254,255] therefore puts the limits $E_0 \gtrsim M_{\rm Pl}$ or $c_i^2 - c^2 \gtrsim -2\times10^{-17}$. In contrast, if $\chi > 0$, or $c_i > c$, ε will grow with energy for $E \gtrsim E_c$ until there is no significant number of target photon density available and the Universe becomes transparent to UHE photons. A clear test of this possibility would be the observation of $\gtrsim 100\,$TeV photons from distances $\gtrsim 100\,$Mpc [256].

In addition, the dispersion relation Eq. (40) implies that a photon signal at energy E will be spread out by

$$\Delta t \simeq \left(\frac{d}{c}\right)\left(\frac{E}{E_0}\right) \simeq 1\left(\frac{d}{100\,\rm Mpc}\right)\left(\frac{E}{\rm TeV}\right)\left(\frac{E_0}{M_{\rm Pl}}\right)^{-1}\ \rm s.$$

The observation of $\gamma-$rays at energies $E \gtrsim 2\,$TeV within $\simeq 300\,$s from the BL Lac object Markarian 421 therefore puts a limit (independent of χ) of $E_0 \gtrsim 4\times10^{16}\,$GeV, whereas the possible observation of $\gamma-$rays at $E \gtrsim 200\,$TeV within $\simeq 200\,$s from a GRB by HEGRA might be sensitive to $E_0 \simeq M_{\rm Pl}$ [257]. For a recent detailed discussion of these limits see Ref. [258].

A related proposal originally due to Kostelecký in the context of CR suggests the electron neutrino to be a tachyon [259]. This would allow the proton in a nucleus of mass $m(A,Z)$ for mass number A and charge Z to decay via $p \to n + e^+ + \nu_e$ above the energy threshold $E_{\rm th} = m(A,Z)[m(A,Z\pm1) + m_e - m(A,Z)]/|m_{\nu_e}|$ which, for a free proton, is $E_{\rm th} \simeq 1.7\times10^{15}/(|m_{\nu_e}|/{\rm eV})\,$eV. Ehrlich [260] claims that by choosing $m_{\nu_e}^2 \simeq -(0.5\,{\rm eV})^2$ it is possible to explain the knee and several other features of the observed CR spectrum, including the high energy end, if certain assumptions about the source distribution are made. The experimental best fit values of $m_{\nu_e}^2$ from tritium beta decay experiments are indeed negative [261], although this is most likely due to unresolved experimental issues. In addition, the values of $|m_{\nu_e}^2|$ from tritium beta decay experiments are typically larger than the value required to fit the knee of the CR spectrum. This scenario also predicts a neutron line around the knee energy [262].

References

1. M. S. Longair, R. A. Sunyaev: Sov. Phys. Usp. **14**, 569 (1972)
   M. T. Ressel, M. S. Turner: Comments Astrophys. **14**, 323 (1990).
2. K. Greisen: Phys. Rev. Lett. **16**, 748 (1966)
3. G. T. Zatsepin, V. A. Kuzmin: Pis'ma Zh. Eksp. Teor. Fiz. **4**, 114 (1966) [JETP. Lett. **4**, 78 (1966)]
4. Review of Particle Physics: Phys. Rev. D **54**, 1 (1996)
5. V. S. Berezinsky, A. Z. Gazizov: Phys. Rev. D **47**, 4206 (1993); ibid., 4217 (1993)
6. S. Lee: Phys. Rev. D **58**, 043004 (1998)
7. Proc. *19th Texas Symposium on Relativistic Astrophysics*, Paris (France), eds. E. Aubourg, et al., Nuc. Phys. B (Proc. Supp.) **80B** (2000)
8. A. Muecke, et al.: Publ. Astron. Soc. Austral. **16** (1999) 160
   A. Muecke, et al.: Comput. Phys. Commun. **124**, 290 (2000)
   A. Muecke, et al.: astro-ph/9905153, to appear in Ref. [7].
9. G. R. Blumenthal: Phys. Rev. D **1**, 1596 (1970)
10. M. J. Chodorowski, A. A. Zdziarski, M. Sikora: Astrophys. J. **400**, 181 (1992)
11. O. E. Kalashev, V. A. Kuzmin, D. V. Semikoz: astro-ph/9911035
12. See, e.g., C. J. Copi, D. N. Schramm, M. S. Turner: Science **267**, 192 (1995)
13. F. W. Stecker: Phys. Rev. **180**, 1264 (1969)
14. J. L. Puget, F. W. Stecker, J. H. Bredekamp: Astrophys. J. **205**, 638 (1976)
15. W. Tkaczyk, J. Wdowczyk, A. W. Wolfendale: J. Phys. A **8**, 1518 (1975)
16. S. Karakula, W. Tkaczyk: Astropart. Phys. **1**, 229 (1993)
17. See, e.g.:
    S. Biller, et al.: Phys. Rev. Lett. **80**, 2992 (1998)
    T. Stanev, A. Franceschini: Astrophys. J. **494**, L159 (1998)
    F. W. Stecker, O. C. de Jager: Astron. Astrophys. **334**, L85 (1998)
    J. R. Primack, R. S. Somerville, J. S. Bullock, J. E. G. Devriendt: astro-ph/0011475, Proc. *International Symposium on Gamma-Ray Astronomy*, Heidelberg, June 2000, eds. F. Aharonian, H. Völk, AIP Conf. Proc.
18. K. Mannheim: Science **279**, 684 (1998)
19. M. A. Malkan, F. W. Stecker: Astrophys. J. **496**, 13 (1998)
20. F. W. Stecker: Phys. Rev. Lett. **80**, 1816 (1998)
21. L. N. Epele, E. Roulet: Phys. Rev. Lett. **81**, 3295 (1998)
    L. N. Epele, E. Roulet: J. High Energy Phys. **10**, 009 (1998)
22. F. W. Stecker: Phys. Rev. Lett. **81**, 3296 (1998)
23. F. W. Stecker, M. H. Salamon: Astrophys. J. **512**, 521 (1999)
24. J. W. Elbert, P. Sommers: Astrophys. J. **441**, 151 (1995)
25. L. A. Anchordoqui, G. E. Romero, J. A. Combi: Phys. Rev. D **60**, 103001 (1999)
26. A. I. Nikishov: Zh. Eksp. Teor. Fiz **41**, 549 (1961) [Sov. Phys. JETP **14**, 393 (1962)]
    J. V. Jelley: Phys. Rev. Lett. **16**, 479 (1966)
    R. J. Gould, G. P. Schréder: Phys. Rev. **155**, 1404 (1967)
    V. Berezinskii: Sov. J. Nucl. Phys **11**, 222 (1970)
27. F. W. Stecker: Astrophys. J. **157**, 507 (1969)
    G. G. Fazio, F. W. Stecker: Nature **226**, 135 (1970)
28. R. J. Protheroe, P. L. Biermann: Astropart. Phys. **6**, 45 (1996); erratum, ibid. **7**, 181 (1997)
29. T. A. Clark, L. W. Brown, J. K. Alexander: Nature **228**, 847 (1970)
30. V. S. Berezinsky: Yad. Fiz. **11**, 339 (1970)

31. S. A. Bonometto: Il Nuovo Cim. Lett. **1**, 677 (1971)
32. M. R. Allcock, J. Wdowczyk: Il Nuovo Cim. **9B**, 315 (1972)
J. Wdowczyk, W. Tkaczyk, C. Adcock, A. W. Wolfendale: J. Phys. A **4**, L37 (1971)
J. Wdowczyk, W. Tkaczyk, A. W. Wolfendale: J. Phys. A **5**, 1419 (1972)
33. F. W. Stecker: Astrophys. Sp. Sc. **20**, 47 (1973)
34. R. J. Gould, Y. Rephaeli: Astrophys. J. **225**, 318 (1978)
35. F. A. Aharonian, P. Bhattacharjee, D. N. Schramm: Phys. Rev. D **46**, 4188 (1992)
36. V. S. Berezinsky, S. V. Bulanov, V. A. Dogiel, V. L. Ginzburg, V. S. Ptuskin: *Astrophysics of Cosmic Rays* (North-Holland, Amsterdam, 1990)
37. A. A. Zdziarski: Astrophys. J. **335**, 786 (1988)
38. F. Halzen, R. J. Protheroe, T. Stanev, H. P. Vankov: Phys. Rev. D **41**, 342 (1990)
39. J. Wdowczyk, A. W. Wolfendale: Astrophys. J. **349**, 35 (1990)
40. A. Chen, J. Dwyer, P. Kaaret: Astrophys. J. **463**, 169 (1996)
P. Sreekumar, et al.: Astrophys. J. **494**, 523 (1998)
P. Sreekumar, F. W. Stecker, S. C. Kappadath: Proc. *4th Compton Symposium*, Williamsburg (April 1997)
41. G. Sigl, K. Jedamzik, D. N. Schramm, V. Berezinsky: Phys. Rev. D **52**, 6682 (1995)
42. R. W. Brown, W. F. Hunt, K. O. Mikaelian, I. J. Muzinich: Phys. Rev. D **8**, 3083 (1973)
43. A. Borsellino: Nuovo Cimento **4**, 112 (1947)
E. Haug: Zeit. Naturforsch. **30a**, 1099 (1975)
A. Mastichiadis: Mon. Not. R. Astron. Soc. **253**, 235 (1991)
44. M. Postma: Phys. Rev. D **64**, 023001 (2001)
See also H. Athar, G.-L. Lin, J.-J. Tseng: Phys. Rev. D **64**, 071302 (2001)
45. R. J. Gould: Astrophys. J. **230**, 967 (1979)
46. R. Svensson, A. A. Zdziarski: Astrophys. J. **349**, 415 (1990)
47. A. A. Zdziarski, R. Svensson: Astrophys. J. **344**, 551 (1989)
48. G. D. Kribs, I. Z. Rothstein: Phys. Rev. D **55**, 4435 (1997)
49. T. Erber: Rev. Mod. Phys. **38**, 626 (1966)
50. T. J. Weiler: Phys. Rev. Lett. **49**, 234 (1982)
T. J. Weiler: Astrophys. J. **285**, 495 (1984)
51. E. Roulet: Phys. Rev. D **47**, 5247 (1993)
52. S. Yoshida: Astropart. Phys. **2**, 187 (1994)
53. S. Yoshida, H. Dai, C. C. H. Jui, P. Sommers: Astrophys. J. **479**, 547 (1997)
54. D. Seckel: Phys. Rev. Lett. **80**, 900 (1998)
55. D. A. Dicus, W. W. Repko: Phys. Rev. Lett. **79**, 569 (1997)
56. M. Harris, J. Wang, V. L. Teplitz: astro-ph/9707113
57. P. Bhattacharjee, C. T. Hill, D. N. Schramm: Phys. Rev. Lett. **69**, 567 (1992)
58. G. Sigl, S. Lee, P. Bhattacharjee, S. Yoshida: Phys. Rev. D **59**, 043504 (1999)
59. See, e.g., M. A. Lawrence, R. J. O. Reid, A. A. Watson: J. Phys. G Nucl. Part. Phys. **17**, 733 (1991) 733, and references therein
see also http://ast.leeds.ac.uk/haverah/hav-home.html
60. D. J. Bird, et al.: Phys. Rev. Lett. **71**, 3401 (1993)
D. J. Bird, et al.: Astrophys. J. **424**, 491 (1994) 491; ibid. **441**, 144 (1995)
61. N. Hayashida, et al.: Phys. Rev. Lett. **73**, 3491 (1994)
S. Yoshida, et al.: Astropart. Phys. **3**, 105 (1995)
M. Takeda, et al.: Phys. Rev. Lett. **81**, 1163 (1998)
see also http://icrsun.icrr.u-tokyo.ac.jp/as/project/agasa.html

62. P. Bhattacharjee, G. Sigl: Phys. Rep. **327**, 109 (2000)
63. R. Cowsik, J. McClelland: Phys. Rev. Lett. **29**, 669 (1972)
R. Cowsik, J. McClelland: Astrophys. J. **180**, 7 (1973)
R. Cowsik, P. Ghosh: Astrophys. J. **317**, 26 (1987)
64. D. Fargion, B. Mele, A. Salis: Astrophys. J. **517**, 725 (1999)
65. T. J. Weiler: Astropart. Phys. **11**, 303 (1999)
66. E. Waxman: astro-ph/9804023
67. S. Yoshida, G. Sigl, S. Lee: Phys. Rev. Lett. **81**, 5505 (1998)
68. G. Gelmini, A. Kusenko: Phys. Rev. Lett. **84**, 1378 (2000)
69. Proc. *24th International Cosmic Ray Conference* (Istituto Nazionale Fisica Nucleare, Rome, Italy, 1995)
70. G. Sigl, S. Lee: in Ref. [69], Vol. **3**, 356.
71. E. Waxman, J. Bahcall: Phys. Rev. D. **59**, 023002 (1999)
72. K. Mannheim, R. J. Protheroe, J. P. Rachen, Phys. Rev. D 63 (2001) 023003
J. P. Rachen, R. J. Protheroe, K. Mannheim, astro-ph/9908031, in Ref. [7]
73. J. Bahcall, E. Waxman: Phys. Rev. D **64**, 023002 (2001)
74. J. J. Blanco-Pillado, R. A. Vázquez, E. Zas: Phys. Rev. D **61**, 123003 (2000)
75. G. Gelmini, A. Kusenko: Phys. Rev. Lett. **82**, 5202 (1999)
76. P. B. Pal, K. Kar: Phys. Lett. B **451**, 136 (1999)
77. I. Affleck, M. Dine: Nucl. Phys. B **249**, 361 (1985)
78. Y. Fukuda, et al. (Super-Kamiokande collaboration): Phys. Rev. Lett. **81**, 1562 (1998)
79. See, e.g.:
F. W. Stecker, C. Done, M. Salamon, P. Sommers: Phys. Rev. Lett. **66**, 2697 (1991)
A. P. Szabo, R. J. Protheroe: Astropart. Phys. **2**, 375 (1994)
F. W. Stecker, M. H. Salamon: Space. Sc. Rev. **75**, 341 (1996)
V. S. Berezinsky: Nucl. Phys. B (Proc. Suppl.) **38B**, 363 (1995)
80. K. Mannheim: Astropart. Phys. **3**, 295 (1995)
81. F. Halzen, E. Zas: Astrophys. J. **488**, 669 (1997)
82. R. Protheroe: Proc. *Accretion Phenomena and Related Outflows*, Vol. **163** of IAU Colloquium, eds. D. Wickramasinghe, G. Bicknell, L. Ferrario (Pub. Astron. Soc. Pac., 1997), p. 585.
83. R. Protheroe: astro-ph/9809144, Proc. *Neutrino 98*, Takayama, 4-9 June 1998.
84. F. W. Stecker: Astrophys. J. **228**, 919 (1979)
85. C. T. Hill, D. N. Schramm: Phys. Lett. B **131**, 247 (1983)
86. Z. Fodor, S. D. Katz: hep-ph/0105064; hep-ph/0105336
87. Proc. *VERITAS Workshop on TeV Astrophysics of Extragalactic Sources*, eds. M. Catanese, J. Quinn, T. Weekes, Astropart. Phys. **11** (1999)
88. K. Mannheim: in Ref. [87], Astropart. Phys. **11**, 49 (1999)
89. J. G. Learned, S. Pakvasa: Astropart. Phys. **3**, 267 (1995)
90. F. Halzen, D. Saltzberg: Phys. Rev. Lett. **81**, 4305 (1998)
91. R. Horvat: Phys. Rev. D **59**, 123003 (1999)
92. P. Keränen, J. Maalampi, J. T. Peltoniemi: Phys. Lett. **B461**, 230 (1999)
93. G. M. Frichter, D. W. McKay, J. P. Ralston: Phys. Rev. Lett. **74**, 1508 (1995)
94. R. Gandhi, C. Quigg, M. H. Reno, I. Sarcevic: Astropart. Phys. **5**, 81 (1996)
R. Gandhi, C. Quigg, M. H. Reno, I. Sarcevic: Phys. Rev. D **58**, 093009 (1998)
95. M. Glück, S. Kretzer, E. Reya: Astropart. Phys. **11**, 327 (1999)
96. M. Derick, et al. (ZEUS collaboration): Phys. Lett. **B316**, 412 (1993)
M. Derick, et al. (ZEUS collaboration): Z. Phys. C **65**, 379 (1995)
I. Abt, et al. (H1 collaboration): Nucl. Phys. **407B**, 515 (1993)
T. Ahmed, et al. (H1 collaboration): Nucl. Phys. **439B**, 471 (1995)

97. G. Gelmini, P. Gondolo, G. Varieschi: Phys. Rev. D **61**, 056011 (2000)
98. J. Kwiecinski, A. D. Martin, A. Stasto: Phys. Rev. D **59**, 093002 (1999)
99. D. A. Dicus, S. Kretzer, W. W. Repko, C. Schmidt: Phys. Lett. **B514**, 103 (2001)
100. M. Kachelriess, M. Plümacher: private communication.
101. J. L. Feng, P. Fisher, F. Wilczek, T. M. Yu: hep-ph/0105067
A. Kusenko, T. Weiler: hep-ph/0106071
102. G. Sigl: Phys. Rev. D **57**, 3786 (1998)
103. V. A. Naumov, L. Perrone: Astropart. Phys. **10**, 239 (1999)
104. P. Jain, J. P. Ralston, G. M. Frichter: Astropart. Phys. **12**, 193 (1999)
105. J. Bordes, et al.: Astropart. Phys. **8**, 135 (1998)
J. Bordes, et al.: Proc. *Beyond the Standard Model. From Theory to Experiment* (Valencia, Spain, 13-17 October 1997), eds. I. Antoniadis, L. E. Ibanez, J. W. F. Valle (World Scientific, Singapore, 1998), p. 328 (hep-ph/9711438)
106. G. Sigl: Proc. *TAUP99*, eds. M. Froissart, J. Dumarchez, D. Vignaud, Paris (France), Nucl. Phys. (Proc. Suppl.) **87B**, p.439 (2000)
107. P. Jain, D. W. McKay, S. Panda, J. P. Ralston: Phys. Lett. **B484**, 267 (2000)
J. P. Ralston, P. Jain, D. W. McKay, S. Panda: hep-ph/0008153
108. G. Burdman, F. Halzen, R. Gandhi: Phys. Lett. **B417**, 107 (1998)
109. G. Domokos, S. Kovesi-Domokos: Phys. Rev. Lett. **82**, 1366 (1999)
110. N. Arkani-Hamed, S. Dimopoulos, G. Dvali: Phys. Lett. **B429**, 263 (1998)
I. Antoniadis, N. Arkani-Hamed, S. Dimopoulos, G. Dvali: Phys. Lett. **B436**, 257 (1998)
N. Arkani-Hamed, S. Dimopoulos, G. Dvali: Phys. Rev. D **59**, 086004 (1999)
111. S. Nussinov, R. Shrock: Phys. Rev. D **59**, 105002 (1999)
S. Nussinov, R. Shrock: Phys. Rev. D **64**, 047702 (2001)
112. M. Kachelrieß M. Plümacher: Phys. Rev. D **62**, 103006 (2000); hep-ph/0109184
113. L. Anchordoqui, et al.: Phys. Rev. D **63**, 124009 (2001)
114. see, e.g., C. Spiering: Nucl. Phys. Proc. Suppl. **91**, 445 (2000)
115. J. W. Cronin: Nucl. Phys. B (Proc. Suppl.) **28B**, 213 (1992)
The Pierre Auger Observatory Design Report (2nd edition), March 1997
see also http://www.auger.org/ and
http://www-lpnhep.in2p3.fr/auger/welcome.html
116. Proc. *25th International Cosmic Ray Conference*, eds.: M. S. Potgieter, et al. (Durban, 1997)
117. Proc. *International Symposium on Extremely High Energy Cosmic Rays: Astrophysics, and Future Observatories*, ed. M. Nagano (Institute for Cosmic Ray Research, Tokyo, 1996)
118. J. F. Ormes, et al.: in Ref. [116], Vol. **5**, 273
Y. Takahashi, et al.: in Ref. [117], p. 310
see also http://lheawww.gsfc.nasa.gov/docs/gamcosray/hecr/OWL/
119. See http://www.ifcai.pa.cnr.it/Ifcai/euso.html
120. J. Linsley: in Ref. [116], Vol. **5**, 381; ibid., 385
P. Attinà, et al.: in Ref. [116], 389
J. Forbes, et al., in Ref. [116], 273
see also http://www.ifcai.pa.cnr.it/Ifcai/airwatch.html
121. C. Tyler, A. Olinto, G. Sigl: Phys. Rev. D. **63**, 055001 (2001)
122. J. Alvarez-Muniz, F. Halzen, T. Han, D. Hooper: hep-ph/0107057
123. W. Rhode, et al.: Astropart. Phys. **4**, 217 (1996)
124. R. M. Baltrusaitis, et al.: Astrophys. J. **281**, L9 (1984)
R. M. Baltrusaitis, et al.: Phys. Rev. D **31**, 2192 (1985)

125. Proc. *26th International Cosmic Ray Conference*, (Utah, 1999)
126. Proc. *First International Workshop on Radio Detection of High-Energy Particles*, Amer. Inst. of Phys., 2001, and at http://www.physics.ucla.edu/~moonemp/radhep/workshop.html
127. P. W. Gorham, K. M. Liewer, C. J. Naudet: astro-ph/9906504, in Ref. [125]. P. W. Gorham, et al.: astro-ph/0102435, in Ref. [126].
128. For general information see http://amanda.berkeley.edu/ see also F. Halzen: New Astron. Rev **42**, 289 (1999) for newest status see, e.g., G. C. Hill, et al. (Amanda Collaboration): astro-ph/0106064, Proceedings of *XXXVIth Recontres de Moriond*, March 2001
129. For general information see http://www.ps.uci.edu/~icecube/workshop.html see also F. Halzen: Am. Astron. Soc. Meeting **192**, # **62 28** (1998) AMANDA collaboration: astro-ph/9906205, Proc. *8th International Workshop on Neutrino Telescopes*, Venice, Feb. 1999.
130. J. J. Blanco-Pillado, R. A. Vázquez, E. Zas: Phys. Rev. Lett. **78**, 3614 (1997) K. S. Capelle, J. W. Cronin, G. Parente, E. Zas: Astropart. Phys. **8**, 321 (1998) A. Letessier-Selvon: astro-ph/0009444
131. See, e.g., P. Lipari: Astropart. Phys. **1**, 195 (1993)
132. See, e.g., R. J. Protheroe, P. A. Johnson: Astropart. Phys. **4**, 253 (1996); erratum ibid. **5**, 215 (1996)
133. see, e.g., S. Cullen, M. Perelstein, M. E. Peskin: Phys. Rev.D **62**, 055012 (2000), and references therein. I. Antoniadis, K. Benakli: hep-ph/0108174
134. S. Dimopoulos, R. Emparan: hep-ph/0108060, and references therein.
135. J. L. Feng, A. D. Shapere: hep-ph/0109106
136. F. Cornet, J. L. Illana, M. Masip: Phys. Rev. Lett. **86**, 4235 (2001)
137. R. Emparan, M. Masip, R. Rattazzi: hep-ph/0109287
138. see also L. Anchordoqui, H. Goldberg: hep-ph/0109242
139. S. Cullen, M. Perelstein: Phys. Rev. Lett. **83**, 268 (1999) V. Barger, T. Han, C. Kao, R.-J. Zhang: Phys. Lett. **B461**, 34 (1999)
140. S. Hannestad, G. Raffelt: Phys. Rev. Lett. **87**, 051301 (2001)
141. S. Hannestad: Phys. Rev. D **64**, 023515 (2001)
142. see, e.g., K. R. Dienes: hep-ph/0108115
143. J. C. Long, H. W. Chan, J. C. Price: Nucl. Phys. **539B**, 23 (1999)
144. D. A. Morris, A. Ringwald: Astropart. Phys. **2**, 43 (1994)
145. H. Goldberg, T. J. Weiler: Phys. Rev. D **59**, 113005 (1999)
146. S. Raby: Phys. Lett. **B422**, 158 (1998) S. Raby, K. Tobe: Nucl. Phys. B **539**, 3 (1999)
147. M. B. Voloshin, L. B. Okun: Sov. J. Nucl. Phys. **43**, 495 (1986)
148. G. R. Farrar: Phys. Rev. Lett. **76**, 4111 (1996)
149. Proc. *Workshop on Observing Giant Cosmic Ray Air Showers from* $> 10^{20}$ *eV Particles from Space*, eds. J. F. Krizmanic, J. F. Ormes, R. E. Streitmatter (AIP Conference Proceedings 433, 1997)
150. D. J. H. Chung, G. R. Farrar, E. W. Kolb: Phys. Rev. D **57**, 4696 (1998) G. R. Farrar: astro-ph/9801020, in Ref. [149].
151. I. F. Albuquerque, et al. (E761 collaboration): Phys. Rev. Lett. **78**, 3252 (1997)
152. J. Adams, et al. (KTeV Collaboration): Phys. Rev. Lett. **79**, 4083 (1997) A. Alavi-Harati, et al. (KTeV collaboration): Phys. Rev. Lett. **83**, 2128 (1999)
153. L. Clavelli: hep-ph/9908342
154. R. Plaga: private communications.

155. I. F. M. Albuquerque, G. F. Farrar, E. W. Kolb: Phys. Rev. D **59**, 015021 (1999)
156. D. S. Gorbunov, G. G. Raffelt, D. V. Semikoz: Phys. Rev. D **64**, 096005 (2001)
157. N. I. Kochelev: JETP Lett. **70**, 491 (1999)
158. S. Lee, A. V. Olinto, G. Sigl: Astrophys. J. **455**, L21 (1995)
159. A. Karle, et al.: Phys. Lett. **B347**, 161 (1995)
F. A. Aharonian et al. (HEGRA collaboration): astro-ph/0109145
160. M. C. Chantell, et al.: Phys. Rev. Lett. **79**, 1805 (1997)
161. E. Waxman, J. Miralda-Escudé: Astrophys. J. **472**, L89 (1996)
162. J. Miralda-Escudé, E. Waxman: Astrophys. J. **462**, L59 (1996)
163. G. Sigl, D. N. Schramm, P. Bhattacharjee: Astropart. Phys. **2**, 401 (1994)
164. P. P. Kronberg: Rep. Prog. Phys. **57**, 325 (1994)
165. J. P. Vallée: Fund. Cosm. Phys. **19**, 1 (1997)
166. T. Stanev: Astrophys. J. **479**, 290 (1997)
167. G. A. Medina Tanco, E. M. de Gouveia Dal Pino, J. E. Horvath: Astrophys. J. **492**, 200 (1998)
168. D. Harari, S. Mollerach, E. Roulet: J. High Energy Phys. **08**, 022 (1999) 022; ibid. **02**, 035 (2000); ibid. **10**, 047 (2000)
169. E.-J. Ahn, G. Medina-Tanco, P. L. Biermann, T. Stanev: astro-ph/9911123
170. P. Billoir, A. Letessier-Selvon: astro-ph/0001427
171. P. Blasi, S. Burles, A. V. Olinto: Astrophys. J. **514**, L79 (1999)
172. G. R. Farrar, T. Piran: Phys. Rev. Lett. **84**, 3527 (2000)
173. D. Ryu, H. Kang, P. L. Biermann: Astron. Astrophys. **335**, 19 (1998)
174. J. A. Eilek: astro-ph/9906485, Proc. *Ringberg Conference on Diffuse Thermal, and Relativistic Plasma in Galaxy Clusters*, Germany (1999)
K. Dolag, M. Bartelmann, H. Lesch: astro-ph/9907199, to appear ibid.
175. G. Sigl, M. Lemoine, P. Biermann: Astropart. Phys. **10**, 141 (1999)
176. M. Lemoine, G. Sigl, P. Biermann: astro-ph/9903124
177. J. Rachen: Proc. *19th Texas Symposium on Relativistic Astrophysics and Cosmology*, Chicago (USA, 1996)
178. F. Halzen, R. V'azques, T. Stanev, H. P. Vankov: Astropart. Phys. **3**, 151 (1995)
179. M. Giller, M. Zielińska: in Ref. [117], p. 382.
180. T. Stanev, et al.: Phys. Rev. Lett. **75**, 3056 (1995)
181. L. J. Kewley, R. W. Clay, B. R. Dawson: Astropart. Phys. **5**, 69 (1996)
182. N. Hayashida, et al.: Phys. Rev. Lett. **77**, 1000 (1996)
183. M. Takeda, et al.: Astrophys. J. **522**, 225 (1999)
Hayashida, et al.: astro-ph/0008102
184. Y. Uchihori, et al.: Astropart. Phys. **13**, 151 (2000)
185. Y. Uchihori, et al.: in Ref. [117], p. 50.
186. E. Waxman, K. B. Fisher, T. Piran: Astrophys. J. **483**, 1 (1997)
187. P. L. Biermann, H. Kang, D. Ryu: in Ref. [117], p. 79
P. L. Biermann, H. Kang, J. P. Rachen, D. Ryu: astro-ph/9709252, Proc. *Moriond Meeting on High Energy Phenomena*, Jan. 1997, Les Arcs.
188. G. A. Medina Tanco: astro-ph/9707054
189. M. Milgrom, V. Usov: Astrophys. J. **449**, L37 (1995)
190. T. Stanev, R. Schaefer, A. Watson: Astropart. Phys. **5**, 75 (1996)
191. C. W. Akerlof, et al.: astro-ph/9706123
192. A. Lindner: hep-ph/9801311, Proc. *International Europhysics Conference on High-Energy Physics (HEP 97)* (Jerusalem, Israel, 19-26 August 1997)
193. G. R. Farrar, P. L. Biermann: Phys. Rev. Lett. **81**, 3579 (1998)
194. C. Hoffman: Phys. Rev. Lett. **83**, 2471 (1999)

195. G. R. Farrar, P. L. Biermann: Phys. Rev. Lett. **83**, 2472 (1999)
196. G. Sigl, D. F. Torres, L. A. Anchordoqui, G. E. Romero: Phys. Rev. D **63**, 081302 (2001)
197. A. Virmani, et al.: astro-ph/0010235; to appear in Astropart. Phys.
198. P. G. Tinyakov, I. I. Tkachev: astro-ph/0102476
199. A. A. Mikhailov: astro-ph/9906302, to appear in Ref. [125].
200. V. S. Berezinsky, S. I. Grigor'eva: Astron. Astrophys. **199**, 1 (1988)
201. F. A. Aharonian, B. L. Kanevsky, V. V. Vardanian: Astrophys. Space Sci. **167**, 93 (1990)
202. J. P. Rachen, P. L. Biermann: Astron. Astrophys. **272**, 161 (1993)
203. J. Geddes, T. C. Quinn, R. M. Wald: Astrophys. J. **459**, 384 (1996)
204. E. Waxman: Astrophys. J. **452**, L1 (1995)
205. L. A. Anchordoqui, M. T. Dova, L. N. Epele, J. D. Swain: Phys. Rev. D **55**, 7356 (1997)
206. S. Yoshida, M. Teshima: Prog. Theor. Phys. **89**, 833 (1993)
207. C. T. Hill, D. N. Schramm: Phys. Rev. D **31**, 564 (1985)
208. F. A. Aharonian, J. W. Cronin: Phys. Rev. D **50**, 1892 (1994)
209. G. Sigl, M. Lemoine, A. Olinto: Phys. Rev. D. **56**, 4470 (1997)
210. G. Sigl, M. Lemoine: Astropart. Phys. **9**, 65 (1998)
211. R. J. Protheroe, T. Stanev: Phys. Rev. Lett. **77**, 3708 (1996); erratum, ibid. **78**, 3420 (1997)
212. R. J. Protheroe, T. Stanev: Mon. Not. R. Astron. Soc. **245**, 453 (1990)
213. For a review see, e.g., T. K. Gaisser, F. Halzen, T. Stanev: Phys. Rep. **258**, 173 (1995)
214. G. Sigl, S. Lee, D. N. Schramm, P. S. Coppi: Phys. Lett. **B392**, 129 (1997)
215. U. F. Wichoski, J. H. MacGibbon, R. H. Brandenberger: hep-ph/9805419
U. F. Wichoski, R. H. Brandenberger, J. H. MacGibbon: hep-ph/9903545, Proc. *2nd Meeting New Worlds in Astroparticle Physics* (Faro, Portugal, 3-5 Sep 1998), eds. A. Mourao, M. Pimenta, P. Sa (World Scientific, Singapore, 1999)
216. J. Wdowczyk, A. W. Wolfendale: Nature **281**, 356 (1979)
M. Giler, J. Wdowczyk, A. W. Wolfendale: J. Phys. G.: Nucl. Phys. **6**, 1561 (1980)
V. S. Berezinskii, S. I. Grigo'eva, V. A. Dogiel, Sov. Phys. JETP **69**, 453 (1989)
217. S. I. Syrovatskii: Sov. Astron. **3**, 22 (1959)
218. P. Blasi, A. V. Olinto: Phys. Rev. D. **59**, 023001 (1999)
219. R. Lampard, R. W. Clay, B. R. Dawson: Astropart. Phys. **7**, 213 (1997)
220. G. A. Medina Tanco, E. M. de Gouveia Dal Pino, J. E. Horvath: Astropart. Phys. **6**, 337 (1997)
221. M. Lemoine, G. Sigl, A. V. Olinto, D. N. Schramm: Astrophys. J. **486**, L115 (1997)
222. G. A. Medina Tanco: Astrophys. J. **495**, L71 (1998)
223. J. D. Barrow, P. G. Ferreira, J. Silk: Phys. Rev. Lett. **78**, 3610 (1997)
224. A. Loeb, A. Kosowsky: Astrophys. J. **469**, 1 (1996)
225. K. Subramanian, J. D. Barrow: Phys. Rev. Lett. **81**, 3575 (1998)
226. G. Sigl, D. N. Schramm, S. Lee, C. T. Hill: Proc. Natl. Acad. Sci. USA **94**, 10501 (1997)
227. S. C. Corbató, et al.: Nucl. Phys. B (Proc. Suppl.) **28B**, 36 (1992)
D. J. Bird, et al., in Ref. [69], Vol. **2**, p.504; Vol. **1**, p.750
M. Al-Seady, et al., in Ref. [117], p. 191
see also http://hires.physics.utah.edu/

228. M. Teshima, et al.: Nucl. Phys. B (Proc. Suppl.) **28B**, 169 (1992)
M. Hayashida, et al.: in Ref. [117], p. 205
see also http://www-ta.icrr.u-tokyo.ac.jp/
229. A. Achterberg, Y. Gallant, C. A. Norman, D. B. Melrose: astro-ph/9907060, submitted to Mon. Not. R. Astron. Soc.
230. T. Stanev, et al.: Phys. Rev. D **62**, 093005 (2000)
231. T. Stanev, D. Seckel, R. Engel: astro-ph/0108338
232. R. H. Kraichnan: Phys. Fluids **8**, 1385 (1965)
233. D. G. Wentzel: Ann. Rev. Astron. Astrophys. **12**, 71 (1974)
234. see also V. S. Ptuskin, et al., Astron. Astrophys. **268**, 726 (1993)
235. F. Casse, M. Lemoine, G. Pelletier: astro-ph/0109223, to appear in Phys. Rev. D
236. G. R. Farrar, T. Piran: astro-ph/0010370
237. C. Isola, M. Lemoine, G. Sigl: astro-ph/0104289, to appear in Phys. Rev. D.
238. G. A. Medina Tanco: Astrophys. J. **510**, L91
G. A. Medina Tanco: astro-ph/9905239, to appear in Ref. [125].
239. S. L. Dubovsky, P. G. Tinyakov, I. I. Tkachev: Phys. Rev. Lett. **85**, 1154 (2000)
Z. Fodor, S. D. Katz: Phys. Rev. D **63**, 023002 (2001)
P. G. Tinyakov, I. I. Tkachev: Pis'ma Zh. Eksp. Teor. Fiz. **74**, 3 (2001) [JETP. Lett. **74**, 1 (2001)]
240. G. A. Medina Tanco: astro-ph/9809219, Proc. *Topics in cosmic-ray astrophysics*, ed. M. A. DuVernois (Nova Scientific, New York, 1999)
241. E. Waxman, K. B. Fisher, T. Piran: Astrophys. J. **483**, 1 (1997)
R. Ugoccioni, L. Teodoro, U. Wichoski: astro-ph/0011171
242. S. Coleman, S. L. Glashow: Phys. Lett. **B405**, 249 (1997)
S. Coleman, S. L. Glashow: Phys. Rev. D **59**, 116008 (1999)
243. L. Gonzalez-Mestres: Nucl. Phys. B (Proc. Suppl.) **48**, 131 (1996)
L. Gonzalez-Mestres: in Ref. [149], p. 148.
244. R. Cowsik, B. V. Sreekantan: Phys. Lett. **B449**, 219 (1999)
245. S. Coleman, S. L. Glashow: hep-ph/9808446
246. H. Sato, T. Tati: Prog. Theor. Phys. **47**, 1788 (1972)
247. D. A. Kirzhnits, V. A. Chechin: Sov. J. Nucl. Phys. **15**, 585 (1972)
248. see, e.g., R. Aloisio, P. Blasi, P. L. Ghia, A. F. Grillo: Phys. Rev. D **62**, 053010 (2000)
249. T. Kifune: Astrophys. J. **518**, L21 (1999)
250. G. Yu. Bogoslovsky, H. F. Goenner: Gen. Rel. Grav. **31**, 1565 (1999)
251. L. Gonzalez-Mestres: hep-ph/9905430, to appear in Ref. [125].
252. A. Halprin, H. B. Kim: Phys. Lett. **B469**, 78 (1999)
253. see J. Ellis, N. E. Mavromatos, D. V. Nanopoulos: Gen. Rel. Grav. **31**, 1257 (1999), and references therein.
254. J. A. Zweerink, et al. (WHIPPLE collaboration): Astrophys. J. **490**, L141 (1997)
255. F. Aharonian, et al. (HEGRA collaboration): Astron. Astrophys. **327**, L5 (1997)
256. W. Kluźniak: astro-ph/9905308, Proc. 8th *International Workshop on Neutrino Telescopes*, ed. Milla Baldo Ceolin (Venice, 1999)
W. Kluźniak: Astropart. Phys. **11**, 117 (1999)
257. see G. Amelino-Camelia, et al.: Nature **393**, 763 (1998), and references therein.
258. J. Ellis, K. Farakos, N. E. Mavromatos, V. Mitsou, D. V. Nanopoulos: Astrophys. J. **535**, 139 (2000)
259. V. A. Kostelecký: Proc. *Topics on Quantum Gravity, and Beyond*, eds. F. Mansouri, J. J. Scanio (World Scientific, Singapore, 1993)
260. R. Ehrlich: Phys. Rev. D **60**, 017302 (1999)
261. C. Caso, et al.: European Phys. J. C **3**, 1 (1998)
262. R. Ehrlich: Phys. Rev. D **60**, 073005 (1999)

Neutrino Cascades: The Byproducts of Propagation of Ultra-High-Energy Neutrinos

Shigeru Yoshida

Institute for Cosmic Ray Research, University of Tokyo, 5-1-5 Kashiwa-no-Ha, Kashiwa City, Chiba 277-8582, Japan

Abstract. Ultra high energy cosmic neutrinos may not propagate unattenuated in space over cosmological distances because they can interact with the relic cosmological neutrino background. This process initiates neutrino cascades which provide a possible signature of the background neutrinos and could give rise to the observed cosmic ray flux beyond the GZK cutoff. We discuss high energy neutrino propagation in space with special focus on generation and development of neutrino cascades. Analytical estimates as well as numerical investigations are presented to show how this byproduct of high energy neutrino propagation can cause various interesting phenomena in the Universe.

1 Why Neutrinos matter?

We have seen that some of the observational results on ultra high energy cosmic rays above 10^{18} eV (UHECRs) seem difficult to reconcile with the consequences of the GZK effect (see the introduction by P. Biermann and G. Sigl). At least the existence of super-GZK particles with energies beyond the GZK threshold energy requires explanation. Moreover, if the event clusters observed by the AGASA experiment are real [1], it may be necessary to introduce completely new ideas to circumvent the GZK effect and explain why there appear to be no nearby astronomical counterparts in the directions of the clusters. The most straightforward idea to circumvent the necessity of nearby sources of the observed UHECRs would be to consider cosmic particles which do NOT interact with the cosmic microwave background (CMB) photons. Among the known particles in our elementary particle catalog, the only candidate which is stable or very long lived and whose existence is firmly established is the neutrino. This is the motivation to study the behavior of ultra high energy (UHE) neutrinos.

Observations of UHE neutrinos would also provide unique informations on the Universe which we could not obtain otherwise because neutrinos can penetrate cosmological distances in the Universe and their trajectories are not deflected because they have no electric charge. They carry information about extremely high energy particle production processes, even in the early Universe.

In the ultra high energy range, the unique feature of cosmic neutrinos is their possible interactions with the relic neutrino background (RNB), which are relics from the Big Bang [2]. Under certain circumstances the Universe becomes opaque to UHE neutrinos due to the interactions. If the energies of UHE neutrinos are high enough, the collision energy can reach the mass of the Z_0 boson, and the cross section rapidly increases due to the resonance effect. The secondary

Martin Lemoine, Günter Sigl (Eds.): LNP 576, pp. 255–274, 2001.

products of muons, taus, and pions emit neutrinos through their decays and further contribute to the neutrino "cascade" [3].This cascade phenomenon could explain the UHECR observations and may even lead to a road to "search" for the RNB, which are firmly believed to exist but extremely difficult to detect. It is, therefore, very interesting to study the detailed behavior of propagation of UHE neutrinos in the RNB field which is at the heart of UHE neutrino astrophysics. In this review we discuss the UHE neutrino propagation in detail for both negligible-mass and massive neutrinos. We start our discussion with a brief introduction to the RNB as a relic of the early Universe. Then we discuss the interactions of UHE neutrinos during their propagation. We also focus on the interesting possibility that the secondary produced particles in the neutrino cascades supply the observed bulk of super-GZK particles. We use natural units in which $\hbar = c = k_{\rm B} = 1$ in this contribution unless otherwise noted.

2 The Cosmic Neutrino Background

In the Standard Hot Big Bang Model the early Universe was radiation dominated. The radiation was in thermal equilibrium in the hot universe. The fundamental laws that govern the evolution of the radiation-dominated Universe are the Einstein equation which couples the radiation energy density to the space-time topology, the energy-momentum conservation law, and entropy conservation. For our purposes, these laws read [4]:

$$\dot{H}(t)^2 + \frac{k}{R(t)^2} = \frac{8\pi G}{3}\rho_{\rm rad}(t), \tag{1}$$

$$H(t) \equiv \frac{\dot{R}(t)}{R(t)}, \tag{2}$$

$$\rho_{\rm rad}(t) = \rho_{\rm rad}(t_0)\left[\frac{R(t_0)}{R(t)}\right]^4 = \rho_0(1+z)^4, \tag{3}$$

$$S = \frac{4}{3}\frac{R(t)^3}{T}\rho_{\rm rad}(t) \propto \left[\frac{R(t)}{T}\right]^3 = \text{constant}. \tag{4}$$

Here $\rho_{\rm rad}$ is the total radiation energy density, z is the redshift value at a given epoch, S is the radiation entropy, and T is the equilibrium temperature. Furthermore, $H(t)$ is the expansion rate of the Universe at cosmic time t, t_0 is the age of the present Universe, $R(t)$ is the scale factor (radius) of the Universe, and G is Newton's cosntant.

Let us start our tour of the history of the Universe when electrons, photons, and neutrinos are in thermal equilibrium. The temperature was $\sim 10^{12}$ Kelvin which is below the muon's mass, but much above the mass of an electron. $e^{\pm}$ production and annihilation with photons and neutrinos occurred so frequently that electrons and neutrinos were a significant part of the thermal radiation. Thermal electrons and neutrinos follow the Fermi distribution while thermal photons follow the Bose distribution. Energy and number density of the thermal equilibrium particles are then given by the energy integral of these distributions,

$$
\begin{aligned}
\rho_\gamma &= \frac{\pi^2}{15(\hbar c)^3}(k_B T)^4 \equiv aT^4\,, \quad a = 7.56\times 10^{-10}\mathrm{erg\ cm^{-3}K^{-4}} \\
n_\gamma &= \frac{2\zeta(3)}{\pi^2}\left(\frac{k_B T}{\hbar c}\right)^3 \equiv 2bT^3\,, \quad b = 10^{13}\mathrm{cm^{-3}K^{-3}} \\
s_\gamma &= \frac{4}{3}\rho_\gamma = \frac{4}{3}aT^4 \\
\rho_e &= \frac{7}{8}aT^4\,, \quad n_e = \frac{3}{2}bT^3\,, \quad s_e = \frac{4}{3}\frac{7}{8}aT^3 = \frac{7}{6}aT^3 \\
\rho_\nu &= \frac{7}{16}aT^4\,, \quad n_\nu = \frac{3}{4}bT^3\,, \quad s_\nu = \frac{4}{3}\frac{7}{16}aT^4 = \frac{7}{12}aT^4
\end{aligned}
\tag{5}
$$

The quantities in the third line are for electrons only (with an equal contribution for positrons) and the neutrino densities refer to one flavor (with an equal contribution for anti-neutrinos), and s is the entropy density. The total energy density of relativistic particles ($e^{\pm}$,ν_e,ν_μ,ν_τ, $\bar{\nu_e}$,$\bar{\nu_\mu}$,$\bar{\nu_\tau}$,γ) is given by

$$
\rho_{\mathrm{rad}} = aT^4\left(1 + 2\times\frac{7}{8} + 6\times\frac{7}{16}\right) = \frac{11}{4}aT^4 + \frac{21}{8}aT^4, \tag{6}
$$

where the second term of the righthand side is the contribution from the neutrinos showing that their energy density is comparable with that of electrons and photons. The entropy of the radiation is given by Eq. (4),

$$
S_{\mathrm{rad}} = \frac{4}{3}\frac{R(t)^3}{T}\rho_{\mathrm{rad}} = \frac{11}{3}a\,[R(t)T]^3 + \frac{7}{2}a\,[R(t)T]^3\,. \tag{7}
$$

Collisions involving neutrinos occur only via weak interactions and their cross section is much smaller than for electromagnetic interactions. Thus as the Universe expands and the density of neutrinos decreases, the neutrino interaction probability becomes too small to keep them in thermal equilibrium. Neutrino decouple and fall out of thermal equilibrium eventually. Decoupling takes place when the interaction rate becomes smaller than the expansion rate of the Universe. The cross section of neutrino interactions is of order $G_{\mathrm{F}}^2 s$ where $G_{\mathrm{F}} = 1.166\times 10^{-5}$ GeV^{-2} is the Fermi coupling constant and s is the Lorentz invariant center of mass (CM) energy squared. Because s is of order $E_\nu^2 \sim T^2$, the total cross section is approximately given by

$$
\sigma_\nu \sim G_{\mathrm{F}}^2 s \sim G_{\mathrm{F}}^2 T^2 \sim 4\times 10^{-44}\left(\frac{T}{10^{10}\mathrm{K}}\right)^2 \mathrm{cm}^2\,. \tag{8}
$$

The number density of the neutrinos is given by Eq. (5) and the interaction mean free time is thus

$$
\tau_\nu \equiv (n_\nu c\sigma_\nu)^{-1} \simeq 50\left(\frac{T}{10^{10}K}\right)^{-5} \mathrm{sec}\,. \tag{9}
$$

The expansion rate of the Universe is given by $H(t)$ as expressed in Eq. (2). Its evolution is given by the Einstein equation, Eq. (1), yielding

$$H(t) \simeq \left[\frac{8\pi G}{3}\rho_{\rm rad}(t)\right]^{1/2} . \tag{10}$$

From Eq. (6),

$$H(t) \sim 5 \times 10^{-1} \left(\frac{T}{10^{10}{\rm K}}\right)^2 \sec^{-1} . \tag{11}$$

The neutrinos decouple when $H(t)\tau_\nu \gtrsim 1$ which leads to

$$\left(\frac{T}{10^{10}K}\right)^3 \lesssim 25 . \tag{12}$$

Thus the neutrino decoupling temperature is

$$T_{\rm dec} \simeq 3 \times 10^{10}\,{\rm K} . \tag{13}$$

From the entropy conservation Eq. (4), the equilibrium temperature must be inversely proportional to $R(t)$ and hence the redshift at neutrino decoupling is $(1 + z_{\rm dec}) \sim (T_{\rm dec}/2.7\,{\rm K})$ [1]$\sim 10^{10}$. Also, according to Eq. (11), neutrino decoupling took place about 1 sec after the Big Bang. Since then, the decoupled neutrinos have been "free" particles playing no role in the evolution of the Universe except for contributing to its energy density which has been decreasing as the Universe further expanded. Their temperature has been redshifted without any significant influence from interactions with other particles. These cold neutrinos now constitute a thermal relic cosmological background (RNB) with temperature T_ν. The neutrino temperature at present is related to $T_{\rm dec}$ by

$$R(t_{\rm dec})T_{\rm dec} = R_0 T_\nu , \tag{14}$$

as derived from Eq. (4). Here $R(t_{\rm dec})$ is the scale factor at decoupling and R_0 is the present scale factor.

The total (conserved) entropy of the electrons and photons in equilibrium is $(11/3)a\,(R(t)T)^3$ as expressed in Eq. (7). When the equilibrium temperature further cooled down to well below the electron mass $\sim 10^9$ K, electron-positron pair production could no longer occur and $e^\pm$ simply annihilated into photons. From the electromagnetically interacting particles only photons survived in the thermal radiation component of the Universe, since then forming the CMB photons in the present Universe. Because the photon entropy is given by $(4/3)a\,[R(t)T]^3$, as given in Eq. (5), whereas at temperatures above the electron mass the electromagnetic entropy is given by $(11/3)a\,[R(t)T]^3$, see Eq. (7), the entropy conservation law, Eq. (4), leads to

$$\frac{11}{3}\left[R(t_{\rm dec})T_{\rm dec}\right]^3 = \frac{4}{3}(R_0 T_0)^3, \tag{15}$$

[1] $(1 + z_{\rm dec}) \simeq (T_{\rm dec}/1.9\,{\rm K})$ is a more accurate estimate taking into account photon heating by $e^\pm$ annihilation as we will see in the next paragraph.

where $T_0 = 2.7$ K is the black body temperature of the CMB photons. From Eq. (14), the neutrino background temperature T_ν is obtained as

$$T_\nu = \left(\frac{4}{11}\right)^{\frac{1}{3}} T_0 \simeq 1.9\,\mathrm{K}\,. \tag{16}$$

The number density of the RNB per flavor is consequently, see Eq. (5),

$$n_\nu + n_{\bar{\nu}} = \frac{3}{2} b T_\nu^3 = \frac{6}{11} b T_0^3 \simeq 108\,\mathrm{cm}^{-3}\,. \tag{17}$$

The present total radiation energy density of the CMB photons and background neutrinos is calculated to be

$$\rho_{\mathrm{rad}} = aT_0^4 + 6 \times \frac{7}{16} aT_\nu^4 = aT^4 + \frac{21}{8}\left(\frac{4}{11}\right)^{\frac{4}{3}} aT^4 \simeq 1.68 aT_0^4\,. \tag{18}$$

The background neutrinos are still responsible for 40 % of the whole radiation energy density in the Universe. It should be remarked that taking into account this contribution, we can now estimate when the radiation density becomes equivalent to that of matter. We find

$$1 + z_{eq} \sim \rho_{\mathrm{matter}}/\rho_{\mathrm{rad}} \sim 2.4 \times 10^4 h^2. \tag{19}$$

Here the Hubble constant was written as $H_0 = 100h$ km sec^{-1}Mpc^{-1}.

3 The Neutrino Dark Matter

Invoking a small mass for the RNB would supply a hot dark matter component. It has been pointed out that its admixture with cold dark matter (CDM) matches better with the data of galaxy group properties such as the number density of clusters [5]. Since the recent neutrino oscillation measurement by the SuperKamiokande detector indeed indicates that neutrinos have masses [6], the RNB may be more likely to form a bulk of nonrelativistic massive neutrinos. The massive neutrino case has even more important ramifications for UHE cosmic neutrinos because their interactions with the massive RNB become especially significant if the relic neutrinos have masses m_ν in the eV range because the Z boson resonance then occurs in the UHE energy range relevant for the highest energy cosmic rays observed.

We can put an upper bound on RNB masses because their energy density is constrained by cosmological considerations if the neutrinos are stable [7]. The energy density of the nonrelativistic neutrinos with masses m_ν is given by

$$\rho_\nu = \sum_{i=\nu_e,\nu_\mu,\nu_\tau} m_\nu^i (n_\nu^i + n_{\bar{\nu}}^i)\,, \tag{20}$$

where the suffix i denotes the neutrino flavor. Comparing the cosmological critical density and the neutrino energy density above gives a constraint on the masses of stable neutrinos as

$$\sum_{i=\nu_e,\nu_\mu,\nu_\tau} m_\nu^i \lesssim 92\Omega_0 h^2\,\mathrm{eV}\,, \tag{21}$$

where Ω_0 is the total cosmological mass density normalized by the critical density of the Universe. Neither h nor Ω_0 is very well known. The age of the Universe t_0 can constrain the allowed region of h and Ω_0, however. The relation between t_0, H_0, and Ω_0 is given as follows:

$$H_0 t_0 = \begin{cases} -\frac{1}{\Omega_0-1} + \frac{\Omega_0}{(\Omega_0-1)^{3/2}} \tan^{-1}\sqrt{\Omega_0 - 1} & \Omega_0 > 1\,, \\ -\frac{1}{\Omega_0-1} - \frac{\Omega_0}{(1-\Omega_0)^{3/2}} \tanh^{-1}\sqrt{1-\Omega_0} & \Omega_0 < 1\,, \\ \frac{2}{3} & \Omega_0 = 1 \end{cases} \tag{22}$$

The age of the Universe can be estimated from measured ages of star clusters which is $\simeq 13 - 17$ Gyr [8]. Requiring $t_0 \gtrsim 13\,\mathrm{Gyr}$, we get from Eq. (22)

$$\Omega_0 h^2 \lesssim 0.4\,. \tag{23}$$

Then Eq. (21) gives the conservative bound

$$\sum_{i=\nu_e,\nu_\mu,\nu_\tau} m_\nu^i \lesssim 37\,\mathrm{eV}\,. \tag{24}$$

The scenario of neutrino dark matter with m_ν in the eV range is not excluded by this limit, nor is it by other limits. Parametrizing all three neutrino masses by m_ν for simplicity, we have

$$\Omega_\nu = 0.03 \left(\frac{m_\nu}{\mathrm{eV}}\right) h^{-2} \tag{25}$$

for the ratio of the massive neutrino density to the cosmological critical density Ω_ν. The cosmological neutrino dark matter with $\sim$ eV masses would thus not be a majority of the total mass in the Universe.

4 Neutrino Cascades: *Massless Neutrinos*

The possibility of producing UHE particles by the annihilation or collapse of topological defects (TDs) such as monopoles, cosmic strings, etc., has been proposed already in the 1980s [9] (see Ref. [10] for a recent detailed review). The maximum energy of the particles such as neutrinos produced by TDs can reach the typical GUT energy scale $\simeq 10^{16}$ GeV. If cosmic neutrino energies indeed exceed 10^{14} GeV even at high redshifts of $z \gtrsim 100$, as expected in the TD scenario, significant interactions of these superhigh energy neutrinos with the RNB should occur because the cross section rapidly increases due to the Z_0 boson resonance.

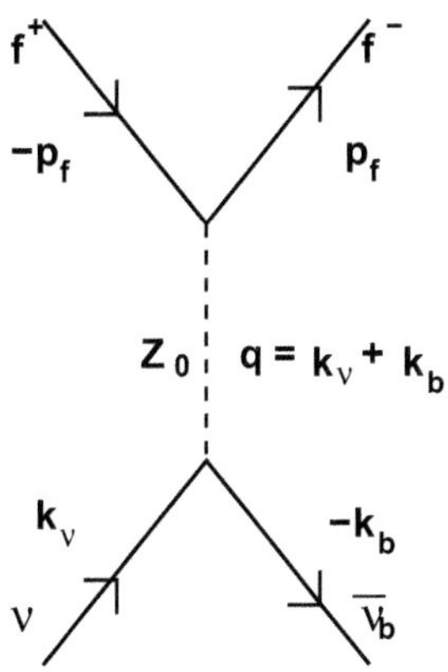

Fig. 1. A diagram that contributes to the neutrino interactions via the Z_0 exchange in the s-channel.

The most important interaction channel for UHE neutrinos is the coupling of neutrinos and anti-neutrinos through Z_0 in the s-channel, which has resonant behavior. Schematically,

$$\begin{aligned}
&\nu_e\bar{\nu}_e \rightarrow e^+e^-, \quad \mu^+\mu^-, \quad \tau^+\tau^-, \quad \nu\bar{\nu}, \quad q\bar{q} \\
&\nu_\mu\bar{\nu}_\mu \rightarrow .... \\
&\nu_\tau\bar{\nu}_\tau \rightarrow ....
\end{aligned}$$

The Feynman diagram of this channel is shown in Fig. 1. In this channel an UHE neutrino with energy E_ν couples a cosmological background anti-neutrino with energy k_b via an exchange of a Z_0 boson producing a fermion pair $f\bar{f}$. The differential cross section is calculated to be

$$\frac{d\sigma}{d\cos\theta^*} = \frac{G_F^2 s}{4\pi}\frac{M_z^4}{(s-M_z^2)^2+M_z^2\Gamma_z^2}\left[g_L^2(1+\cos\theta^*)^2+g_R^2(1-\cos\theta^*)^2\right], \tag{26}$$

where θ^* is the rotation angle of the collision in the CM system as schematically illustrated in Fig. 3. Furthermore, M_z is the mass of the Z_0, Γ_z is the decay width of the Z_0, and g_L and g_R are the left-handed and right-handed coupling constants, respectively. The squared CM energy s can be written in the UHECR lab system as

$$s = 2E_\nu k_b(1+\cos\theta), \tag{27}$$

where θ is the collision angle between the RNB and UHE neutrino in the lab system.

The t-channel $W^\pm$ exchange also produces leptons. Its Feynman diagram is shown in Fig. 2. This interaction is the most important channel for the coupling of neutrinos with different flavors and becomes significant at energies beyond and below the Z_0-peak.

Muons and tauons produced by these processes emit neutrinos through their decays. Quarks produced by s-channel Z_0 exchange fragment and produce jets

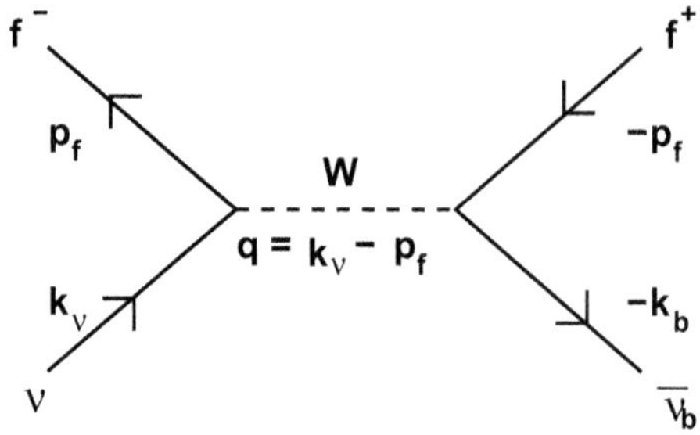

Fig. 2. A diagram that contributes to the neutrino interactions via the $W^{\pm}$ exchange in the t-channel.

of hadrons. Most of the hadrons in the jets are pions which emit secondary UHE neutrinos. The neutrino cascade hence develops during UHE neutrino propagation. The profile of the cascade is calculated by the appropriate transport equations in a manner analogous to the approach presented in Sect. 8.1 of the contribution by G. Sigl in this volume. Adapted to the case of neutrinos these equations describe the rate of change of the neutrino number density per energy $N_i(E_{\nu,i}, z)$ at redshift z and read [3]

$$\frac{dN_i}{dL}(E_{\nu,i}, z) = \sum_{j=\nu_e,\nu_\mu,\nu_\tau} \int_{E_{\nu,i}}^{E_{\nu,i}^{max}} dE'_{\nu,j} N_j(E'_{\nu,j}, z) \frac{dN_{j\to i}(E'_{\nu,j}, E_{\nu,i}, z)}{dE_{\nu,i} dL} - \frac{N_i(E_{\nu,i}, z)}{\lambda_i(E_{\nu,i}, z)} + \frac{\partial}{\partial E_{\nu,i}} \left[H_0 (1+z)^{\frac{3}{2}} E_{\nu,i} N_i(E_{\nu,i}, z) \right] , \quad (28)$$

for $i = \nu_e, \nu_\mu, \nu_\tau$. Here λ_i is the interaction length of a neutrino of flavor i, and $dN_{j\to i}/dE_{\nu,i}dL$ is the number of neutrinos with flavor i produced from a primary neutrino with flavor j per unit length per unit energy of the produced. The collision term is given by

$$\frac{dN_{j\to i}}{dE_{\nu,i} dL} = \int dE_{\rm rec} \frac{dn}{dE_{\nu,i}}(E_{\nu,i}, E_{\rm rec}) \int ds \frac{d\sigma}{dE_{\rm rec}}(s) \frac{1}{4\pi} \int d\Omega (1+\cos\theta) \frac{dn_{\nu_b}}{dE_{\nu_b}} \frac{dE_{\nu_b}}{ds} . \quad (29)$$

Here ν_b denotes the background neutrino, and $E_{\rm rec}$ is its recoil energy, i.e. the energy transfered by the primary neutrino in the collision with the RNB. The RNB energy distribution dn_{ν_b}/dE_{ν_b} is given by the Fermi distribution. Because the black-body temperature of the RNB changes with redshift as Eq. (14), the dependence on the cosmological evolution scales as

$$\frac{dN_{j\to i}}{dE_{\nu,i} dL}(E_{\nu,i}, E_{\nu,j}, z) = (1+z)^2 \frac{dN_{j\to i}}{dE_{\nu,i} dL}(E_{\nu,i}/(1+z), E_{\nu,j}/(1+z), 0) . \quad (30)$$

This scaling implies that collisions occur more frequently and with a lower threshold energy at high redshifts. From Eq. (27), we find that the Z bosons are resonantly produced by neutrinos of energy

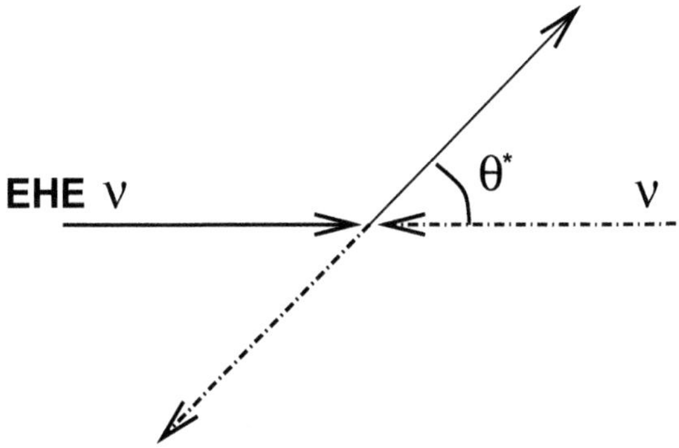

Fig. 3. Collision of an UHE neutrino and a neutrino of the RNB in the CM system.

$$E_{\rm res} \simeq 4 \times 10^{25}(1+z)^{-1}\left(\frac{k_b}{10^{-4}\,{\rm eV}}\right)\,{\rm eV} \tag{31}$$

which thus scales as $(1+z)^{-1}$.

In Eq. (29) $dn/dE_{\nu,i}$ is the energy distribution of the secondary neutrinos produced by the interaction. For elastic collisions, these will be delta-functions, and for the produced muons and taus, they can be derived from the decay matrix elements. When quarks are produced, they fragment ("hadronize") and produce jets of hadrons which emit neutrinos through their decays, so the distribution $dn/dE_{\nu,i}$ for the hadrons can be obtained by convolution of the hadronic fragmentation spectrum with the parton decay spectrum. The differential cross section $d\sigma/dE^{\rm rec}$ can be obtained from the corresponding quantity $d\sigma/d\cos\theta^*$ in the CM system. We have

$$E^{\rm rec} = \frac{E_{\nu,i}}{2}(1+\cos\theta^*), \tag{32}$$

The collision in the CM system is illustrated in Fig. 3. It is convenient to express the equations above in dimensionless energies,

$$\frac{dN_{j\to i}}{d\eta_{\nu,i}dL} = \int\limits_{\eta_{\nu,i}}^{1} \frac{d\eta_{\rm rec}}{\eta_{\rm rec}} \frac{dn}{d(\frac{\eta_{\nu,i}}{\eta_{\rm rec}})} \int ds \frac{d\sigma}{d\eta_{\rm rec}}(s)\frac{1}{4\pi}\int d\Omega(1+\cos\theta)\frac{dn_{\nu_b}}{ds}\,, \tag{33}$$

where the dimensionless $\eta \equiv E/E_{\nu,j}$ is normalized by the primary neutrino energy $E_{\nu,j}$. As seen in Eq. (32), $\eta_{\rm rec}$ is equal to $(1+\cos\theta^*)/2$. For example, the contribution to ν_μ production via pion decay in the hadronic jets generated by $\nu\nu$ collisions by s-channel Z_0 exchange is given by

$$\frac{dN_{j\to\mu}}{d\eta_{\nu_\mu}dL} = \int\limits_{\frac{\eta_{\nu_\mu}}{(1-r_\pi)}}^{1} \frac{dx}{x}\frac{1}{1-r_\pi}\int\limits_{x}^{1}\frac{d\eta_{\rm rec}}{\eta_{\rm rec}}\frac{dn_h}{d(\frac{x}{\eta_{\rm rec}})}\int ds\frac{d\sigma}{d\eta_{\rm rec}}(s)\frac{1}{4\pi}\int d\Omega(1+\cos\theta)\frac{dn_{\nu_b}}{ds}, \tag{34}$$

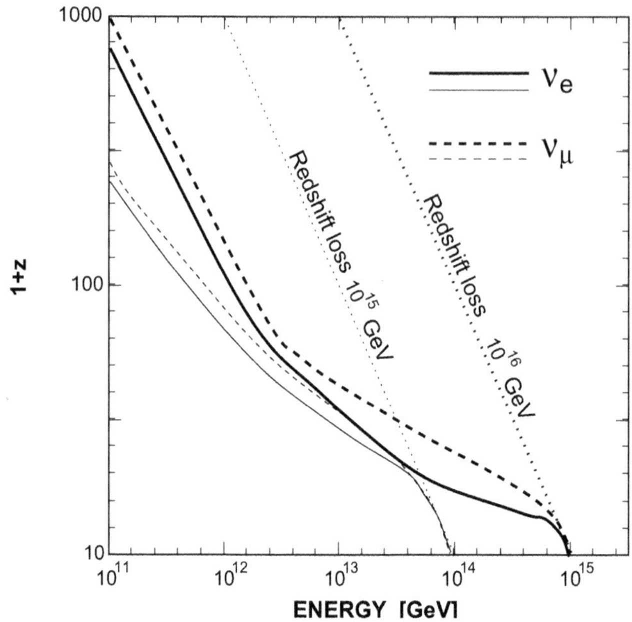

Fig. 4. The neutrino horizon of the Universe as a function of present-day neutrino energies [3]. The primary energies are 10^{15} GeV (thin lines) and 10^{16} GeV (thick lines). The solid lines show electron-neutrinos and the dashed lines show muon-neutrinos. The dotted lines show the upper bounds of the horizon taking into account only the energy loss due to redshift.

where $r_\pi = m_\mu^2/M_\pi^2$ and dn_h/dy is the hadronic fragmentation spectrum with $y \equiv E_\pi/E_{\rm jet}$.

To illustrate the effect of the neutrino cascade on the propagation of UHE neutrinos, let us determine the maximum redshift up to which neutrinos are not attenuated in their propagation (referred to as the "neutrino horizon of the Universe"). For a monochromatic primary energy spectrum at a given epoch z_a represented by $dN_i/dE_{\nu,i}(z_a) = N_0\delta(E_{\nu,i} - E_0)$, the present-day energy distribution of neutrinos after propagation, $dN_i/dE_{\nu,i}(0)$, is calculated by the transport equation Eq. (28). The effective cutoff energy k is defined by

$$\int_k^{E_0} dE_{\nu,i} \frac{dN_i}{dE_{\nu,i}}(0) = \frac{1}{e} N_0 \,. \tag{35}$$

We consider the redshift z_a as the horizon of the Universe for neutrinos with present-day energy k for a primary energy E_0. Fig. 4 shows the curves of the horizon as a function of the present-day neutrino energies. The primary energies are 10^{15} GeV and 10^{16} GeV, respectively. The upper bounds of the horizon determined from the redshift energy loss only are also shown for comparison. It is found that the energy loss effect due to the interactions with the RNB contracts the horizon. This means that the distant sources at very high redshift epoch do not contribute to the bulk of UHE neutrinos. These UHE neutrinos are *absorbed.* However, the bulk of the secondary neutrinos moderate the absorption effect for the propagation from a high redshift epoch when the neutrino cascade develops significantly as expected from the scaling law Eq. (30). As seen in Fig. 4, the cascading effect expands the horizon for neutrinos below 10^{12} GeV compared with that obtained by extrapolation from the higher energy region in which the cascade does not develop significantly during propagation. Twice as many muon neutrinos are produced as electron neutrinos by the pion decay processes, and the cascade expands the horizon for muon neutrinos further. The expansion of the horizon due to the cascade leads to an enhancement of the superhigh energy neutrino flux.

5 Neutrino Cascades: *Massive Neutrinos*

If neutrinos are massive and the RNB is non-relativistic, the cascading is initiated at much lower energies because the squared CM energy s is now given by

$$s \simeq 2E_\nu m_\nu \,, \tag{36}$$

and the hadronic decay of Z bosons resonantly produced occurs with neutrinos of energy

$$E_{\rm res} = M_z^2/2m_\nu = 4\times 10^{21}\left(\frac{m_\nu}{1\mathrm{eV}}\right)^{-1}\ \mathrm{eV}, \tag{37}$$

which, for neutrino masses in the eV range, is four orders of magnitude lower than for the massless case expressed by Eq. (31). Thus cascading would be more likely to take place.

Let us see how the energy spectrum of UHE neutrinos would be modified after their propagation in the massive RNB field. The collision term of the transport equation, Eq. (33) simplifies in case of massive neutrinos because the RNB has now a non-relativistic approximately monochromatic energy distribution. We find [11]

$$\frac{dN_{j\to i}}{d\eta_{\nu,i}dL} = (1+z)^3 \int_{\eta_{\nu,i}}^{1} \frac{d\eta_{\rm rec}}{\eta_{\rm rec}} \frac{dn}{d(\frac{\eta_{\nu,i}}{\eta_{\rm rec}})} n_{\nu_b} \left.\frac{d\sigma}{d\eta_{\rm rec}}\right|_{s=2m_\nu E_\nu}, \tag{38}$$

where the number density of the RNB, n_{ν_b}, is given by Eq. (17), $\eta_{\rm rec} \equiv E_{\rm rec}/E_{\nu,j}$, and $\eta_{\nu,i} \equiv E_{\nu,i}/E_{\nu,j}$. The mean collision free path is obtained to be

$$\lambda(E_\nu, z) \simeq (1+z)^{-3}\left[n_0\sigma|_{s=2m_\nu E_\nu}\right]^{-1}, \tag{39}$$

and is shown in Fig. 5. It exhibits a clear resonance structure around $E_{\rm res}$. Numerical integration of the transport equation then shows the energy distribution of UHE neutrinos after propagation. The simple case when a single source emits UHE neutrinos with a E^{-2} spectrum is shown in Fig. 6. The sharp dip appears because of the Z_0 resonance behavior, as described above. We also find a slight

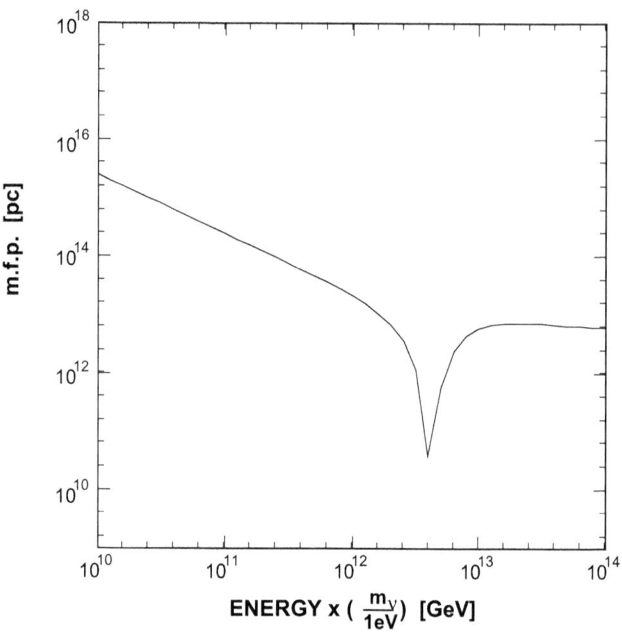

Fig. 5. Interaction length of UHE neutrinos in the massive RNB field at zero redshift.

enhancement of the flux in the energy region below the dip. This enhancement created by the bulk of the secondary neutrinos is drastically increased when UHE neutrinos are emitted at very high redshift epochs because of the $(1+z)^3$ dependence of the interaction length as expressed in Eq. (39). Figure 7 shows the UHE neutrino spectrum in a TD scenario when the topological defects distributed up to $z \gtrsim 1000$ decay into UHE neutrinos exclusively [12]. It is found that the spectrum is strongly modified below E$= 3 \times 10^{19}$ eV due to the secondary neutrinos produced in the cascades. This case provides the most prominent signatures of the massive background neutrinos via UHE neutrino fluxes, which is one of the "visible" effects of the existence of the cosmological relic neutrinos.

6 The Neutrino Cascades and Super-GZK Particles

We have seen that collisions of UHE cosmic neutrinos with the RNB could initiate neutrino cascades. It should be remarked that the cascades also produce electrons and photons via pion and muon decay processes, and via the direct production of leptons by $\nu\nu$ collisions. Moreover, the hadronic jets contain a small fraction of nucleons. These nucleonic and electromagnetic particles may constitute the observed population of UHECRs with energies beyond the GZK cutoff. This is called the "Z-burst" scenario [13], the attempt to explain the observation without invoking new physics beyond the standard model except neutrino mass. The energy distribution of the produced secondary nucleons, photons, and

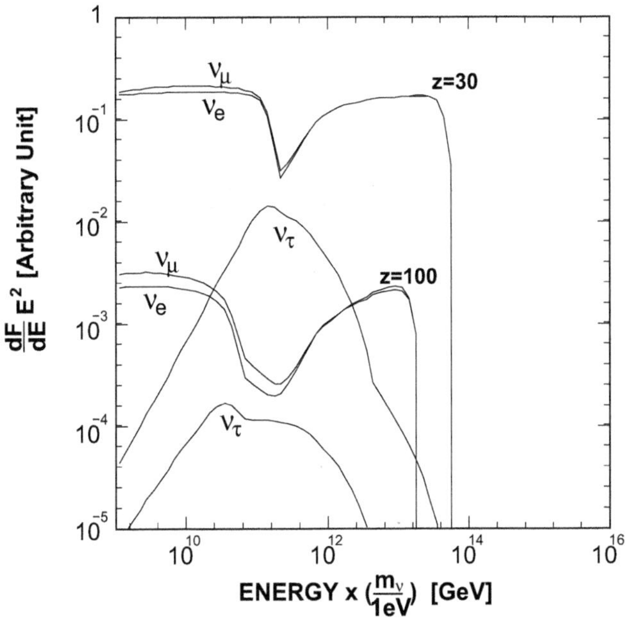

Fig. 6. The energy spectra of UHE neutrinos propagating in the massive cosmological background neutrino field. The primary spectra are assumed to be $\propto E^{-2}$. All ν_τ shown here are secondaries produced in the neutrino cascade.

electrons can be calculated from an expanded version of the transport equation, Eq. (28) which includes propagation of the nucleonic and electromagnetic components [14] (see also Sect. 8.1 of the contribution by G. Sigl in this volume). Figure 8 shows energy distributions of particles in the neutrino cascade after 1 Mpc propagation of an electron neutrino of original energy $E = E_{\mathrm{res}}$. The ν_μ, ν_τ, e, γ, and nucleons are produced as secondaries in the cascades. These distributions were determined quite accurately because all the interactions involved in the cascades occur in the well-measured LEP[2] energy range. For example, the proton energy distribution drawn in the figure was in fact measured with great accuracy by studying e^+e^- collisions at the Z boson resonance energy by several experiments at CERN. Thus the calculation is quite solid in terms of particle physics. The produced protons, electrons, and photons are subject to the GZK effect and lose energy forming a "low" energy population of cosmic rays below the cutoff. The super-GZK component can be generated by the particle production in the cascade which occurs in our neighborhood where the propagation distances are too short for the particles to lose a significant fraction of their energies by the GZK mechanism. An illustration of this "Z-burst" mechanism is

[2] The Large Electron Positron Collider at CERN. The world largest $e^\pm$ collider began its operation in the summer of 1989 and for six years the collision energy of the electrons and positrons was tuned exactly to the value needed to produce the Z_0.

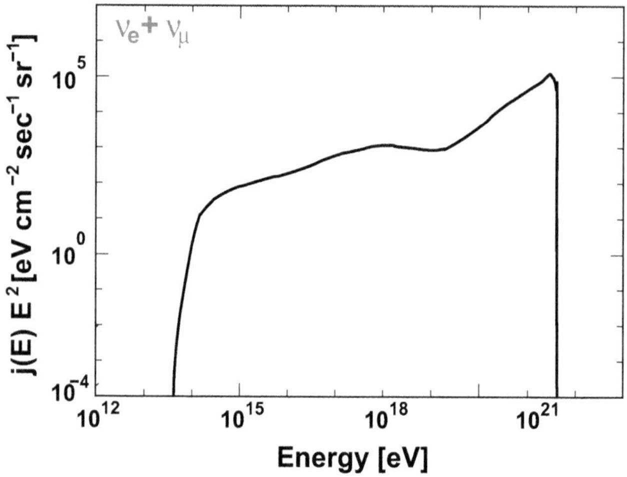

Fig. 7. Neutrino energy spectra in a top-down scenario where topological defects emit particles of mass $\simeq 10^{13}$ GeV which exclusively decay into electron or tau neutrinos. The normalization is only indicative (see Ref. [12] for more details.

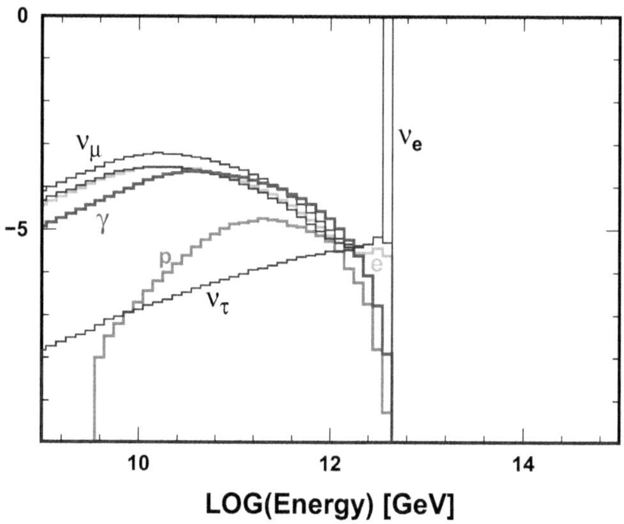

Fig. 8. Energy distribution of particles in the neutrino cascade after propagation of 1 Mpc. Primary input spectrum is monochromatic energy distribution of $E_{\rm res}$ electron neutrinos.

shown in Fig. 9. In fact, nearby particle production may be higher than average because the massive background neutrinos are expected to cluster by gravitational interactions. The Fermi distribution with a velocity dispersion v limits the overdensity to

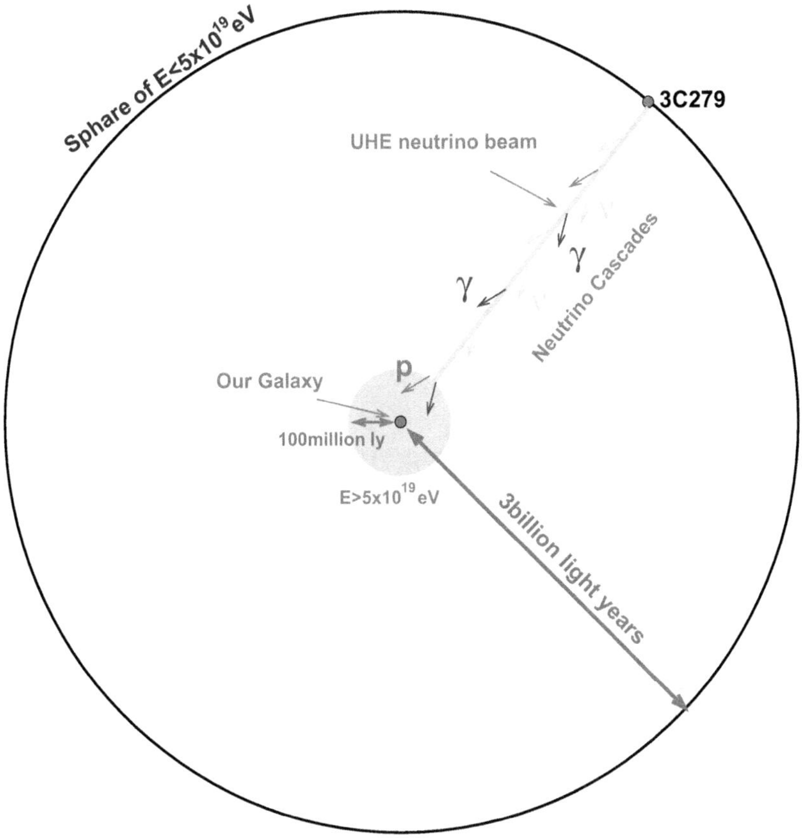

Fig. 9. Sketch of the "Z-burst" scenario (see text for details)

$$f_\nu \lesssim \frac{v^3 m_\nu^3}{(2\pi)^{3/2}\bar{n_\nu}} \simeq 330 \left(\frac{v}{500\,\mathrm{km\,sec^{-1}}}\right)^3 \left(\frac{m_\nu}{\mathrm{eV}}\right)^3 \tag{40}$$

over the uniform RNB. If clustering occurs on a scale l_ν comparable to the local Supercluster (~ 5 Mpc), nucleons and $\gamma-$rays around 10^{20} eV produced in the neutrino cluster can make significant contributions to the UHECR flux. The required intensity of primary UHE neutrinos to account for the observed super GZK particle flux can be approximately estimated as follows: Based on Eq. (38) we find that the number of produced $\gamma-$rays per unit length per UHE neutrino is roughly given by

$$\frac{dN_{\nu\to\gamma}}{d\eta_\gamma dL} = \frac{4}{3}\int\limits_{\eta_\gamma}^{1}\frac{dy}{y}\int\limits_{y}^{1}\frac{d\eta_{\mathrm{rec}}}{\eta_{\mathrm{rec}}}\frac{dn_h}{d\left(\frac{y}{\eta_{\mathrm{rec}}}\right)} n_{\nu_b}\left.\frac{d\sigma}{d\eta_{\mathrm{rec}}}\right|_{s=2m_\nu E_\nu}, \tag{41}$$

where $y \equiv E_\pi/E_\nu$, $\eta_{\mathrm{rec}} \equiv E_q/E_\nu$, $\eta_\gamma \equiv E_\gamma/E_\nu$. The hadron fragmentation spectrum dn_h/dx can be approximated by the power law form $\sim ax^{-\delta}$ ($\delta \sim$

1.4) in the range of interest here. The factor of 4/3 approximately accounts for contributions of both γ's from π^0 and $e^{\pm}$ from $\pi^{\pm}$. The cross section of the s channel Z_0 exchange is given by Eq. (26). We find

$$\begin{aligned}\frac{d\sigma}{d\eta_{\rm rec}} &= \frac{2G_{\rm F}^2}{\pi} M_z^4 \frac{s}{(s-M_z^2)^2 + M_z^2\Gamma_z^2}\left[g_L^2\eta_{\rm rec}^2 + g_R^2(1-\eta_{\rm rec})^2\right] \\ &\equiv \frac{2G_{\rm F}^2}{\pi} M_z^2 F(s)\left[g_L^2\eta_{\rm rec}^2 + g_R^2(1-\eta_{\rm rec})^2\right] \end{aligned} \tag{42}$$

$$\sigma = \frac{2G_{\rm F}^2}{3\pi} M_z^2 F(s)(g_L^2 + g_R^2) = \sigma_{\rm res}\Gamma_z^2 F(s)\,. \tag{43}$$

As g_L is much larger than g_R for hadronic Z_0 decay, all the integrations in Eq. (41) can be approximated to give

$$\frac{dN_{\nu\to\gamma}}{d\eta_\gamma dL} = \frac{4}{\delta(\delta+2)} n_{\nu_b}\sigma|_{s=2m_\nu E_\nu} a\eta_\gamma^{-\delta}\,. \tag{44}$$

Then the number of photons per unit length is obtained as

$$\frac{dN_\gamma}{dE_\gamma dL} = \int dE_\nu \frac{dN_{\nu\to\gamma}}{d\eta_\gamma dL}\frac{dN_\nu}{dE_\nu}\frac{1}{E_\nu}\,, \tag{45}$$

where dN_ν/dE_ν is the primary energy spectrum of UHE neutrinos which follows the power law $\sim E_\nu^{-\gamma}$. Using the fact that the power law index of the fragmentation spectrum δ is close to that of the UHE neutrino spectrum γ, we finally obtain

$$\frac{dN_\gamma}{dE_\gamma dL} \simeq \frac{4}{\delta(\delta+2)}\sqrt{2}\pi\frac{\Gamma_z}{M_z}\frac{1}{\lambda_z}\left.\frac{dN_\nu}{dE_\nu}\right|_{E_\nu=E_{\rm res}}\cdot\left.\frac{dn_h}{dx}\right|_{x=E_\gamma/E_{\rm res}}\,, \tag{46}$$

where $\lambda_z = (n_{\nu_b}\sigma_{\rm res})^{-1}$ is the interaction length of UHE neutrinos at the Z boson resonance energy $E_{\rm res}$. Then the UHE γ−ray flux produced within the neutrino dark matter cluster over the distance scale l_ν is obtained by

$$\frac{dN_\gamma}{dE_\gamma} \simeq \frac{4}{\delta(\delta+2)}\sqrt{2}\pi\frac{\Gamma_z}{M_z}\frac{f_\nu l_\nu}{\lambda_z}\left.\frac{dN_\nu}{dE_\nu}\right|_{E_\nu=E_{\rm res}}\cdot\left.\frac{dn_h}{dx}\right|_{x=E_\gamma/E_{\rm res}}\,. \tag{47}$$

If we approximate dn_h/dx by the Hill formula [15]

$$\frac{dn_h}{dx} \sim 0.3\frac{15}{16}x^{-1.5}(1-x)^2\,, \tag{48}$$

we get

$$\left.\frac{dN_\gamma}{dE_\gamma}\right|_{E_\gamma=10^{20}\rm eV} = 0.26\left(\frac{f_\nu}{300}\right)\left(\frac{l_\nu}{5\rm Mpc}\right)\left.\frac{dN_\nu}{dE_\nu}\right|_{E_\nu=E_{\rm res}=4\times10^{21}\rm eV} \tag{49}$$

if $m_\nu \simeq 1$ eV. The produced photon intensity should be comparable to the observed UHECR flux given by

$$E_\gamma^2\left.\frac{dN_\gamma}{dE_\gamma}\right|_{E_\gamma=10^{20}\rm eV} \simeq 1\rm eV\ cm^{-2}sr^{-1}sec^{-1}\,. \tag{50}$$

Then we obtain that the required neutrino energy intensity at $E_{\rm res}$ is

$$E_{\rm res}^2 \left.\frac{dN_\nu}{dE_{\rm res}}\right|_{E_{\rm res}=4\times10^{21}{\rm eV}} \simeq 6\times10^3\left(\frac{f_\nu}{300}\right)^{-1}\left(\frac{l_\nu}{5{\rm Mpc}}\right)^{-1} {\rm eV\ cm^{-2}sr^{-1}sec^{-1}}. \tag{51}$$

An observational upper bound on the UHE neutrino flux was obtained by the Fly's Eye measurement around 10^{18} eV. Recent observations have slightly improved this limit and have extended it up to a few 10^{21} eV [16]. Roughly, this limit can be written as

$$E_\nu^2\frac{dN_\nu}{dE_\nu} \lesssim 3\times10^4 {\rm eV\ cm^{-2}sr^{-1}sec^{-1}} \quad {\rm for}\ 10^{17}\,{\rm eV} \lesssim E_\nu \lesssim 10^{21}\,{\rm eV}\,. \tag{52}$$

Future experiments will have considerably higher sensitivities, see Fig. 7 in the contribution by G. Sigl in this volume.

Consistency of the required intensity of UHE neutrinos with existing upper limits therefore requires an overdensity $f_\nu \gtrsim 60$ over several Mpc. This is at least consistent with the phase space bound Eq. (40) as long as $m_\nu \gtrsim 1\,$eV.

It is also necessary to compute the differential spectrum of the secondary particles to see whether they are consistent with the observed UHECR spectrum and the diffuse $\gamma-$ray flux measured at lower energies. Thus solving the transport equations is inevitable for an accurate evaluation of the consequences of this scenario. In Fig. 10 we show numerically calculated spectra for a typical case where a homogeneous distribution of sources radiating UHE neutrinos with a constant differential spectrum $\propto E^{-1}$ (assuming flavor ratios resulting from pion decay in the absence of neutrino mixing) and a luminosity per comoving volume that scales as $(1+z)^3$ between $z=0$ and $z=3$ [14]. And an intermediate strength for the (poorly known) universal radio background relevant for UHE $\gamma-$ray interactions and an extragalactic magnetic field of 10^{-9} Gauss were assumed. The neutrino parameters were chosen as $m_{\nu_e}=m_{\nu_\mu}=m_{\nu_\tau}=1$ eV, $f_\nu\simeq20$, and $l_\nu=5$ Mpc. It can be seen that for this parameter combination the predicted fluxes are consistent with the measurement of the diffuse $\gamma-$ray flux by EGRET in the GeV range, and with upper limits on neutrino fluxes, Eq. (52), the latter in accord with the analytical estimate given above in Eqs. (51) and (52). This also shows that the UHE part of the secondary $\gamma-$rays and protons possibly constitute a hard component of the observed UHECRs without a pronounced GZK cutoff.

The electromagnetic component produced in the neutrino cascades at cosmological distances forms a diffuse $\gamma-$ray flux in the MeV-GeV region as their energies are recycled down due to the interactions with the CMB photons and synchrotron cooling (for a detailed discussion of electromagnetic cascades see the contribution by G. Sigl in this volume). The ratio of this diffuse component to the UHECR component produced in the local neutrino overdense region is indirectly proportional to the product of the overdensity f_ν and the length scale of neutrino clustering l_ν. This follows from the fact that the diffuse GeV $\gamma-$ray flux is insensitive to neutrino clustering whereas the produced UHECR flux is

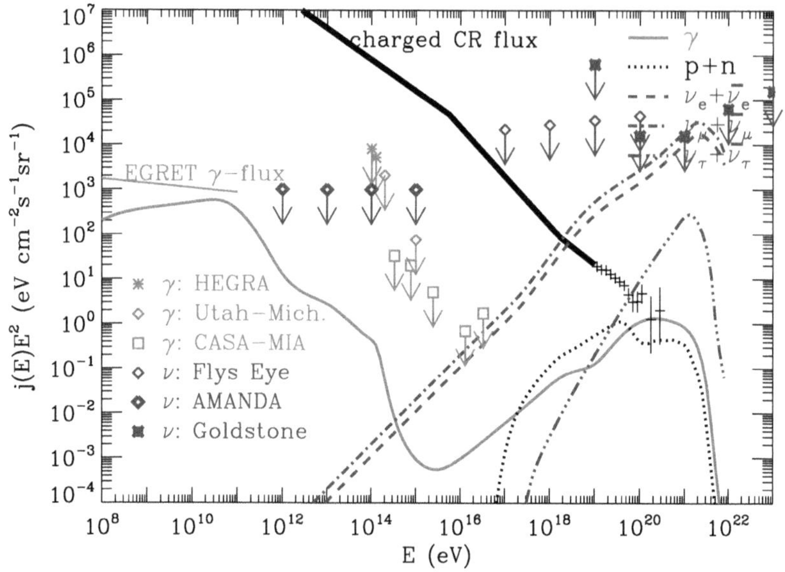

Fig. 10. Energy spectra of nucleons, $\gamma-$rays and neutrinos for the scenario described in the text. 1 sigma error bars are the combined data from the Haverah Park [17], Fly's Eye [18] and the AGASA [19] experiments above 10^{19} eV. Also shown are piecewise power law fits to the observed charged CR flux below 10^{19} eV, the EGRET measurement of the diffuse $\gamma-$ray flux between 30 MeV and 100 GeV [20]. Experimental upper limits on neutrino fluxes are shown from the Fly's Eye experiment [21], the Goldstone radio telescope [16], and the AMANDA neutrino telescope [22]. Upper limits on the $\gamma-$ray flux below 10^{17} eV and on neutrino fluxes above 10^{17} eV (except for AMANDA) from various experiments are as indicated (see Ref. [10] for more details)

given by Eq. (47). Consequently the EGRET bound in the GeV region leads to a lower bound of $f_\nu \gtrsim 20$ [14]. Otherwise the secondary produced electromagnetic component would give rise to a diffuse $\gamma-$ray flux that is higher than the EGRET limit.

The results obtained lead to the following relevant points in the scenario where the cascades initiated by UHE cosmic neutrino beams are at the origin of UHECRs in the highest energy region:

- Neutrino cascades could contribute to the observed cosmic ray flux above 3×10^{19} eV, regardless of the nature of the neutrino sources if the maximum neutrino energy reaches to the Z_0 boson pole region and the massive background neutrinos are clustered on the Supercluster scale. Neutrino emitters at cosmological distances can be responsible for the UHECR flux we are observing. Thus this mechanism is indeed a loophole of the GZK mechanism.

- Super-GZK particles should consist of protons and $\gamma-$rays according to this scenario.
- The intensity of the diffuse MeV-GeV photon background produced by the neutrino cascade is below the EGRET bound as long as the background neutrinos are sufficiently strongly clustered in our local neighborhood, satisfying $(f_\nu/20)(l_\nu/5\,\mathrm{Mpc}) \gtrsim 1$. A similar bound follows from the requirement that the required primary neutrino flux be smaller than existing upper limits, Eq. (52). Figure 10 demonstrates this limiting case.
- If the UHE neutrino emitters involved in the model radiate not only neutrinos but also $\gamma-$rays with comparable power, this bound strengthens to $(f_\nu/10^3)(l_\nu/5\,\mathrm{Mpc}) \gtrsim 1$. This requires neutrino masses in the eV range in order to satisfy the phase space bound Eq. (40).
- How to generate UHE neutrinos with $E \sim E_{\mathrm{res}}$ remains a problem. If these neutrinos are produced from the conventional mechanism invoking photopion production by accelerated protons, the protons must be accelerated up to $\sim 20E_{\mathrm{res}} \sim 10^{23}(m_\nu/1\mathrm{eV})^{-1}$ eV which seems difficult to realize in the usual astronomical environments (see contribution by G. Pelletier in this volume). Furthermore, the sources would need a dense photon target to absorb these protons and to produce a sufficiently high neutrino luminosity. The optical depth for protons in the sources must be high also because otherwise the observed nucleon flux below the GZK cutoff would be comparable to the neutrino flux, and thus inconsistent with the observed cosmic ray flux in the ankle region (see Fig. 10).
- The required energy luminosity of UHE neutrinos in this model is

$$L_\nu \simeq \frac{4\pi}{c} H_0 I_\nu = 2.4 10^{46} h \left(\frac{I_\nu}{1.2 10^4 \ \mathrm{eV/cm^2 sec\ sr}}\right)\left(\frac{f_\nu}{300}\right)^{-1} \ \mathrm{erg/Mpc^3 yr}. \tag{53}$$

The required primary proton luminosity to produce UHE neutrinos via photopion production at the source is thus $L_p \sim 10^{47}$ erg/Mpc^{3}yr. This is about three orders of magnitude lower than the total luminosity of the Universe, but extremely high. Among the existing astronomical objects, the Gamma Ray Bursts (GRBs) may be a candidate in terms of the energetics, if the total energy of a fireball is $E^{\mathrm{iso}}(\Delta\Omega/4\pi) \sim 10^{56}(\Delta\Omega/4\pi)$ erg as suggested in the proton-synchrotron model of GRBs [23], where $\Delta\Omega/4\pi$ is the beaming factor. The energy luminosity of GRBs is estimated to be

$$L_{\mathrm{GRB}} = 7 \times 10^{47} \left(\frac{E^{\mathrm{iso}}}{10^{56}\mathrm{erg}}\right)\left(\frac{R^{\mathrm{iso}}}{7 \times 10^{-9}\mathrm{Mpc^{-3}yr^{-1}}}\right) \ \mathrm{erg/Mpc^3yr}. \tag{54}$$

This may be large enough to supply L_p required in the model. Production of neutrinos with $E \gtrsim E_{\mathrm{res}}$ in GRBs faces difficulties, however. Acceleration of protons to $\sim 20E_{\mathrm{res}}$ and significant synchrotron cooling of photoproduced pions [24] would be challenging for UHE neutrino production. See the contribution by E. Waxman in this volume for a detailed prediction of the UHE neutrino flux from GRBs.

Acknowledgments

The author wishes to thank the School organizers for their warm hospitality at this meeting. He also wishes to acknowledge his collaborators to work on the UHE neutrino issues, especially Günter Sigl and Pijushpani Bhattacharjee. He is grateful to Eli Waxman for his several useful comments on the neutrino production models with the Gamma Ray Bursts. This works was partly supported by the Inamori Foundation at Kyoto.

References

1. M. Takeda, et al.: Astrophys. J. **522**, 225 (1999)
2. T. J. Weiler: Phys. Rev. Lett. **49**, 234 (1982)
   E. Roulet: Phys. Rev. D **47**, 5247 (1993)
3. S. Yoshida: Astropart. Phys. **2**, 187 (1994)
4. S. Weinberg: *Gravitation and Cosmology* (Wiley, New York, 1972)
5. R. W. Strickland, D. N. Schramm: Astrophys. J. **481**, 571 (1997)
6. Y. Fukuda, et al.: Phys. Rev. Lett. **81**, 1562 (1998)
7. S. A. Bludman: Phys. Rev. D **45**, 4720 (1992)
8. A. Sandage, C. Cacciari: Astrophys. J. **350**, 645 (1990)
9. P. Bhattacharjee, C. T. Hill, and D. N. Schramm: Phys. Rev. Lett. **69** (1992) 567.
10. P. Bhattacharjee and G. Sigl: Phys. Rep. **327**, 109 (2000).
11. S. Yoshida, H. Dai, C. H. Jui, P. Sommers: Astrophys. J. **479**, 547 (1997)
12. G. Sigl, S. Lee, P. Bhattacharjee, S. Yoshida: Phys. Rev. D **59**, 043504 (1998)
13. D. Fargion, B. Mele, A. Salis: Astrophys. J. **517**, 725 (1999)
    T. J. Weiler: Astropart. Phys. **3**, 303 (1999)
14. S. Yoshida, G. Sigl, S. Lee: Phys. Rev. Lett. **81**, 5505 (1998)
15. C. T. Hill: Nucl. Phys. **224B** (1983) 469.
16. P. W. Gorham, K. M. Liewer, C. J. Naudet: astro-ph/9906504, Proc. *26th International Cosmic Ray Conference*, (Utah, 1999)
    P. W. Gorham, et al.: astro-ph/0102435, Proc. *First International Workshop on Radio Detection of High-Energy Particles*, Amer. Inst. of Phys., 2001, and at http://www.physics.ucla.edu/~moonemp/radhep/workshop.html
17. M. A. Lawrence, R. J. O. Reid, A. A. Watson: J. Phys. G Nucl. Part. Phys **17**, 733 (1991)
18. D. J. Bird, et al.: Phys. Rev. Lett. **71**, 3401 (1993)
19. M. Takeda, et al.: Phys. Rev. Lett. **81**, 1163 (1998)
20. P. Sreekumar, et al.: Astrophys. J. **494**, 523 (1998)
21. R. M. Baltrusaitis, et al.: Phys. Rev. D **31**, 2192 (1985)
22. For general information see http://amanda.berkeley.edu/
    see also F. Halzen: New Astron. Rev **42**, 289 (1999)
    for newest status see, e.g., G. C. Hill, et al. (Amanda Collaboration): astro-ph/0106064
23. T. Totani: Mon. Not. R. Astron. Soc. **307**, L41 (1999)
24. E. Waxman, J. N. Bahcall: Phys. Rev. D **59**, 023002 (1999)

Extreme-Energy Cosmic Rays: Hints to New Physics Beyond the Standard Model?

Pijushpani Bhattacharjee[1] and Günter Sigl[2]

[1] Indian Institute of Astrophysics, Bangalore-560 034. INDIA. E-mail: pijush@iiap.ernet.in
[2] Institut d'Astrophysique de Paris, CNRS, 98bis Boulevard Arago, 75014 Paris, France. E-mail: sigl@iap.fr

Abstract. The observed extreme energy cosmic ray (EECR) events above 10^{20} eV are difficult to explain within the standard scenario of origin of cosmic rays in which charged particles are accelerated in magnetized shocks in powerful astrophysical sources and their interaction is governed by the Standard Model of particle physics. Several ideas involving possible new physics beyond the Standard Model have been suggested in order to explain these events. Here we review some of the major proposed new physics solutions of EECR origin and discuss how the up-coming and future EECR experiments will be able to probe some possible forms of physics beyond the Standard Model otherwise not possible in terrestrial accelerators.

1 Introduction

The origin of the observed cosmic ray (CR) events of Extremely High Energy (EHE)[1] – those with energy $\gtrsim 10^{20}$ eV [1,2,3,4,5,6] – is one of the major unsolved problems in contemporary astrophysics [7,8]. About 20 of these Extreme Energy Cosmic Ray (EECR) events have been reported in published literature so far, with more expected in the near future from the ongoing experiments. The highest energy event reported so far – the one detected by the Fly's Eye group [5] – is at energy $3^{+0.36}_{-0.54} \times 10^{20}$ eV.

The energies and nature of the primary EECR particles are inferred from the properties of the extensive air showers (EAS) of secondary particles initiated by the primary EECR particles in the Earth's atmosphere. Because of the relatively small number of EECR events detected so far, the nature of the primary EECR particles is not known with certainty. The current data are consistent with EECR primaries being mainly nucleons, although photon primaries cannot be ruled out at this time. The flux of EECR at $\sim 10^{20}$ eV is $\lesssim 1\text{particle}/\,\text{km}^2/\text{century}$ which exemplifies the difficulty in detecting these events and necessitates construction of ground-based detectors with large area coverage such as the Auger [9], HiRes [10], Telescope Array [11], and the proposed space-based detectors such as OWL/AirWatch [12,13] and EUSO [14] for studying the nature and origin of these particles. The present data, while not sufficient to measure the spectrum

[1] We shall use the abbreviation EHE to specifically denote energies $E \gtrsim 10^{20}$ eV, while the abbreviation UHE for "Ultra-High Energy" will sometimes be used to denote $E \gtrsim 1\text{EeV} \equiv 10^{18}$ eV. Clearly UHE includes EHE but not vice versa.

Martin Lemoine, Günter Sigl (Eds.): LNP 576, pp. 275–299, 2001.

of the EECR accurately, already give indications of a spectrum above 10^{20} eV that seems to be significantly harder than the one below it (see Fig. 1), probably signifying a new component of the spectrum above 10^{20} eV different in origin than the one below it.

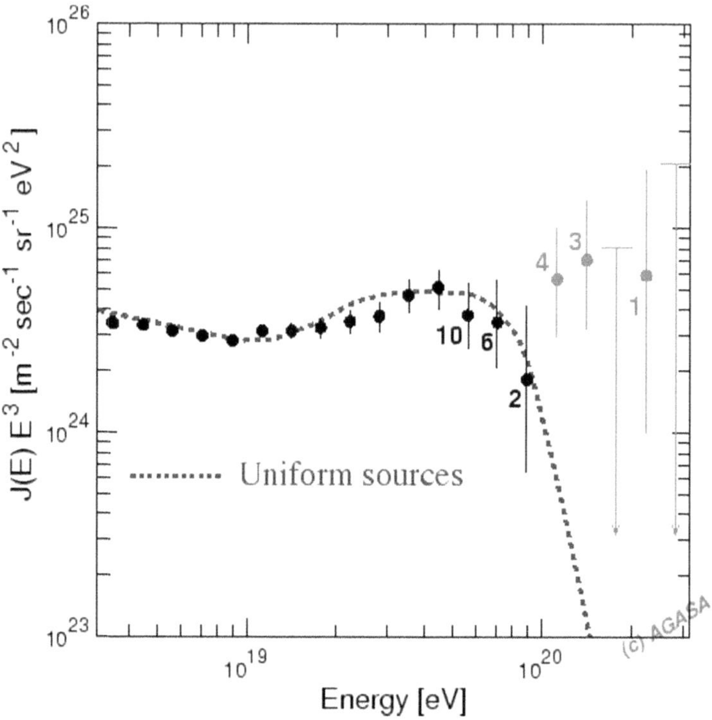

Fig. 1. Energy spectrum of UHECR measured by the AGASA experiment. The dashed curve represents the spectrum expected for extragalactic sources distributed uniformly in the Universe. The numbers attached to the data points are the number of events observed in the corresponding energy bins (From Ref. [6])

The existence of these EECR events at the detected flux level poses a serious challenge for the conventional scenario of origin of CR in which charged CR particles are accelerated in moving magnetized shocks in powerful astrophysical objects and their propagation in the Universe is governed by the physics of the Standard Model (SM) of particle physics. The associated problems have been widely discussed in literature. In addition to the articles in this volume, a number of recent reviews discuss the relevant issues in detail. The detailed review in [15] contains an extensive list of references to original literature as well as to earlier reviews and monographs on the subjects of cosmic rays in general and UHE CR in particular. In addition see, e.g., the reviews in Ref. [16,17,18,19,20,21,7,8]. For reviews of various experiments and experimental issues, see, e.g., Ref. [22] and references therein.

Briefly, there are two basic difficulties encountered in explaining the existence of EECR events within the conventional scenario:

First, the problem of energetics: It is extremely difficult [23,24] to accelerate particles to energies above 10^{20} eV even in the most powerful known astrophysical objects by means of the standard diffusive shock acceleration mechanism [25].

Second, the absence of candidate astrophysical sources: Too small a number of suitably powerful candidate sources are found in our cosmological neighborhood within a distance of $\lesssim$100 Mpc, the limiting source distance beyond which the flux of nucleons, nuclei or even photons of requisite energies would be severely attenuated due to their interaction with the cosmic background radiation fields during their propagation from the source to Earth.

The sources of the EECR particles are widely believed to be extragalactic because the arrival directions of the observed EECR events do not show any significant anisotropy associated with the Galactic disk that would be expected if these particles were accelerated in Galactic sources. Indeed, the observed large-scale isotropy of arrival directions of the EECR events, and the fact that these energetic particles are commonly believed not to be bent significantly by the regular extragalactic as well as Galactic magnetic fields (for scenarios entertaining strong bending see contributions by G. Medina Tanco, by P. Biermann et al., and by G. Sigl in this volume), call for *several* (rather than a few) extragalactic sources of these particles.

There are, however, strong restrictions on the distance of the sources of the EECR particles from Earth. As first pointed out by Greisen and independently by Zatsepin and Kuzmin [26], nucleons above a threshold energy of $\sim 6 \times 10^{19}$ eV produce pions by interacting with the cosmic microwave background (CMB) photons. The mean free path for this photo-pion production process on CMB for nucleons above $\sim 10^{20}$ eV is roughly energy independent and ~ 6 Mpc, and the nucleon losses on average about one-fifth of its energy in each interaction. This means that after traveling a distance of D Mpc, the energy of the nucleon is reduced by a factor of $\sim \exp(-0.22D/6)$. Thus a nucleon of initial energy well above 10^{20} eV will have its energy reduced by about one order of magnitude after traveling a distance of ~ 60 Mpc. More detailed calculations show [27] that the observed energy of a nucleon at Earth, from a source at a distance $\gtrsim 100$ Mpc, will always be less than 10^{20} eV , irrespective of the energy ($\geq 10^{20}$ eV) at the source. Heavy nuclei as well as photons above 10^{20} eV are also not immuned to the destructive effects of background radiation fields: Nuclei are photo-dissociated by CMB as well the infrared background photons [28], while photons are absorbed due to e^+e^- pair production off the Universal Radio Background (URB) photons [29], both on length scales $\lesssim 10 - 20$ Mpc. For detailed discussions and review of propagation of EECR particles, see, e.g., Ref. [15] and the contribution by G. Sigl in this volume. It is thus clear that if the EECR particles are "standard" particles such as nucleons, nuclei and/or photons, then their sources must lie in our cosmological neighborhood within a distance of $\lesssim 100\,\mathrm{Mpc}$. However, most of the powerful extragalactic objects, such as some radio galaxies with powerful jets and/or hotspots and gamma ray bursts (GRBs), which may marginally be able to accelerate particles to energies beyond 10^{20} eV – albeit with optimistic assumptions on various parameters (see contributions by G. Pelletier and by E. Waxman in this volume)– are generally too far away at distances $\gg 100\,\mathrm{Mpc}$. The few cosmologically nearby powerful radio galaxies, such as M87 at $\sim 18\,\mathrm{Mpc}$ or Centaurus A at $\sim 3.4\,\mathrm{Mpc}$, are not in the direction of any of the observed EECR events, and one needs rather strong and special configuration of the extragalactic magnetic fields to bend the trajectories of EHE charged particles from these objects to reproduce the observed large-scale isotropic distribution of the EECR events.

The absence of any suitable sources within the "GZK limiting distance" of $\sim 100\,\mathrm{Mpc}$ should imply a steep decline, generally referred to as the "GZK cutoff", of the EECR spectrum beginning somewhere in the energy region of 5×10^{19} eV. The GZK cutoff is expected if the extragalactic sources of EECR, with typical source particle spectra as expected in standard shock acceleration mechanism, are uniformly distributed in the Universe. The apparent absence of such a cutoff, as indicated by the significant number of detected events above 10^{20} eV (see Fig. 1), is, therefore, a major puzzle.

Actually, the GZK cutoff may or may not be a *complete* cutoff – rather it is more likely to be a GZK "dip" preceded in energy by a "pile-up" and followed by a "recovery". The "strengths" of these various predicted features of the UHE CR spectrum depend on several parameters such as the maximum energy and the spectral index of the CR particles at the source, the inhomogeneity of the distribution of the sources, maximum distance (redshift) of the sources, as well as possible evolution (with cosmological epoch) of the production rate of the CR particles; see, e.g., Ref. [30,31,32,33]. The present EECR data (see Fig. 1) are rather too sparse to allow one to tell, with any high degree of statistical confidence, whether the predicted pile-up and dip structures are present in the spectrum. But with more data expected from up-coming experiments, confirmation of the presence or absence of one or more of these features in the EECR spectrum and the measurements of the parameters characterizing the strengths of these features may be possible, which may have important implications for distinguishing between various proposed solutions of the EECR problem (see below).

The difficulties of explaining the origin of EECR within the conventional scenario have prompted several suggestions of possible "new physics" solutions of the EECR enigma. Almost all of these proposed solutions involve some kinds of physics beyond the SM in one form or another. These suggestions generally fall into two broad classes:

In one class are proposals which attempt to *evade* the GZK distance limit on source locations or avoid the GZK cutoff effect on the spectrum by postulating new physics. Among these are suggestions involving small violation of Lorentz invariance, supersymmetry, a small neutrino mass, new interactions of neutrinos with matter, and so on.

The other class of proposals addresses the problem of energetics itself by doing away with the question of acceleration of particles in astrophysical objects. Instead, the EECR particles are hypothesized to arise simply from decay of some supermassive particles (of mass $> 10^{21}$ eV) originating from fundamental processes in the early Universe. This class of proposals generally goes by the name of *top-down* scenario as opposed to the *bottom-up* scenario in which particles are accelerated from lower energies to the requisite EECR energies in suitable astrophysical environments. In some models of the top-down scenario, the relevant massive particles may be decaying at large cosmological distances beyond the normal GZK limit in which case some GZK limit evading mechanism may have to be invoked in addition.

In this article we review some of the major proposals of these new physics solutions, emphasizing how the highest energy end of the cosmic ray spectrum can be used as a probe of possible new fundamental physics.

In Sect. 2, we discuss possible new physics ways of avoiding the GZK cutoff and source distance limits. The top-down scenario that obviates the need to accelerate particles is discussed in Sect. 3, with conclusions presented in Sect. 4. In this contribution, we use natural units in which $\hbar = c = k_{\mathrm{B}} = 1$ unless otherwise noted.

2 Avoiding the GZK Cutoff and Source Distance Limit

If one assumes that the EECR particles are somehow accelerated in astrophysical objects to the requisite energies or are created in a top-down mechanism, then there are several possible new physics ways of eliminating or extending the GZK upper limit on the distance of the EECR sources. In this section we summarize some of the major suggestions in this regard. Some of this is also discussed in Sect. 4 of the contribution by G. Sigl in this volume, especially the possibility of new interactions associated with neutrinos which are not covered in detail in the present contribution. We will focus here mostly on the qualitative features of these scenarios.

2.1 Violation of Lorentz Invariance

It has been pointed out by a number of authors [34,35] that the GZK effect may be eliminated altogether by allowing violation of Lorentz invariance (VLI) by a tiny amount that is consistent with all current experiments. At a purely theoretical level, several quantum gravity models including some based on string theories do in fact predict non-trivial modifications of space-time symmetries that also imply VLI at extremely short distances (or equivalently at extremely high energies); see e.g., Ref. [36] and references therein. These theories are, however, not yet in forms definite enough to allow precise quantitative predictions of the exact form of the possible VLI. Current formulations of the effects of a possible VLI on high energy particle interactions relevant in the context of EECR, therefore, adopt a phenomenological approach in which the form of the possible VLI is parametrized in various ways. VLI generally implies the existence of a universal preferred frame which is usually identified with the frame that is comoving with the expansion of the Universe, in which the CMB is isotropic.

A direct way of introducing VLI is through a modification of the standard *dispersion relation*, $E^2 - p^2 = m^2$, between energy E and momentum $p = |\boldsymbol{p}|$ of particles, m being the invariant mass of the particle. Currently there is no unique way of parametrizing the possible modification of this relation in a Lorentz non-invariant theory. As a purely illustrative example of how certain forms of the modified dispersion relation can allow one to completely evade the GZK limit, consider the modified dispersion relation [37]

$$E^2 - p^2 \simeq m^2 - \frac{p^3}{M}\,, \tag{1}$$

where M is some large mass scale, such that the standard Lorentz invariant dispersion relation is recovered in the limit $M \to \infty$.

Now, consider the GZK photo-pion production process in which a nucleon of energy E, momentum p and mass m_N collides head-on with a CMB photon of energy ϵ producing a pion and a recoiling nucleon. The threshold initial momentum of the nucleon for this process according to standard Lorentz invariant kinematics is

$$p_{\rm th,0} = (m_\pi^2 + 2m_\pi m_N)/4\epsilon\,, \tag{2}$$

where m_π and m_N are the pion and nucleon masses, respectively. Assuming exact energy-momentum conservation but using the modified dispersion relation given above, it is easy to show after some tedious but straight forward algebra that, in the ultra-relativistic regime $m \ll p \ll M$, and neglecting sub-leading terms, the new nucleon threshold momentum $p_{\rm th}$ under the modified dispersion relation (1) satisfies the equation [37]

$$-\,\alpha x^3 + x - 1 = 0\,, \tag{3}$$

where $x = p_{\rm th}/p_{\rm th,0}$, and

$$\alpha = \frac{2p_{\rm th,0}^3}{(m_\pi^2 + 2m_\pi m_N)M}\,\frac{m_\pi m_N}{(m_\pi + m_N)^2}\,. \tag{4}$$

It is easy to check that, for a fixed $M = M_{\rm Pl}$, say, $M_{\rm Pl}$ being the Planck mass, there is no real positive solution of (3), implying that the GZK process does not take place and consequently the GZK cutoff effect disappears completely. It can also be shown [37] that the same modified dispersion relation (1) also forbids the absorption of high energy gamma rays through e^+e^- pair production on the infrared, microwave or radio backgrounds. Thus EHE nucleons and/or photons will be able to reach Earth from any distance. On the other hand, if future EECR data confirm the presence of a GZK cutoff at some energy then that would imply a lower limit on the mass scale M at which the VLI becomes important, thus probing specific Lorentz non-invariant theories.

It is to be mentioned here that if VLI is due to modification of the space-time structure expected in some quantum gravity theory, for example, then the strict energy-momentum conservation assumed in the above discussion, which requires spacetime translation invariance, is not guaranteed in general, and then the calculation of the modified particle interaction thresholds becomes highly non-trivial and non-obvious. Also, it is possible that a Lorentz non-invariant theory while giving a modified dispersion relation also imposes additional kinematical structures such as a modified law of addition of momenta. Indeed, [36] gives an example of a so-called κ-Minkowski non-commutative space-time in which the modified dispersion relation has the same form as in (1) but there is also a modified momentum addition rule which compensates for the effect of the modified dispersion relation on the particle interaction thresholds discussed above leaving the threshold momentum unaffected and consequently the GZK problem unsolved.

It should also be mentioned that VLI does not necessarily have to be associated with quantum gravity and can even exist at the level of electrodynamics [35,38]. Coleman and Glashow (CG) [35], for example, have studied a Lorentz non-invariant gauge theory by explicitly adding a Lorentz non-invariant term in the Lagrangian. In their analysis one possible parametrization of VLI is that the maximum attainable speed is different for different particles, with the energy-momentum relation for the particle a of rest mass m_a being $E_a^2 = m_a^2 c_a^4 + p_a^2 c_a^2$, where c_a is the limiting speed for the particle a. In this formulation, CG have demonstrated how the GZK effect can be evaded for a certain range of values of $(c_\pi - c_N)$.

There are several other fascinating effects of allowing a small VLI, some of which are relevant for the question of origin and propagation of EECR, and the resulting constraints on VLI parameters from cosmic ray observations are often more stringent than the corresponding laboratory limits; for more details, see Ref. [35] and [15] and the contribution by G. Sigl in this volume for review and further references.

2.2 Supersymmetric Particles as EECR Primaries

One possible way to increase the GZK limit on the source distance is to postulate a new as yet undiscovered particle species as the EECR primaries which, according to (2), would have a higher "GZK threshold" for interaction with the CMB if they were more massive than nucleons. In conformity with the assumed hadronic character of the observed EECR shower events, the new particle species would still have to be hadronic.

As one possible realization of this idea, certain supersymmetric particles have been suggested as possible candidates for the EECR primaries [39]. The particular scenario of [39] involves a light and stable (or at least quasi-stable) *neutral* particle (to avoid deflection by the intergalactic magnetic field), with a mass between 0.1 and 1 GeV [40]. The suggested primary EECR candidate is the lightest gluino-containing baryonic bound state, $uds\tilde{g}$, denoted by S^0, which could be long-lived or stable. The kinematical threshold for the photo-pion production process $S^0 + \gamma_{\rm CMB} \to S^0 + \pi$ would be higher than that for nucleon by a factor $\simeq m_{S^0}/m_N$, where m_{S^0} and m_N are the masses of S^0 and the nucleon, respectively. Furthermore, as in the case of the nucleon, for which the total photo-pion production cross section is dominated by the lowest-lying pion-nucleon resonance Δ (of mass 1232 MeV) which occurs close to the threshold and at which the cross section peaks, the cross section for photo-pion production by S^0 may be expected to peak at the invariant mass $m_{S^{0*}}$ of the lowest-lying resonance, S^{0*}, of S^0. Thus the cross section for $\gamma_{\rm CMB} - S^0$ interaction would be expected to peak at an S^0 energy higher by a factor $(m_{S^0}/m_N)(m_{S^{0*}} - m_{S^0})/(m_\Delta - m_N)$ with respect to that for the case of nucleon. It is expected that $(m_{S^{0*}} - m_{S^0})/(m_\Delta - m_N) \gtrsim 2$. As a result of this, as well as a smaller interaction cross section of S^0 with photons and a smaller fractional energy loss of S^0 relative to those in the case of nucleons, the effective GZK threshold is higher by factors of a few, and sources of primary S^0 particles above $10^{19.5}$ eV could be 15-30 times further away compared to the case of nucleon primaries. The existence of the EECR events was, therefore, proposed [39] as a signal] of supersymmetry. Indeed, Farrar and Biermann [41] reported a possible correlation between the arrival directions of the five highest energy CR events and the directional locations of some compact radio quasars at redshifts between 0.3 and 2.2, as might be expected if these quasars were sources of massive neutral particles. However, with the present data, such "evidence" for directional correlation remains a subject of debate [42].

A specific difficulty of this model within the general bottom-up acceleration scenario is the fact that, of course, the neutral S^0 cannot be accelerated, but rather has to be produced as a secondary of an accelerated proton interacting with the ambient matter. As a consequence, protons must be accelerated to at least 10^{21} eV at the source in order for the secondary S^0 particles to explain the EECR events, which may be difficult. Furthermore, this scenario of a neutral supersymmetric particle in the mass range required to be consistent with the shower characteristics of the observed EECR events seems to be difficult to reconcile with constraints from accelerator experiments [43]. For more details on the nature of extensive air showers generated by gluino-containing hadrons and the resulting constraints on their masses, see the recent work Ref. [44].

2.3 Massive Neutrinos and Z-Burst

The only particle in the SM that can propagate unattenuated with energies above 10^{20} eV from sources at distances $\gg 100$ Mpc is the neutrino; however, in the SM, the probability of neutrinos to directly initiate the observed EECR air shower events is at least a factor of $\sim 10^{-6}$ smaller than the corresponding probability in the case of nucleons. However, as first suggested in [45], neutrinos of sufficiently high energy from cosmologically distant ($\gg 100$ Mpc) sources can *indirectly* give rise to the observed EECR events.

The idea hinges on allowing one of the most conservative deviations from the SM, namely, that neutrinos have small masses in the eV range, which seems to be supported by recent experimental evidence [46] of atmospheric neutrino flavor oscillations. If some flavor of neutrino is assumed to have a small mass

$m_\nu \sim 1\,\text{eV}$, and if there are sources capable of producing neutrinos of sufficiently high energy ($\gtrsim 10^{22}\,\text{eV}$), then interaction of those neutrinos with the neutrinos (ν_b) constituting the cosmic thermal relic neutrino background (RNB) can excite the Z boson resonance, $\nu + \bar{\nu}_b \to Z$, at the EHE neutrino energy $E_{\nu,\text{res}} = (M_Z^2/2m_\nu) \simeq 4 \times 10^{21}(\,\text{eV}/m_\nu)\,\text{eV}$, where $M_Z = 91\,\text{GeV}$ is the Z-boson mass. The decay of each Z (rest-frame life time $\sim 3 \times 10^{-25}\,\text{sec}$) into $q\bar{q}$, the branching ratio for which is $\sim 70\%$, and the subsequent hadronization of the quarks would produce about one nucleon-antinucleon pair, 10 neutral pions and 17 charged pions [47] with neutral pions further decaying into photons and charged pions into neutrinos, electrons and positrons. It has been suggested [45] that the resulting EHE nucleons and photons from the decay of the Z bosons produced *within the GZK distance limit of* $\sim 100\,\text{Mpc}$ *from Earth* could be candidates for the observed EECR events. In this so-called Z-burst scenario, since the final decay products of the Z are dominated by photons and neutrinos, the EECR events are predicted to be mainly photons (like in the top-down scenario in general; see below) rather than nucleons.

Note that for massless neutrinos, the required EHE neutrino energy would be much higher: $E_{\nu,\text{res}}(m_\nu = 0) \simeq 8 \times 10^{24}(4.8 \times 10^{-4}\,\text{eV}/\epsilon_{\nu,b})\,\text{eV}$, where $\epsilon_{\nu,b} \simeq 3T_\nu$ is the typical energy of the relic neutrino, $T_\nu \simeq 1.9\text{K} \simeq 1.6 \times 10^{-4}\,\text{eV}$ being the effective temperature of the relic neutrino background. Such high energy neutrinos are unlikely to be produced in any astrophysical sources.

The invariant energy-averaged cross section for the process $\nu + \bar{\nu}_b \to Z$, defined as $\langle\sigma\rangle \equiv \int ds\sigma(s)/M_Z^2$, with s the square of the energy in the center of momentum frame, is $\langle\sigma\rangle \simeq 4.2 \times 10^{-32}\,\text{cm}^2$. The relative energy width of the Z resonance at FWHM is $\sim$3%. So only EHE neutrinos with energy in a small range around the resonant energy $E_{\nu,\text{res}}$ are involved in producing the Z's. Because the target background neutrinos (of mass in the eV range) are essentially nonrelativistic, the produced Z-boson has the energy $E_Z \simeq E_{\nu,\text{res}}$. The average nucleon energy in the Z decay is $\langle E_N\rangle \sim E_{\nu,\text{res}}/30 \sim 1.3(\,\text{eV}/m_\nu)\times 10^{20}\,\text{eV}$ while the average photon energy $\langle E_\gamma\rangle \sim 0.5\langle E_N\rangle$, since the total particle multiplicity in the Z decay is about 30 and each pion decays into two photons. For $m_\nu \lesssim 0.1\,\text{eV}$, the produced nucleons and photons can be well above the GZK cutoff and can in principle explain the observed EECR events.

Detailed calculations have been done examining the viability of and constraints on the Z-burst scenario; see e.g., Ref. [48,49,50,51]. The major constraints on the scenario are discussed below.

The probability for resonant annihilation of a EHE neutrino with a relic (anti)neutrino of small but finite mass producing Z bosons within a distance $D_{\text{GZK}} \lesssim 100\,\text{Mpc}$ is rather small, $\sim 2.5 \times 10^{-4}$, for a uniformly distributed neutrino background (see, e.g., Ref. [21]). The massive neutrinos would, however, be expected to cluster, and depending on the length-scale and strength of the clustering, the above probability can be somewhat larger, though perhaps not larger than about 1% [21]. In general, because of the relatively small probability of the process, a rather large EHE neutrino flux is required in order to successfully explain the EECR events. This neutrino flux, when extrapolated to lower energies of order 10^{17} eV with a spectrum going as E^{-2} expected from typical astrophysical sources, generally conflict with the limit on neutrino flux [52] at $> 10^{17}$ eV obtained from non-observation of horizontal air showers that could be initiated by the neutrinos. The conflict can be avoided [50] if the source neutrino spectrum is rather hard, $dN_\nu/dE_\nu \propto E_\nu^{-\gamma}$ with spectral index $\gamma \lesssim 1.2$ [50]. Such hard spectra of neutrinos are, however, not usually expected from astrophysical sources.

It has also been pointed out [48] that in the Z-burst scenario of EECR origin, in addition to the requirement of relatively hard spectrum neutrino sources, significant local neutrino clustering is required to avoid generating a diffuse background of 30 MeV to 100 GeV photons in excess of that measured by the EGRET experiment [53]. This comes about in the following way: While the contribution to the observed EECR would come only from Z-bursts occurring in our cosmological neighborhood within $\lesssim 100\,$Mpc, the accompanying electromagnetic (EM) energy injected into the Universe by the sources that produce the EHE neutrinos as well as the EM component of the Z-bursts themselves at large cosmological distances ($\gg 100\,$Mpc) would cascade down to lower energies through the process of EM cascading in the cosmological radiation background fields (see, e.g., Ref. [15] and contribution by G. Sigl in this volume for a review of the cosmological EM cascading process), and would thus give rise to a diffuse gamma ray background peaking at around 10 GeV. A relatively lower flux of Z-burst-initiating EHE neutrino flux, as would obtain in the case of a local clustering of the relic neutrinos relative to the no-clustering case, would yield a correspondingly lower level of the diffuse gamma ray background in the MeV–GeV region. The analysis of [48] shows that in order for the Z-burst scenario of EECR origin to be consistent with the EGRET bound on the diffuse gamma ray flux in the 30 MeV–100 GeV region, the relic neutrino overdensity f_ν over a length scale l_ν has to satisfy $f_\nu \gtrsim 10^3 (l_\nu/5\,\mathrm{Mpc})^{-1}$, if the total photon luminosity of the sources is comparable to their total neutrino luminosity, as would be expected in most source models.

Furthermore, if the EHE neutrinos causing the Z-bursts are produced in astrophysical sources, where they would presumably be produced through interaction of accelerated protons with the dense matter and radiation in the source, then those sources must be such as to trap the accelerated protons within the sources because otherwise the observable proton flux below the GZK cutoff would be comparable to the neutrino flux [54,55] in contradiction with observation. In other words, in order for the Z-burst mechanism to contribute significantly to the observed EECR flux, the existence of a new class of high energy neutrino sources, possibly unrelated to the sources of $> 10^{19}$ eV cosmic rays, may have to be invoked. Moreover, it has been argued [55] that the energy generation rate of these EHE neutrino sources would have to be comparable to the total photon luminosity of the Universe.

In this context, it has been pointed out [56] that a degenerate relic neutrino background with a finite neutrino chemical potential (implying an asymmetry between ν and $\bar{\nu}$), produced, for example, through neutrino flavor oscillation in the early Universe, would allow a much larger density of the Fermi-degenerate relic neutrinos than is predicted in the standard big bang model, and would consequently increase the neutrino annihilation- and thus Z boson production probability. The authors of [56] have argued that for $m_\nu \simeq 0.07$ eV, the value suggested by the Super-Kamiokande experiment [46], and for a relic neutrino density parameter $\Omega_\nu \simeq 0.01$, the resulting requirement on the source EHE neutrino flux (in order to explain the observed EECR flux) implies energy generation rate of the EHE neutrino sources well below the total photon luminosity of the Universe.

If the EECR events are indeed due to the Z-burst mechanism, then it offers the exciting possibility of *determining* the mass of the heaviest neutrino, as pointed out in [57] and studied in more details in [51], by fitting the predictions of the Z-burst scenario to the observed UHE CR data. The neutrino mass so determined [51] from the present data is consistent with the value indicated by the Super-Kamiokande experiment [46]. Note, however, that for such neutrino masses, the sources are required to almost exclusively produce neutrinos at least up to 10^{22} eV for the Z-burst mechanism to work. Such high energies are rather difficult to obtain within conventional bottom-up models, but are easily obtained in top-down models (see below), making the Z-burst scenario more likely to play a role in the latter.

2.4 New Neutrino Interactions above the Electroweak Scale

The main idea here is that the neutrino-nucleon cross section at center of mass (CM) energies $\sqrt{s} \simeq (2m_N E_\nu) \simeq 430(E_\nu/10^{20}\,\mathrm{eV})^{1/2}$ TeV, relevant for interactions in the detector medium, could be much larger than predicted by the Standard Model, possibly close to hadronic cross sections. This is not in conflict with existing accelerator data which reach only up to CM energies of a few hundred GeV. At the same time propagation of neutrinos would not be influenced because the relevant CM energies are $\simeq 450(\varepsilon/\mathrm{eV})^{1/2}(E_\nu/10^{20}\,\mathrm{eV})^{1/2}$ GeV, and thus in the well known regime even for massive RNB neutrinos of energy $\varepsilon \sim 1\,\mathrm{eV}$. The possibility of non-standard neutrino-nucleon cross sections beyond a TeV is discussed in more detail in Sect.4.1 of the contribution by G. Sigl in this volume.

3 Avoiding the Acceleration Problem: The Top–Down Scenario

As already mentioned in the Introduction, it is extremely difficult to accelerate particles to energies beyond 10^{20} eV by the standard diffusive shock acceleration mechanism in known astrophysical objects. An alternative possibility is that the enormous energies of the EECR particles are not due to any acceleration process; instead, they could arise simply from decay of very massive particles of mass $\gg 10^{20}$ eV. Two possible realizations of this top-down scenario have been suggested, both of which require physics beyond SM. Below we discuss them briefly; for detailed review and references to original literature, see Ref. [15].

3.1 EECR from Decays of Metastable Superheavy Relic Particles

It has been suggested [58,59] that EECR may be produced from the decay of some metastable superheavy relic particles (MSRPs) of mass $m_X \gtrsim 10^{12}$ GeV and lifetime larger than or comparable to the age of the Universe. The long but finite lifetime of MSRPs could be due to slow decay of the MSRPs through non-perturbative instanton effects or through quantum gravity effects, for example. The MSRP "X" particles would typically decay into quarks and leptons. The hadronization of the quarks produces a photon- and neutrino dominated spectrum of particles with energy up to m_X. The EECR are hypothesized to be mainly the photons from these MSRP decays.

There are no MSRP candidates within the SM. Possible candidates for MSRPs and their possible decay mechanisms giving them sufficiently long lifetime have been discussed in the context of specific particle physics models beyond SM, by a number of authors; see, e.g., Ref. [60,61]. Several non-thermal mechanisms of production of MSRPs in the post-inflationary epoch in the early Universe have also been studied; see, e.g, Ref. [62] for a review of these mechanisms. Under certain circumstances MSRPs can exist in the Universe with sufficient abundance so as to act as non-thermal superheavy dark matter.

Obviously, the flux of EECR produced by this mechanism depends on the abundance as well as the lifetime of the MSRPs, neither of which is known with much confidence. An interesting aspect of this scenario is that, like the cold dark matter (CDM) particles, the MSRPs would gravitationally cluster, in particular, on the scale of the Galactic Halo (GH). The flux of EECR photons and nucleons will, therefore, be dominated by the contribution of MSRP decay within the GH, which would naturally explain the *absence* of the GZK cutoff (see, however, below), since the size of the GH is much less than the GZK distance limit. Because of the general isotropic distribution of the MSRPs within the GH, the scenario also naturally explains the observed isotropy of the EECR. There will, however, be a small anisotropy associated with the off-center location of the solar system in the GH [63], which will hopefully be detectable by the up-coming detector such as the Pierre Auger, providing an important test of the scenario. It has also been pointed out [64] that the model of decaying MSRPs in the GH (we will denote it by GHMSRP hereafter) can also explain the possible small-scale anisotropies (clustering) of the UHE CR events reported recently [65][2], provided the MSRP distribution in the GH is suitably clumped. Recently, the possibility that EECR may result from *annihilation* (rather than decay) of the MSRP X particles in the GH has also been investigated [66]. The required annihilation cross section, however, turns out to be uncomfortably large, much larger than the unitarity limit.

Another important aspect of the top-down scenario in general, including the MSRP decay model as well as the topological defect model discussed below, is that the injection spectrum of the EECR particles is mainly determined by the spectrum of the hadronization products of the quarks (and/or gluons) from the

[2] The statistical significance of the reported small-scale clustering is, however, not very strong with the present data.

decay of the massive X particles, which in turn is determined by QCD. However, since reliable measurements of QCD hadronization spectra of quarks/gluons are available from current accelerator data only for center-of-mass energy up to $\sim 100\,\mathrm{GeV} \ll m_X$, large extrapolations of the measured hadronization spectra are required to predict the spectra of particles from X particles decay. This is beset with uncertainties associated with possible but currently unknown new physics effects beyond SM (such as supersymmetry, for example). Recently, the effects of incorporating supersymmetry in the hadronization spectra of X decay products have been studied in some details both semi-analytically [67] (by numerically solving the so-called DGLAP evolution equation for QCD jet fragmentation functions) as well through Monte Carlo simulations [68]. There are, however, significant differences between the results of these studies, thus reflecting the present degree of theoretical uncertainties in reliably determining the expected spectra of the X decay products.

In spite of the uncertainties discussed above, one (hopefully) robust feature of the spectra of the X particle decay products is that these spectra are generally expected to be significantly harder than the generic particle spectra predicted in the bottom-up (shock acceleration) scenario; see, e.g., Ref. [15] for more discussion on this point. Since the spectra of EECR nucleons and photons are relatively unaffected by cosmological propagation effects (since the EECR particles would have to originate at rather close-by distances $\lesssim 100\,\mathrm{Mpc}$), their observed spectra would essentially mimic their injection spectra, which in the top-down model, as already mentioned, is determined essentially by QCD. Thus, should the top-down scenario of EECR origin be confirmed by future experiments, the measured spectra of the EECR particles in those experiments would be a probe of QCD at energies well beyond those currently accessible in particle colliders.

Coming back to the GHMSRP model, it may be mentioned that this model predicts a *complete* absence of the GZK cutoff, whereas the current EECR data (see, e.g., Fig. 1) may be interpreted (albeit without strong statistical significance) to indicate the presence of only a GZK "dip" (rather than a complete cutoff) followed by a "recovery". Should such features be confirmed by future data, they would seem to be more consistent with an extragalactic top-down model than with the Galactic Halo based top-down model [69].

3.2 EECR from Collapse or Annihilation of Cosmic Topological Defects

The X particles of the top-down model discussed above were required to be metastable with lifetime comparable to the age of the Universe so that they decay in the present epoch in order to produce the observed EECR particles. An alternative source of the supermassive X particles is possible in many Grand Unified Theories (GUTs) which allow formation of cosmic Topological Defects (TDs) such as magnetic monopoles, and cosmic strings during symmetry-breaking phase transitions in the early Universe (see Ref. [70] for a review on TDs). In this case, the candidate X particles (for a top-down model of EECR) are the superheavy gauge bosons, higgs bosons, fermions, etc., of the underlying spontaneously broken gauge theory, which can have mass m_X as high as a typical GUT mass scale of $\sim 10^{16}$ GeV. TDs essentially represent topologically non-trivial configurations of the classical fields representing these X particles. Although in most GUT models the *free* X particles themselves have extremely short lifetime, the X particles are prevented from decaying as long as they are "trapped" inside TDs. On the other hand, the X particles can be emitted continuously (on cosmological time scales) from collapsing and/or annihilating TDs. Once released from TDs, the X particles would decay essentially instantaneously. The decaying X particles released from TDs collapsing and/or annihilating in the present cosmological epoch can be the sources of the EECR particles which can have energies up to $m_X \sim 10^{16}$ GeV.

The injection rate of X particles into the Universe due to a wide variety of processes involving TDs can be written, on dimensional grounds, in the form [32]

$$\dot{n}_X(t) = \kappa m_X^p t^{-4+p}\,, \tag{5}$$

where κ and p are dimensionless constants whose values depend on the specific process involving specific kinds of TDs [32]. This form is expected to be valid for any TD system for which there is no intrinsic time and energy scales involved other than the Hubble time t and mass scale m_X. This is the case in situations in which the TDs under consideration evolve in a scale-independent way such that the energy density of the TD network always scales in time with – and thereby remains a fixed fraction of – the total energy density of the Universe. This scaling is indeed a property of evolution of most of the interesting kinds of TDs, for if it were not the case, then the energy density in the form of TDs would either dominate the total energy density of the Universe at some epoch or would become negligible[3].

Processes involving specific TDs, such as collapsing cosmic string loops ($p = 1$), monopole-antimonopole annihilation ($p = 1$), collapsing necklaces (closed loops of cosmic strings with monopole "beads" on them) ($p = 1$), current-saturated superconducting cosmic string loops (generally $p < 1$), vorton decay ($p = 2$), and so on, have been studied in the literature (see Ref. [15] for references)[4]. Again, a photon and neutrino dominated injection spectrum, determined mainly by QCD, is predicted, while the absolute flux depends on the specific TD model and is rather uncertain.

Unlike the Galactic Halo MSRP decay case, however, the extragalactic TD scenario is significantly constrained by the measured flux of the diffuse γ–ray background in the several MeV to several GeV region, in addition to being constrained by the observed EECR flux. This is because, in most TD models, the X particle production from TDs and their decay occur not only in the present

[3] It is easy to see that similar arguments imply that the injection rate of X particles due to decay of MSRPs discussed in the previous section must also be expressible in the form of (5), in general, with $p = 2$.

[4] See also the new possibilities raised for the cosmic string model by the recent work of [71].

epoch at small cosmological distances, but also at earlier epochs (or equivalently at large cosmological distances). While those X particles produced and decaying at small redshifts ($z \ll 1$) would give rise to the observed EECR, the mainly EM energy injected at ultrahigh energies at larger redshifts cascades down to lower energies (in the MeV–GeV region) in the present epoch due to the development of EM cascades on the background radiation fields. Whether or not the resulting gamma ray background in the MeV–GeV region satisfies the observational constraints depends significantly on the strength of the URB, the strength of the extragalactic magnetic field (EGMF), the mass m_X of the X particles released from the TDs, and so on [15]. In addition, it turns out that TD processes with $p < 1$ (some processes of X particle production from superconducting cosmic strings, for example) generally lead to unacceptably high rate of energy injection in the early cosmological epochs, which would cause excessive ^{4}He photo-disintegration and CMB distortion [72], and are, therefore, currently unfavored in the context of EECR.

Figure 2 shows an example of the spectra of nucleons, photons, and neutrinos expected in a typical extragalactic TD model in which the TD sources are uniformly distributed in the Universe and produce X particles at a rate $\propto t^{-3}$, i.e., with $p = 1$, where t is the cosmic time. The absolute flux is normalized in the standard way by maximizing the likelihood of explaining the observed UHECR data, and the resulting spectra are then seen to be consistent with all other observational bounds such as those from gamma ray observations and upper limits at lower energies.

The normalization procedure to the EECR flux described above imposes the constraint $Q^0_{\rm EECR} \lesssim 10^{-22}\,{\rm eV\,cm^{-3}\,sec^{-1}}$ within a factor of a few [85,49,86] for the total energy release rate Q_0 from TDs at the current epoch. In most TD models, because of the unknown values of the parameters involved, it is currently not possible to calculate the exact value of Q_0 from first principles, although it has been shown that the required values of Q_0 (in order to explain the EECR flux) mentioned above are quite possible for certain kinds of TDs. Some cosmic string simulations and the necklace scenario suggest that defects may lose most of their energy in the form of X particles and estimates of this rate have been given [87,88]. If that is the case, the constraint on $Q^0_{\rm EECR}$ translates via Eq. (5) into a limit on the mass m_X of the X particle and hence on the symmetry breaking scale: $m_X \lesssim 10^{13}\,{\rm GeV}$ [89]. Independently of whether or not this scenario explains EECR, the EGRET measurement of the diffuse GeV $\gamma-$ray background leads to a similar bound, $Q^0_{\rm EM} \lesssim 2.2\times10^{-23}\,h(3p-1)\,{\rm eV\,cm^{-3}\,sec^{-1}}$, which leaves the bound on m_X practically unchanged. Furthermore, constraints from limits on CMB distortions and light element abundances from ^{4}He-photodisintegration are comparable to the bound from the directly observed diffuse GeV $\gamma-$rays [72]. That these crude normalizations lead to values of η in the right range suggests that defect models require less fine tuning than decay rates in scenarios of metastable massive dark matter.

An important point to note from Fig. 2 is that, although the injection spectrum in the TD scenario is always dominated by photons over nucleons, the final

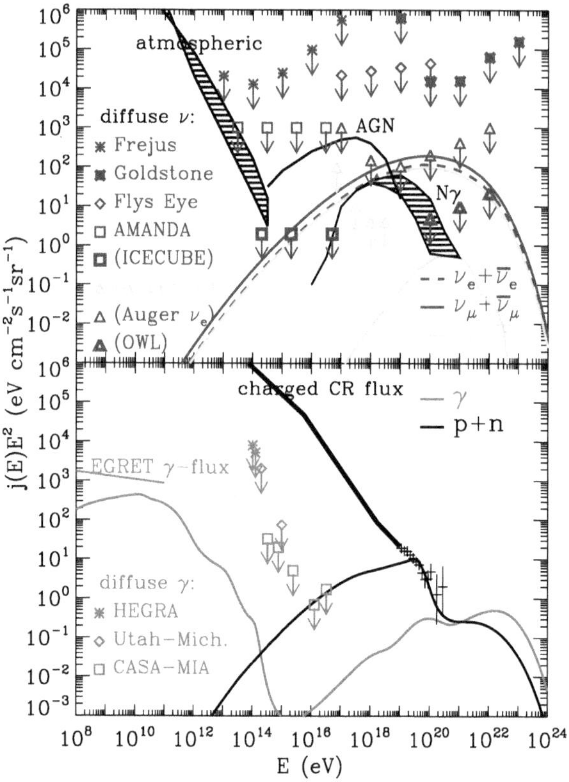

Fig. 2. All particle spectra for a top-down model involving the decay into two quarks of non-relativistic X particles of mass 10^{16} GeV, released from homogeneously distributed topological defects. Lower panel: The fluxes of the "visible" particles, nucleons and γ–rays. 1 sigma error bars are the combined data from the Haverah Park [3], the Fly's Eye [5], and the AGASA [6] experiments above 10^{19} eV. Also shown are piecewise power law fits to the observed charged CR flux below 10^{19} eV, the measurement of the diffuse γ–ray flux between 30 MeV and 100 GeV by the EGRET instrument [53], as well as upper limits on the diffuse γ–ray flux at higher energies from the HEGRA [73], the Utah-Michigan [74], and the CASA-MIA [75] experiments, as indicated (see Ref. [15] for more details). Upper panel: Neutrino fluxes. Shown are experimental neutrino flux limits from the Frejus underground detector [76], the Fly's Eye [52], the Goldstone radio telescope [79], and the Antarctic Muon and Neutrino Detector Array (AMANDA) neutrino telescope [80], as well as projected neutrino flux sensitivities of ICECUBE, the planned kilometer scale extension of AMANDA [81], the Pierre Auger Project [82] (for electron and tau neutrinos separately) and the proposed space based OWL [12] concept. For comparison also shown are the atmospheric neutrino background [83] (hatched region marked "atmospheric"), and neutrino flux predictions for a model of active galactic nuclei optically thick to nucleons ("AGN"), and for "cosmogenic" UHECR interactions with the CMB [84] ("$N\gamma$", dashed range indicating typical uncertainties for moderate source evolution). The top-down fluxes are shown for electron-, muon, and tau-neutrinos separately, assuming no (lower ν_τ-curve) and maximal $\nu_\mu - \nu_\tau$ mixing (upper ν_τ-curve, which would then equal the ν_μ-flux), respectively

evolved spectrum in the EHE region, for the assumed strengths of the URB and the EGMF, is still dominated by nucleons over photons. The situation may, however, well be reversed for other assumptions on URB and EGMF strengths; see Ref. [15] for examples. We emphasize this point here because it is often naively stated in the literature that the TD scenario *always* predicts a photon dominated evolved EECR spectrum, and so the TD scenario would be ruled out in the event of confirmation of a nucleon dominated EECR spectrum by future data. Clearly, this is not true because, in the TD scenario, the predicted composition of the EECR depends significantly on the strengths of the URB and EGMF which are rather uncertain at this time.

In some TD models — as in the case of monopole-antimonopole annihilation through formation and subsequent collapse of metastable monopole-antimonopole bound states called monopolonia, or in the case of collapsing necklaces — the relevant TDs would be clustered in the GH, giving rise to predicted EECR spectra having properties similar to those in the MSRP decay model discussed above, in which case the EECR would always be photon dominated because the propagation and evolution of the spectra would then be unaffected by URB and EGMF.

3.3 Neutrino Fluxes in the Top–Down Scenario

Perhaps, the most important aspect of the extragalactic top-down models in general is the predicted dominant EHE neutrino flux whose possible detection in the up-coming experiments would provide a clear signature of the top-down scenario. As discussed in Sect. 3.1, in top-down scenarios most of the energy is released in the form of EM particles and neutrinos. If the X particles decay into a quark and a lepton, the quark hadronizes mostly into pions and the ratio of energy release into the neutrino versus EM channel is $r \simeq 0.3$.

In the absence of neutrino oscillations the electron neutrino and anti-neutrino fluxes are about a factor of 2 smaller than the muon neutrino and anti-neutrino fluxes, whereas the τ−neutrino flux is in general negligible. In contrast, if the interpretation of the atmospheric neutrino deficit in terms of nearly maximal mixing of muon and τ−neutrinos proves correct, the muon neutrino fluxes would be maximally mixed with the τ−neutrino fluxes. Figure 2 shows that the TD flux component clearly dominates above $\sim 10^{19}$ eV.

In order to translate neutrino fluxes into event rates, one has to fold in the interaction cross sections with matter, i.e. with nucleons and nuclei. At UHEs these cross sections are not directly accessible to laboratory measurements. Resulting uncertainties therefore translate directly to bounds on neutrino fluxes derived from, for example, the non-detection of UHE atmospheric muons produced in charged current interactions. In the following, we will assume the estimate

$$\sigma_{\nu N}(E) \simeq 2.36 \times 10^{-32} \left(\frac{E}{10^{19}\,\mathrm{eV}}\right)^{0.363} \mathrm{cm}^2 \quad (10^{16}\,\mathrm{eV} \lesssim E \lesssim 10^{21}\,\mathrm{eV})\,. \quad (6)$$

based on the Standard Model for the charged current muon-neutrino-nucleon cross section $\sigma_{\nu N}$ [90] if not indicated otherwise.

For an (energy dependent) ice or water equivalent acceptance $A(E)$ (in units of volume times solid angle), one can obtain an approximate expected rate of UHE muons produced by neutrinos with energy $> E$, $R(E)$, by multiplying $A(E)\sigma_{\nu N}(E)n_{\mathrm{H_2O}}$ (where $n_{\mathrm{H_2O}}$ is the nucleon density in water) with the integral muon neutrino flux $\simeq Ej_{\nu_\mu}$. This can be used to derive upper limits on diffuse neutrino fluxes from a non-detection of muon induced events. Figure 2 shows bounds obtained from several experiments: The Frejus experiment derived upper bounds for $E \gtrsim 10^{12}$ eV from their non-detection of almost horizontal muons with an energy loss inside the detector of more than 140 MeV per radiation length [76]. The AMANDA neutrino telescope has established an upper limit in the TeV-PeV range [80]. The Fly's Eye experiment derived upper bounds for the energy range between $\sim 10^{17}$ eV and $\sim 10^{20}$ eV [52] from the non-observation of deeply penetrating particles. The NASA Goldstone radio telescope has put an upper limit from the non-observation of pulsed radio emission from cascades induced by neutrinos above $\simeq 10^{20}$ eV in the lunar regolith (this quite strongly depends on systematic effects and the limit shown in Fig. 2 is an optimistic estimate). The AKENO group has published an upper bound on the rate of near-horizontal, muon-poor air showers [91] (not shown in Fig. 2). Horizontal air showerscreated by electrons, muons or tau leptons that are in turn produced by charged current reactions of electron, muon or tau neutrinos within the atmosphere have recently also been pointed out as an important method to constrain or measure UHE neutrino fluxes [82] with next generation detectors.

Clearly, the TD model shown in Fig. 2 is not only consistent with observed "visible" particle fluxes, but also with all existing neutrino flux limits within 2-3 orders of magnitude. What, then, are the prospects of detecting UHE neutrino fluxes predicted by TD models? In a 1 km^3 2π sr size detector, the scenario from Fig. 2, for example, predicts a muon-neutrino event rate of $\simeq 0.08\,\mathrm{yr}^{-1}$, and an electron neutrino event rate of $\simeq 0.05\,\mathrm{yr}^{-1}$ above 10^{19} eV, where "backgrounds" from conventional sources should be negligible. Further, the muon-neutrino event rate above 1 PeV should be $\simeq 0.6\,\mathrm{yr}^{-1}$, which could be interesting if conventional sources produce neutrinos at a much smaller flux level. Moreover, the neutrino flux around 10^{17} eV could have a slight enhancement due to neutrinos from muons produced by interactions of UHE photons and electrons with the CMB at high redshift [92], an effect that has not been taken into account in Figs. 2 and 3. Of course, above $\simeq 100$ TeV, instruments using ice or water as detector medium, have to look at downward going muon and electron events due to neutrino absorption in the Earth. However, $\tau-$neutrinos obliterate this Earth shadowing effect due to their regeneration from τ decays [93]. The presence of $\tau-$neutrinos, for example, due to mixing with muon neutrinos, as suggested by recent experimental results from Super-Kamiokande, can therefore lead to an increased upward going event rate [94]. $\tau-$neutrinos skimming the Earth at small angles below the horizon can also lead to an increase of sensitivity of fluorescence and ground array detectors [95,96,97].

For detectors based on the fluorescence technique such as the HiRes [10] and the Telescope Array [11] (see Sect. 1), the sensitivity to UHE neutrinos is often expressed in terms of an effective aperture $a(E)$ which is related to $A(E)$ by $a(E) = A(E)\sigma_{\nu N}(E)n_{\mathrm{H_2O}}$. For the cross section of Eq. (6), the apertures given in Ref. [10] for the HiRes correspond to $A(E) \simeq 3\,\mathrm{km}^3 \times 2\pi\,\mathrm{sr}$ for $E \gtrsim 10^{19}\,\mathrm{eV}$ for muon neutrinos. The expected acceptance of the ground array component of the Pierre Auger project for horizontal UHE neutrino induced events is $A(10^{19}\,\mathrm{eV}) \simeq 20\,\mathrm{km}^3\,\mathrm{sr}$ and $A(10^{23}\,\mathrm{eV}) \simeq 200\,\mathrm{km}^3\,\mathrm{sr}$ [82], with a duty cycle close to 100%. We conclude that detection of neutrino fluxes predicted by scenarios such as the scenario shown in Fig. 2 requires running a detector of acceptance $\gtrsim 10\,\mathrm{km}^3 \times 2\pi\,\mathrm{sr}$ over a period of a few years. Apart from optical detection in air, water, or ice, other methods such as acoustical and radio detection [98] (see, e.g., the RICE project [99] for the latter) or even detection from space [12,14,13] appear to be interesting possibilities for detection concepts operating at such scales. For example, the space based OWL/AirWatch satellite concept would have an aperture of $\simeq 3 \times 10^6\,\mathrm{km}^2\,\mathrm{sr}$ in the atmosphere, corresponding to $A(E) \simeq 6 \times 10^4\,\mathrm{km}^3\,\mathrm{sr}$ for $E \gtrsim 10^{20}\,\mathrm{eV}$, with a duty cycle of $\simeq 0.08$ [12]. The backgrounds seem to be in general negligible [100,101]. As indicated by the numbers above and by the projected sensitivities shown in Fig. 2, the Pierre Auger Project and especially the space based AirWatch type projects should be capable of detecting typical TD neutrino fluxes. This applies to any detector of acceptance $\gtrsim 100\,\mathrm{km}^3\,\mathrm{sr}$. Furthermore, a 100 day search with a radio telescope of the NASA Goldstone type for pulsed radio emission from cascades induced by neutrinos or cosmic rays in the lunar regolith could reach a sensitivity comparable to or better than the Pierre Auger sensitivity above $\sim 10^{19}\,\mathrm{eV}$ [79].

A more model independent estimate [86] for the average event rate $R(E)$ can be made if the underlying scenario is consistent with observational nucleon and $\gamma-$ray fluxes and the bulk of the energy is released above the pair production threshold on the CMB at $\simeq 3 \times 10^{14}\,\mathrm{eV}$. Let us assume that the ratio of energy injected into the neutrino versus EM channel is a constant r. As discussed above, cascading effectively reprocesses most of the injected EM energy into low energy photons whose spectrum peaks at $\simeq 10\,\mathrm{GeV}$ [102]. Since the ratio r remains roughly unchanged during propagation, the height of the corresponding peak in the neutrino spectrum should roughly be r times the height of the low-energy diffuse $\gamma-$ray peak, i.e., we have the condition $\max_E\left[E^2 j_{\nu_\mu}(E)\right] \simeq r \max_E\left[E^2 j_\gamma(E)\right]$. Imposing the observational upper limit on the diffuse $\gamma-$ray flux around 10 GeV shown in Fig. 2, $\max_E\left[E^2 j_{\nu_\mu}(E)\right] \lesssim 2 \times 10^3 r\,\mathrm{eV cm^{-2} sec^{-1} sr^{-1}}$, then bounds the average diffuse neutrino rate above pair production threshold on the CMB, giving

$$R(E) \lesssim 0.34\,r\left[\frac{A(E)}{1\,\mathrm{km}^3 \times 2\pi\,\mathrm{sr}}\right]\left(\frac{E}{10^{19}\,\mathrm{eV}}\right)^{-0.6}\,\mathrm{yr}^{-1} \quad (E \gtrsim 10^{15}\,\mathrm{eV}), \qquad (7)$$

assuming the Standard Model cross section Eq. (6). Comparing this with the flux bounds shown in Fig. 2 results in an upper bound on r. For example, the Fly's Eye bound translates into $r \lesssim 20(E/10^{19}\,\mathrm{eV})^{0.1}$. We stress again that TD models are not subject to the Waxman Bahcall bound which applies only to neutrinos produced as secondaries of primary charged CRs (see Sect. 4.1 in the contribution by G. Sigl in this volume for more details). In contrast, in the top-down scenarios the nucleons produced are considerably less abundant than and are not the primaries of injected γ−rays and neutrinos.

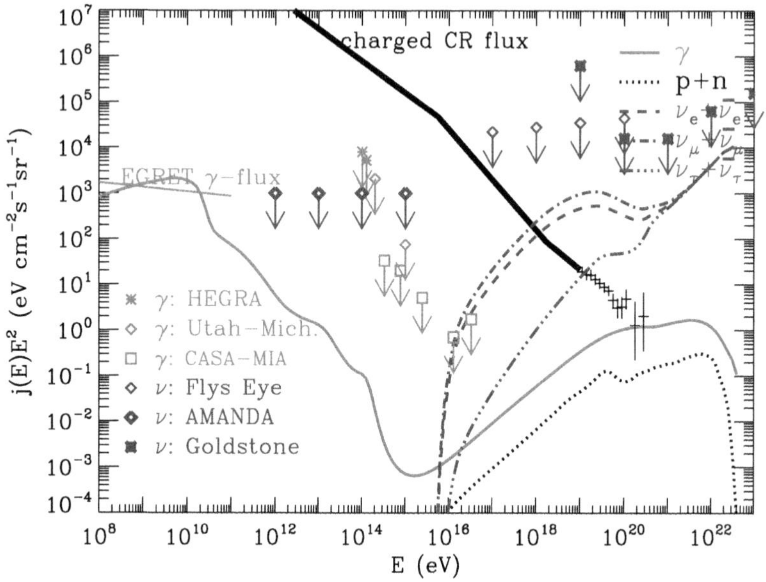

Fig. 3. Flux predictions for a TD model characterized by $p = 1$, $m_X = 10^{14}$ GeV, with X particles exclusively decaying into neutrino-antineutrino pairs of all flavors (with equal branching ratio), assuming neutrino masses $m_{\nu_e} = 0.1$ eV, $m_{\nu_\mu} = m_{\nu_\tau} = 1$ eV. For neutrino clustering, an overdensity of $\simeq 50$ over a scale of $l_\nu \simeq 5$ Mpc was assumed. The calculation assumed an intermediate URB estimate from Ref. [103] and an EGMF $\ll 10^{-11}$ G. Flux upper limits are as in Fig. 2

In typical TD models such as the one discussed above where primary neutrinos are produced by pion decay, $r \simeq 0.3$. However, in TD scenarios with $r \gg 1$ neutrino fluxes are only limited by the condition that the *secondary* γ−ray flux produced by neutrino interactions with the RNB be below the experimental limits. In this case the observed EECR flux would be produced by the Z-burst mechanism discussed in Sect. 3.1. An example for such a scenario is given by X particles exclusively decaying into neutrinos (although this is not very likely in most particle physics models, but see Ref. [49] and Fig. 3 for a scenario involving topological defects and Ref. [104] for a scenario involving decaying superheavy relic particles, both of which explain the observed EECR events as secondaries of neutrinos interacting with the RNB). Such scenarios predict appreciable event rates above $\sim 10^{19}$ eV in a km^3 scale detector, but require unrealistically strong clustering of relic neutrinos (a homogeneous relic neutrino overdensity would make the EGRET constraint only more severe because neutrino interactions beyond $\sim 50\,\mathrm{Mpc}$ contribute to the diffuse GeV γ−ray background but not to the UHECR flux). A detection would thus open the exciting possibility to establish an experimental lower limit on r. Being based solely on energy conservation, Eq. (7) holds regardless of whether or not the underlying TD mechanism explains the observed EECR events.

The transient neutrino event rate could be much higher than Eq. (7) in the direction to discrete sources which emit particles in bursts. Corresponding pulses in the EHE nucleon and γ−ray fluxes would only occur for sources nearer than $\simeq 100\,\mathrm{Mpc}$ and, in case of protons, would be delayed and dispersed by deflection in Galactic and extragalactic magnetic fields [105,106]. The recent observation of a possible clustering of UHECR above $\simeq 4\times10^{19}\,\mathrm{eV}$ by the AGASA experiment [65] might suggest sources which burst on a time scale $t_b \ll 1\,\mathrm{yr}$. A burst fluence of $\simeq r\left[A(E)/1\,\mathrm{km}^3\times 2\pi\,\mathrm{sr}\right](E/10^{19}\,\mathrm{eV})^{-0.6}$ neutrino induced events within a time t_b could then be expected. Associated pulses could also be observable in the GeV − TeV γ−ray flux if the EGMF is smaller than $\simeq 10^{-15}\,\mathrm{G}$ in a significant fraction of extragalactic space [107].

In contrast to roughly homogeneous sources and/or mechanisms with branching ratios $r \gg 1$, in scenarios involving clustered sources such as metastable superheavy relic particles decaying with $r \sim 1$, the neutrino flux is comparable to (not significantly larger than) the UHE photon plus nucleon fluxes and thus comparable to the universal cosmogenic flux marked "$N\gamma$" in Fig. 2. This can be understood because the neutrino flux is dominated by the extragalactic contribution which scales with the extragalactic nucleon and γ−ray contribution in exactly the same way as in the unclustered case, whereas the extragalactic contribution to the "visible" flux to be normalized to the EECR data is much smaller in the clustered case. The resulting neutrino fluxes in these scenarios would thus be much harder to detect even with next generation experiments.

4 Conclusion

The solution of the EECR enigma seems to require some kind of new physics beyond the Standard Model, either to solve the problem of energetics or to solve the problem of absence of sufficiently powerful identifiable astrophysical sources in the nearby Universe. The future in this subject appears promising and exciting because several on-going as well as up-coming and proposed large EECR detectors will have the potential to probe some forms of possible new physics beyond the Standard Model suggested in this context.

References

1. J. Linsley: Phys. Rev. Lett. **10**, 146 (1963); Proc. 8th *International Cosmic Ray Conference*, **4**, 295 (1963)
2. R. G. Brownlee, et al.: Can. J. Phys. **46**, S259 (1968)
   M. M. Winn, et al.: J. Phys. G **12**, 653 (1986)
   See also http://www.physics.usyd.edu.au/hienergy/sugar.html

3. M. A. Lawrence, R. J. O. Reid, A. A. Watson: J. Phys. G Nucl. Part. Phys. **17**, 733 (1991) and references therein
See also http://ast.leeds.ac.uk/haverah/hav-home.html
4. B. N. Afnasiev: Proc. *International Symposium on Extremely High Energy Cosmic Rays: Astrophysics and Future Observatories*, ed. by M. Nagano (Institute for Cosmic Ray Research, Tokyo 1996) p. 32
5. D. J. Bird, et al.: Phys. Rev. Lett. **71**, 3401 (1993); Astrophys. J. **424**, 491 (1994); ibid. **441**, 144 (1995)
6. N. Hayashida, et al.: Phys. Rev. Lett. **73**, 3401 (1994)
S. Yoshida, et al.: Astropart. Phys. **3**, 105 (1995)
M. Takeda, et al.: Phys. Rev. Lett. **81**, 1163 (1998)
See also http://www-akeno.icrr.u-tokyo.ac.jp/AGASA/
7. J. W. Cronin: Proc. *Unsolved Problems in Astrophysics*, ed. by J. N. Bahcall and J. P. Ostriker (Princeton University Press, Princeton 1997) p. 325
8. J. W. Cronin: Rev. Mod. Phys. **71**, S165 (1999)
9. J. W. Cronin: Nucl. Phys. B (Proc. Suppl.) **28B**, 213 (1992)
See also http://www.auger.org/ and
http://www-lpnhep.in2p3.fr/auger/welcome.html
10. S. C. Corbató, et al.: Nucl. Phys. B (Proc. Suppl.) **28B**, 36 (1992)
D. J. Bird, et al.: Proc. *24th International Cosmic Ray Conference* (Rome 1995) **2**, 504 (1995); **1**, 750
See also http://hires.physics.utah.edu/
11. M. Teshima, et al.: Nucl. Phys. B (Proc. Suppl.) **28B**, 169 (1992)
See also http://www-ta.icrr.u-tokyo.ac.jp/
12. D. B. Cline, F. W. Stecker: *OWL/AirWatch science white paper*, astro-ph/0003459
See also http://lheawww.gsfc.nasa.gov/docs/gamcosray/hecr/OWL/
13. J. Linsley: Proc. *25th International Cosmic Ray Conference*, Durban **5**, 381 (1997)
See also http://www.ifcai.pa.cnr.it/Ifcai/airwatch.html
14. See http://www.ifcai.pa.cnr.it/Ifcai/euso.html
15. P. Bhattacharjee, G. Sigl: Phys. Rep. **327**, 109 (2000)
16. G. Sigl: Science **291**, 73 (2001)
17. A. V. Olinto: Phys. Rep. **333**, 329 (2000); astro-ph/0011106
18. M. Nagano, A. A. Watson: Rev. Mod. Phys. **72**, 689 (2000)
19. P. L. Biermann: J. Phys. G: Nucl. Part. Phys. **23**, 1 (1997)
20. V. Berezinsky: Nucl. Phys. B (Proc. Suppl.) **70B**, 419 (1999)
21. T. Weiler: hep-ph/0103023
22. S. Yoshida, H. Dai: J. Phys. G **24**, 905 (1998)
23. A. M. Hillas: Ann. Rev. Astron. Astrophys. **22**, 425 (1984)
24. C. A. Norman, D. B. Melrose, A. Achterberg: Astrophys. J. **454**, 60 (1995)
25. L. O'C. Drury: Rep. Prog. Phys. **46**, 973 (1983)
R. D. Blandford, D. Eichler: Phys. Rep. **154**, 1 (1987)
F. C. Jones, D. C. Ellison: Sp. Sc. Rev. **58**, 259 (1991)
F. Takahara: Proc. *Physics and Astrophysics of Neutrinos*, ed. M. Fukugita, A. Suzuki (Springer-Verlag, Tokyo 1994) p. 900
J. G. Kirk, P. Duffy: J. Phys. G: Nucl. Part. Phys. **25**, R163 (1999)
26. K. Greisen: Phys. Rev. Lett. **16**, 748 (1966)
G. T. Zatsepin, V. A. Kuzmin: Pis'ma Zh. Eksp. Teor. Fiz. **4**, 114 (1966) [JETP. Lett. **4**, 78 (1966)]

27. F. A. Aharonian, J. W. Cronin: Phys. Rev. D **50**, 1892 (1994)
28. F. W. Stecker: Phys. Rev. **180**, 1264 (1969)
J. L. Puget, F. W. Stecker, J. H. Bredekamp: Astrophys. J. **205**, 638 (1976)
L. N. Epele, E. Roulet: Phys. Rev. Lett. **81**, 3295 (1998)
F. W. Stecker: Phys. Rev. Lett. **81**, 3296 (1998)
29. A. I. Nikishov: Zh. Eksp. Teor. Fiz **41**, 549 (1961) [Sov. Phys. JETP **41**, 393 (1962)]
J. V. Jelley: Phys. Rev. Lett. **16**, 479 (1966)
R. J. Gould, G. P. Schréder: Phys. Rev. **155**, 1404 (1967)
V. Berezinskii: Sov. J. Nucl. Phys **11**, 222 (1970)
30. C. T. Hill, D. N. Schramm: Phys. Rev. D **31**, 564 (1985)
31. V. S. Berezinsky, S. V. Bulanov, V. A. Dogiel, V. L. Ginzburg, V. S. Ptuskin: *Astrophysics of Cosmic Rays* (North-Holland, Amsterdam 1990)
32. P. Bhattacharjee, C. T. Hill, D. N. Schramm: Phys. Rev. Lett. **69**, 567 (1992)
33. M. Blanton, P. Blasi, A. Olinto: Astropart. Phys. **15**, 275 (2001)
34. H. Sato, T. Tati: Prog. Theor. Phys. **47**, 1788 (1972)
D. A. Kirzhnits, V. A. Chechin: Sov. J. Nucl. Phys. **15**, 585 (1972)
L. Gonzalez-Mestres: Nucl. Phys. B (Proc. Suppl.) **48B**, 131 (1996)
35. S. Coleman, S. L. Glashow: Phys. Lett. **B405**, 249 (1997); Phys. Rev. D **59**, 116008 (1999)
36. G. Amelino-Camelia, T. Piran: Phys. Lett. **B497**, 265 (2001)
37. R. Aloisio, P. Blasi, P. Ghia, A. Grillo: Phys. Rev. D **62**, 053010 (2000)
38. R. Jackiw, V. A. Kostelecky: Phys. Rev. Lett. **82**, 3572 (1999)
39. D. J. H. Chung, G. R. Farrar, E. W. Kolb: Phys. Rev. D **57**, 4696 (1998)
G. R. Farrar: astro-ph/9801020
40. G. R. Farrar: Phys. Rev. Lett. **76**, 4111 (1996)
41. G. R. Farrar, P. L. Biermann: Phys. Rev. Lett. **81**, 3579 (1998)
42. C. M. Hoffman: Phys. Rev. Lett. **83**, 2471 (1999)
G. R. Farrar, P. L. Biermann: ibid. **83**, 2472 (1999)
G. Sigl, D. F. Torres, L. A. Anchordoqui, G. E. Romero: Phys. Rev. D **63**, 081302 (2001)
A. Virmani, et al.: astro-ph/0010235; to appear in Astropart. Phys.
43. I. F. Albuquerque, et al. (E761 collaboration): Phys. Rev. Lett. **78**, 3252 (1997)
A. Alavi-Harati, et al. (KTeV collaboration): Phys. Rev. Lett. **83**, 2128 (1999)
44. V. Berezinsky, M. Kachelrieß , S. Ostapchenko: astro-ph/0109026
45. T. J. Weiler: Astropart. Phys. **11**, 303 (1999)
D. Fargion, B. Mele, A. Salis: Astrophys. J. **517**, 725 (1999)
46. Y. Fukuda, et al. (Super-Kamiokande collaboration): Phys. Rev. Lett. **81**, 1562 (1998)
47. Review of Particle Properties: Euro. Phys. J. **C15**, 1 (2000)
48. S. Yoshida, G. Sigl, S. Lee: Phys. Rev. Lett. **81**, 5505 (1998)
49. G. Sigl, S. Lee, P. Bhattacharjee, S. Yoshida: Phys. Rev. D **59**, 043504 (1999)
50. J. J. Blanco-Pillado, R. A. Vázquez, E. Zas: Phys. Rev. D **61**, 123003 (2000)
51. Z. Fodor, S. D. Katz, A. Ringwald: hep-ph/0105064
52. R. M. Baltrusaitis, et al.: Astrophys. J. **281**, L9 (1984); Phys. Rev. D **31**, 2192 (1985)
53. P. Sreekumar, et al.: Astrophys. J. **494**, 523 (1998)
54. G. Sigl, S. Lee: Proc. *24th International Cosmic Ray Conference*, Rome, **3**, 356 (1995)
55. E. Waxman: astro-ph/9804023

56. G. Gelmini, A. Kusenko: Phys. Rev. Lett. **82**, 5202 (1999)
57. H. Päs, T. Weiler: Phys. Rev. D **63**, 113015 (2001)
58. V. A. Kuzmin, V. A. Rubakov: Phys. Atom. Nucl. **61**, 1028 (1998) [Yad. Fiz. **61**, 1122 (1998)]
59. V. Berezinsky, M. Kachelrieß , A. Vilenkin: Phys. Rev. Lett. **79**, 4302 (1997)
60. J. Ellis, J. L. Lopez, D. V. Nanopolous: Phys. Lett. **B247**, 257 (1990)
J. Ellis, G. Gelmini, J. L. Lopez, D. V. Nanopolous, S. Sarkar: Nucl. Phys. **373B**, 399 (1992)
K. Benakli, J. Ellis, D. V. Nanopolous: Phys. Rev. D **59**, 047301 (1999)
61. S. Chang, C. Corianò, A. E. Faraggi: Nucl. Phys. **477B**, 65 (1996)
P. H. Frampton, B. Keszthelyi, Y. J. Ng: Int. J. Mod. Phys. D **8**, 117 (1999)
K. Hamaguchi, Y. Nomura, T. Yanagida: Phys. Rev. D **58**, 103503 (1998); Phys. Rev. D **59**, 063507 (1999)
C. P. Burgess, L. E. Ibá~nez, F. Quevedo: Phys. Lett. **B447**, 257 (1999)
K. Hamaguchi, K. -I. Izawa, Y. Nomura, T. Yanagida: Phys. Rev. D **60**, 125009 (1999)
G. Dvali: Phys. Lett. **B459**, 489 (1999)
K. Hagiwara, Y. Uehara: Phys. Lett. B **517**, 383 (2001)
62. V. Kuzmin and I. Tkachev: Phys. Rep. **320**, 199 (1999)
63. S. L. Dubovsky, P. G. Tinyakov: Sov. Phys. JETP. Lett. **68**, 107 (1998)
A. Benson, A. Smialkowski, A. Wolfendale: Astropart. Phys. **10**, 313 (1999)
G. A. Medina Tanco, A. A. Watson: Astropart. Phys. **12**, 25 (1999)
V. S. Berezinsky, A. A. Mikhailov: Phys. Lett. **B449**, 237 (1999)
T. Wibig, A. W. Wolfendale: J. Phys. G **26**, 1033 (2000)
W. Evans, F. Ferrer, S. Sarkar: astro-ph/0103085; to appear in Astropart. Phys.
64. V. Berezinsky: Nucl. Phys. Proc. Suppl. **87**, 387 (2000) [hep-ph/0001163]
P. Blasi, R. K. Sheth: Phys. Lett. **B486**, 233 (2000)
65. N. Hayashida, et al.: Phys. Rev. Lett. **77**, 1000 (1996)
Y. Uchihori, et al.: Astropart. Phys. **13**, 151 (2000)
66. P. Blasi, R. Dick, E. W. Kolb: astro-ph/0105232
67. S. Sarkar, R. Toldrà: hep-ph/0108098
68. V. Berezinsky, M. Kachelrieß: Phys. Rev. D **63**, 034007 (2001)
69. Z. Fodor, S. D. Katz: Phys. Rev. Lett. **86**, 3224 (2001)
70. A. Vilenkin, E. P. S. Shellard: *Cosmic Strings and other Topological Defects* (Cambridge Univ. Press, Cambridge 1994)
71. X. Siemens, K. D. Olum: Nucl. Phys. B **611**, 125 (2001)
72. G. Sigl, K. Jedamzik, D. N. Schramm, V. Berezinsky: Phys. Rev. D **52**, 6682 (1995)
73. A. Karle, et al.: Phys. Lett. **B347**, 161 (1995)
74. J. Matthews, et al.: Astrophys. J. **375**, 202 (1991)
75. M. C. Chantell, et al.: Phys. Rev. Lett. **79**, 1805 (1997)
76. W. Rhode, et al.: Astropart. Phys. **4**, 217 (1996)
77. Proc. *26th International Cosmic Ray Conference*, (Utah, 1999)
78. Proc. *First International Workshop on Radio Detection of High-Energy Particles*, Amer. Inst. of Phys., 2001, and at
http://www.physics.ucla.edu/~moonemp/radhep/workshop.html
79. P. W. Gorham, K. M. Liewer, C. J. Naudet: astro-ph/9906504, in Ref. [77]
P. W. Gorham, et al.: astro-ph/0102435, in Ref. [78].
80. For general information see http://amanda.berkeley.edu/
see also F. Halzen: New Astron. Rev **42**, 289 (1999)

for newest status see, e.g., G. C. Hill, et al. (Amanda Collaboration): astro-ph/0106064
81. For general information see http://www.ps.uci.edu/~icecube/workshop.html
see also F. Halzen: Am. Astron. Soc. Meeting **192**, # 62 28 (1998)
AMANDA collaboration: astro-ph/9906205, Proc. *8th Int. Workshop on Neutrino Telescopes* (Venice, Feb 1999)
82. J. J. Blanco-Pillado, R. A. Vázquez, E. Zas: Phys. Rev. Lett. **78**, 3614 (1997)
K. S. Capelle, J. W. Cronin, G. Parente, E. Zas: Astropart. Phys. **8**, 321 (1998)
A. Letessier-Selvon: astro-ph/0009444
83. See, e.g., P. Lipari: Astropart. Phys. **1**, 195 (1993)
84. See, e.g., R. J. Protheroe, P. A. Johnson: Astropart. Phys. **4**, 253 (1996); erratum ibid. **5**, 215 (1996)
85. R. J. Protheroe, T. Stanev: Phys. Rev. Lett. **77**, 3708 (1996); erratum, ibid. **78**, 3420 (1997)
86. G. Sigl, S. Lee, D. N. Schramm, P. S. Coppi: Phys. Lett. **B392**, 129 (1997)
87. G. R. Vincent, N. D. Antunes, M. Hindmarsh: Phys. Rev. Lett. **80**, 2277 (1998)
G. R. Vincent, M. Hindmarsh, M. Sakellariadou: Phys. Rev. D **56**, 637 (1997)
88. V. Berezinsky, A. Vilenkin: Phys. Rev. Lett. **79**, 5202 (1997)
89. U. F. Wichoski, J. H. MacGibbon, R. H. Brandenberger: hep-ph/9805419
U. F. Wichoski, R. H. Brandenberger, J. H. MacGibbon: hep-ph/9903545, Proc. *2nd Meeting New Worlds in Astroparticle Physics* (Faro, Portugal, 3-5 Sep 1998), eds. A. Mourao, M. Pimenta and P. Sa (World Scientific, Singapore 1999)
P. Bhattacharjee, Q. Shafi, F. W. Stecker: Phys. Rev. Lett. **80**, 3698 (1998)
90. R. Gandhi, C. Quigg, M. H. Reno, I. Sarcevic: Astropart. Phys. **5**, 81 (1996)
R. Gandhi, C. Quigg, M. H. Reno, I. Sarcevic: Phys. Rev. D **58**, 093009 (1998)
91. M. Nagano, et al.: J. Phys. G **12**, 69 (1986)
92. M. Postma: Phys. Rev. D **64**, 023001 (2001)
See also H. Athar, G.-L. Lin, J.-J. Tseng: Phys. Rev. D **64**, 071302 (2001)
93. F. Halzen, D. Saltzberg: Phys. Rev. Lett. **81**, 4305 (1998)
94. K. Mannheim: Astropart. Phys. **11**, 49 (1999)
95. D. Fargion: astro-ph/0002453, astro-ph/0101565
96. X. Bertou, et al.: astro-ph/0104452; to appear in Astropart. Phys.
97. J. L. Feng, P. Fisher, F. Wilczek, T. M. Yu: hep-ph/0105067
A. Kusenko, T. Weiler: hep-ph/0106071
98. T. K. Gaisser, F. Halzen, T. Stanev: Phys. Rep. **258**, 173 (1995)
99. For general information see http://kuhep4.phsx.ukans.edu/~iceman/index.html
100. S. Yoshida, H. Dai, C. C. H. Jui, P. Sommers: Astrophys. J. **479**, 547 (1997)
101. P. B. Price: Astropart. Phys. **5**, 43 (1996)
102. P. S. Coppi, F. A. Aharonian: Astrophys. J. **487**, L9 (1997)
103. R. J. Protheroe, P. L. Biermann: Astropart. Phys. **6**, 45 (1996)
104. G. Gelmini, A. Kusenko: Phys. Rev. Lett. **84**, 1378 (2000)
G. Gelmini: hep-ph/0005263, Proc. *4th International Symposium on Sources and Detection of Dark Matter in the Universe*, (Marina del Rey, February 2000)
105. E. Waxman, J. Miralda-Escudé: Astrophys. J. **472**, L89 (1996)
106. M. Lemoine, G. Sigl, A. V. Olinto, D. N. Schramm: Astrophys. J. **486**, L115 (1997)
107. E. Waxman, P. S. Coppi: Astrophys. J. **464**, L75 (1996)

Summary of the School: A Critical View on the Origin of the Ultra-High-Energy Cosmic Rays

Murat Boratav and Antoine Letessier-Selvon

Universités Paris 6 & Paris 7, Laboratoire de Physique Nucléaire et de Hautes Energies (CNRS/IN2P3), 4 Place Jussieu, F-75005 Paris, France

Abstract. The Meudon UHECR2000 school gathered a large number of experts in the various fields in relation with the sources of ultra high energy cosmic rays as well as with the experiments and projects aiming at their observation. Many of the attendants were the young physicists who, no doubt, will "unravel" the threads of the ultra-high energy cosmic ray puzzle with the next generation of experiments.

The authors of this article were asked by the organizers of the School to write a summary of what was said and discussed during these three full days of lectures and debates. We shall do so by giving a short account of the various contributions.

1 Introduction – Overview and Open Questions

The origin of ultra-high energy cosmic rays (UHECR) is one of the outstanding puzzles of modern astrophysics. Although some authors (including some among the contributors to this volume) assert that there is nothing mysterious about them, the innumerable articles written on the subject during these last few years, the many models or theories advocated in favor of the mechanism that produces them and the exciting but controversial debates going on about their nature and origin are proof that nothing decisive has yet been said about the UHECR. The problematics boils down to a few basic questions. Are there astrophysical engines capable of accelerating particles to ZeV ($= 10^{21}$ eV) energies? Are the sources at cosmological distances, and if yes why is the Greisen-Zatsepin-Kuzmin (GZK) cutoff violated? If they are in our neighborhood, why don't we see some counterparts of the necessarily remarkable accelerating mechanism? If the UHECR are results of a top-down decay of some supermassive particle, how did such particles survive from the Big Bang to the present epoch? If the sources are a few and point-like, why do their images seem to be isotropic with no obvious correlation with the local luminous mass distribution? If they are diffuse, why do we see several multiplets in the incoming directions not compatible with a chance coincidence? And so on.

Any of the questions listed above finds one or several answers provided by one or several authors. However none of these answers finds a consensus of opinion among the community. Moreover there is no single model answering all the questions without calling on a few "small miracles". No doubt we will have to await the next generation of experiments (Pierre Auger Observatory, Telescope Array, EUSO...) and their high quality and high statistics data to decide on

Martin Lemoine, Günter Sigl (Eds.): LNP 576, pp. 300–317, 2001.

which paths should be abandoned and which should be explored further to have (hopefully) a definitive answer.

The object of this school was to give an overview of some of these paths and to bring the defenders of various models, experimentalists and theorists, to study together some new ideas to be gone into thoroughly. The opening article of this volume by P. Biermann and G. Sigl is a compact and comprehensive review of what we know and what we don't on cosmic rays over a large energy range (from the TeV to the ZeV).

2 Nature and Propagation of UHECR

Physical effects governing the propagation of the cosmic rays are of paramount importance in what we observe when we detect them. Energy loss processes, propagation distances, deviations by galactic or extragalactic magnetic fields, interactions with Earth's environment (geomagnetic fields, atmosphere) are as decisive on the parameters of the detected cosmic rays as the mechanism which produced them. This is why we prefer to start by an overview of the propagation phenomena and especially of the detailed contribution by G. Sigl. In the following, we use the definition given in this article for the acronym "UHECR", namely cosmic rays with detected energies above 1 EeV (10^{18} eV), and limit our comments to them.

The chemical composition of the UHECR is unknown. However, the number of stable particles which can propagate over cosmological distances is quite limited: heavy or light atomic nuclei, photons and neutrinos. Electrons are not considered as potential UHECR because they radiate most of their energy while crossing the cosmic magnetic fields. To these we should add a number of exotic particles or interaction which will be briefly mentioned below.

2.1 Nucleons and Nuclei: The GZK Cutoff

Light or heavy atomic nuclei (with special emphasis on protons and the most stable iron) are the favorite candidates for UHECR, at least for astrophysical acceleration mechanisms. There is indeed some (weak) experimental indications that one can interpret, in the relevant energy range, as showing a composition which shifts from dominantly heavy (Fe) to dominantly light (p) nuclei (see [1] for a review). However this result is not statistically compelling and its interpretation is to some extent model dependent. A stronger empirical argument is that in an astrophysical acceleration process (Fermi or shock acceleration, unipolar mechanisms ...), all the other particles (conventional or exotic) are necessarily secondaries, therefore less likely candidates in a situation where we have a strong energy crisis.

The dominant energy loss process for UHE protons is the photo-production of pions when the proton reaches the threshold energy (a few tens of EeV) in its interactions with the 2.7 K cosmic microwave background (CMB) (see Sect. 2 of the contribution by G. Sigl). A relevant parameter to be considered is the energy attenuation length, $l_E = l\eta$ where l is the proton interaction length and η is the inelasticity, i.e. the fraction of the initial energy transferred to the final state leading particle. At the threshold energy, the attenuation length is less than 10 Mpc. The conclusion is that unless the proton starts with extravagant energies, its source must lie in a sphere of a few tens of Mpc at most: this is the so-called GZK cutoff. Other processes such as e^+e^- pair production, inverse Compton scattering on the CMB or energy loss due to cosmic expansion (redshift) are negligible above the cutoff. A recent article[2] gives a more detailed account of the proton energy losses and concludes that a local (less than 30 Mpc) over density of sources of at least a factor of 30 (with respect to a uniform distribution) is needed to suppress the GZK cutoff.

Neutrons, because of their β-decay, have ranges much shorter than protons (less than 1 Mpc at 100 EeV) and cannot be accelerated directly by electromagnetic processes. Therefore they are usually not considered as likely UHECR candidates.

Heavy nuclei are interesting since they are easier to accelerate and sometimes necessary ingredients when the sources are at the limit of the energy budget: neutron stars [3] or galactic $\gamma-$ray bursts [4]. They loose energy by photodisintegration. The giant dipole resonance can be reached by their interaction on the CMB or infrared (IR) background depending on the energy of the nucleus. The IR background is not very well known but based on the most recent measurements and simulations, it is likely that up to hundreds of EeV the attenuation length is mainly due to interactions with the CMB, in which case it would be comparable to that of the protons.

2.2 UHE Photons

Photons in the UHE range are of big interest as they would be (together with neutrinos) an almost inescapable signature of superheavy particle decays. In such a scenario, at the source, they are expected to dominate over ordinary hadrons by about a factor of ten. Of course they can also be the secondary products of the interaction of protons with the CMB (in which case they are usually called "GZK photons or neutrinos"). High energy photons traveling through the Universe produce e^+e^- pairs when colliding with the Infrared/Optical (IR/O),CMB, or Universal Radio Background (URB) photons. The threshold energy for pair production on CMB photons is around 3×10^{14} eV with interaction lengths down to $\simeq 10$ kpc. However, one should keep in mind that the "interaction length" is not the right parameter to measure the photon's propagation properties. The electron (positron) of the final state pair can produce a next generation photon by inverse Compton scattering starting an iterative electromagnetic cascading process. Following the leading particle of each generation, one ends up with attenuation lengths much larger than the interaction length. The attenuation length at the pair production threshold is quite small (tens of kpc at most) and remains below 100 Mpc even at the highest energies (see Fig. 4, in the contribution by G. Sigl) where the (poorly known) URB interactions take over.

This cascading process is very important in the sense that it will produce, through successive collisions on the various photon backgrounds, lower and lower energy cascades and pile up in the form of a diffuse photon background below TeV energies with a typical power law spectrum of index $\alpha = 1.5$. The measurements of the diffuse γ−ray background in the $10^7 - 10^{11}$ eV range done for example by EGRET [5] will impose limits on the photon production fluxes of top-down mechanisms and consequently on the abundance of topological defects (TDs) or relic superheavy particles.

2.3 Neutrinos

UHE neutrinos, like photons, are reliable signatures of top-down mechanisms. Moreover, their propagation being governed mainly by their interactions with the relic neutrino background (RNB), they can come from almost arbitrarily remote sources. We shall say more later on the possibility of their detection by ground based detectors. In the contributions by G. Sigl and S. Yoshida, a detailed discussion is presented on the neutrino interaction and propagation properties for various incident neutrino energies, neutrino masses and species. A few simple conclusions can be mentioned.

- The RNB has a temperature of $T_\nu = 1.9\,\mathrm{K}$ and a density of $n_\nu + n_{\bar{\nu}} \simeq 115\,\mathrm{cm}^{-3}$ per neutrino flavor.
- At all energies, neutrino-hadron interactions are always negligible (within the Standard Model); above the W production threshold $\nu\gamma$ interactions should however be taken into account.
- In the UHE range with $E_\nu \lesssim 10^{24}$ eV (i.e. grand unification energies) and reasonable RNB neutrino masses ($\lesssim 1$ eV), the $\nu\nu$ interactions are well described by the Standard Model (SM). Under such conditions, the interaction length of a 100 EeV neutrino substantially exceeds the size of the universe.
- The recently proposed "Z-burst" scenario [6] (UHECR produced in the decays of Z° produced by UHE neutrino-RNB interactions) which circumvents the GZK cutoff is most likely relevant only for top-down models (several tens of ZeV energies needed at production).

2.4 New Particles or Interactions

A series of other possibilities have been envisaged to explain the apparent absence of a cutoff at the highest energies. The neutrino obviously escapes the GZK crisis and can reach Earth from sources gigaparsecs away. However the SM interactions are not compatible with what we observe on the UHECR behavior penetrating the atmosphere. The size at ground level of all observed UHECR showers as well as the position of the shower maximum for a few of them seen by the Fluorescence Detector (FD) technique are all compatible with hadronic or electromagnetic interactions with the atmosphere.

An answer to this experimental contradiction would be that some new mechanism beyond the SM enhances the neutrino - nucleon cross sections and brings them close to hadronic values. Two such mechanisms were envisaged: a new broken SU(3) gauge symmetry or graviton exchanging interactions in the framework of theories with n additional compact dimensions. In the case of an SU(3) flavor symmetry the neutrino - nucleon cross section could have hadronic size, but even if the future detectors such as Pierre Auger, the IceCube or EUSO provide the high statistics that are missing at present, the hadron vs neutrino hypotheses will not be easy to discriminate (e.g. by looking at the energy dependence of the cross sections) unless we have a good knowledge of the fluxes at the source. In the extra dimension models the cross sections are in general too small to make the neutrino a primary candidate of the observed showers, but interesting constraints on its size could be derived from the observation of deeply penetrating air showers.

A supersymmetric solution was also envisaged which moves the incident cosmic ray's interaction threshold with the CMB to much higher values and makes the propagation over distances larger than a gigaparsec possible (Sect. 4.2 of the contribution by G. Sigl). The working hypothesis in this case is the existence of a baryonic bound state including a light gluino and called the S°. A mass larger than the proton (hence a larger threshold energy) and $\gamma - S^\circ$ cross sections lower than hadronic ones give to such a hypothetic particle the possibility to travel over distances up to 30 times larger than a proton. However, stringent limits coming from the accelerator data and the fact that the S° has to be a secondary product of e.g. a proton accelerated to much higher -multi ZeV- energies are constraints difficult to circumvent for such a model.

Some other non standard physics hypotheses may also allow the evading of the GZK cutoff by protons. One such possibility is to postulate a yet undetected violation of the Lorentz invariance, e.g. by the fact that the maximum attainable velocity for a particle would not be the universal c but some value depending on the particle species. Consequences of such models (substantial increase of the cutoff energies, transparency of the universe to UHE photons) and a few other similar attempts are described in Sect. 9 of the contribution by G. Sigl.

2.5 Galactic and Extragalactic Magnetic Fields – Source/Image Relationship

Magnetic fields intersected by the cosmic rays on their journey from the source to the detector are essential ingredients of UHECR phenomenology. As emphasized in the contribution of G. Medina Tanco, magnetic fields and UHECR will have to be tackled together.

Our present knowledge of cosmic magnetic fields is mainly based on Faraday rotation or Zeeman splitting measures. The common creed is of an average value for the extragalactic magnetic field (EGMF) intensity much lower than a nanogauss and galactic magnetic fields (GMF) of the order of a microgauss with an exponential decrease in the halo. With such a model, UHECR astronomy should become possible with protons (and of course neutral particles). The magnetic rigidity of a proton of 100 EeV is such that the image of a source 50 Mpc away should be blurred only with an angular size of a few degrees. Analytic formulae and detailed discussions are given in Sect. 5.2 of the contribution of G. Sigl, and in Sect. 1 of the contribution by G. Medina Tanco.

However, and since the quoted values are based on a limited number of measurements and some theoretical prejudice, things may be much more complicated than that. Several contributions to this volume deal with this problematic (G. Sigl, Sect. 5.2 and 8.2; P. Biermann *et al.*; G. Medina Tanco). We will attempt a concise report on those very detailed discussions.

Up to a few PeV (knee region) cosmic rays are generally associated with a galactic origin (Sect. 1 of the contribution by G. Medina Tanco), and especially supernova remnants (SNR), because their confinement by GMF as well as their acceleration by the first order Fermi mechanism to such energies are possible. At higher energies, the situation is unclear, in particular beyond the ankle. There, the incoming directions seem isotropic, the nature (therefore the charge) of the cosmic rays is mostly unknown (although a bulk composition in photons, neutrinos or heavy nuclei seems disfavored), and all that can be said is that UHECR do probably not originate in the galactic disk, even though galactic, rapidly rotating young neutron stars have been considered by some authors as a possible source for UHECR (see the contribution by B. Rudak, especially Sect. 7).

There seems to be a soft consensus on a GMF model similar to what is found in many spiral galaxies - at least for the few hundreds of parsecs thick region of the galactic disk (Sect. 2.1 of the contribution by G. Medina Tanco): a regular component of about $2\,\mu$G with at least one reversal in the disk superimposed to a random component, with a total field intensity of $\sim 5\,\mu$G probably increasing towards the center. With such a model, and in the range of a few EeV energies, protons undergo quite small deviations (less than $1°$, same order as the experimental reconstruction uncertainties) except in the direction of the galactic center. Iron nuclei, on the other hand, are likely to be more or less isotropised, except when coming from the direction of the galactic anticenter (see Fig. 3 in the contribution by G. Medina Tanco). Other models have also been envisaged and cannot be excluded on the basis of experimental data, such as the one presented in Sect. 1.3 of the contribution by P. Biermann et al., and based on a galactic wind model where the field has a dominant component $B_\phi \sim \sin\theta/r$ decreasing slowly away from the disk. Such a model has the property [7] of tunelling UHECR in the direction of the galactic pole, and should be testable when large statistics will be available (properties of caustics, north-south asymmetries) (see Figs. 15 and 16 in the contribution of G. Medina Tanco).

The Faraday Rotation measures that are the basis of the often assumed models of EGMF (field intensity of less than a nG, coherence lengths of the order of 1 Mpc) are actually a convolution of the integral of the field intensity transverse to the direction of propagation and the electron column density along the line of sight. One can envisage extreme models compatible with such measurements where, in some directions, the UHECR can encounter structures with field intensities ranging up to the μG level. In a laminar-type model where the sources are embedded inside a thin slab of high value fields, the directional information is totally lost, the reconstructed energy spectrum and the observed fluxes being strongly dependent on the position of the detector (the Earth) with respect to the slab (Sect. 2.3 in the contribution by G. Medina Tanco; Sect. 8.2 in the contribution by G. Sigl). In such a model, and even with relatively local sources ($d < 20$ Mpc, e.g. the Local Supercluster), the often assumed rectilinear propagation for all UHE protons actually would occur only at the highest values of the energy (> 200 EeV): the association between source and image will not be possible without very large statistics. Another model (see Sect. 2.4 of the contribution by G. Medina Tanco) correlates the EGMF with the matter distribution (cells) with high field values over small regions of high matter density (galaxies). In this case, the reconstructed UHECR directions will have strong visible correlations with the large scale structures such as the Supergalactic plane or the Virgo cluster (if these structures include the sources).

One of the most intriguing aspects of the UHECR puzzle is the simultaneous observation of an isotropic distribution of the source images on a large scale and the existence of multiplets (two or three events superimposed in direction within the measurement errors). Recent analyses seem to prove that a chance coincidence cannot explain the multiplets (as an example, the analysis done in [8] on the AGASA and Yakutsk events estimates a chance probability of about 10^{-6}). Moreover, the events in the multiplets seem to show no clear correlation between the time of arrival and energy. These observations raise a series of questions: is a "burst type" production mechanism compatible with the last property?; can we find a reasonable scenario with astrophysical point sources capable of producing the data?; same question with a top-down scenario? The answer to such problems calls for complicated models of charged particle optics, magnetic field configurations, source distributions, but also delicate methods of statistical analysis. Many aspects of this issue are treated in Sect. 8.2 of the contribution by G. Sigl and Sect. 3 of the contribution by G. Medina Tanco. A terse conclusion would be that, given the number of parameters on which one can build any model, no clear conclusion can be reached without improving the statistics at least by a factor of 10. The experimental results leave both the top-down and bottom-up families of models open.

3 Extensive Air Showers: Phenomenology and Detection Techniques

3.1 General Properties

Extensive Air Showers (EAS) are the particle cascade following the interaction of a cosmic ray particle with an atom of the upper atmosphere. The atmosphere acts on an incident cosmic ray as a calorimeter with variable density, a vertical thickness of 26 radiation lengths and 11 interaction lengths. Because of their very low flux, cosmic rays at the highest energies (above the PeV range) cannot be detected directly before they interact with Earth's atmosphere (i.e. with balloon or satellite borne detectors). The necessarily large aperture detectors are therefore ground based and they have to reconstruct the properties of the primary cosmic ray (nature, energy, direction) indirectly by measuring the parameters of the EAS.

Several techniques can be used to do so. The detection of the direct Čerenkov emission by the charged secondaries in the EAS is the basis of $\gamma-$ray astronomy. This technique cannot be used for a full-sky coverage in search of rare events, since it needs the Čerenkov telescope to be oriented in the direction of a point source. A technique presently under development is to look for the radio or acoustic waves generated by the cosmic ray shower in various media (air, water, ice, geological pure salt structures or even the Moon's superficial crust) [9]. This detection technique may soon be operational in a full-size detector for physics. Another method, several decades old, is based on the detection of the scintillation (or fluorescence) light generated by the charged secondaries (mainly electrons) in the EAS by a system of mirrors and phototubes (see the contribution by S. Yoshida on the fluorescence technique). It is currently referred to as the "Fly's Eye" technique from the name of the first detector built by a team of the University of Utah in the early eighties [10]. Such a "fluorescence detector" (FD) sees the longitudinal development of the EAS and measures its energy like a calorimeter by the amount of UV light deposited in the atmosphere (excitation of the nitrogen molecules by the shower electrons). Finally the most frequent detection technique is based on an idea first used by Pierre Auger in the late thirties. It consists in sampling the secondaries of the EAS that reach the ground by a network of particle detectors (scintillators, water Čerenkov tanks, muon calorimeters), as explained in detail in the contribution by P. Billoir. The properties of the primary cosmic ray are deduced from the lateral distribution of the secondaries in a plane section of the EAS. The parameters of such a ground array (altitude, instrumented area, spacing between the detector stations) must be adapted to the energy range aimed for. A detector using a combination of two (or more) of these techniques (e.g. fluorescence telescopes with a ground array) is called *hybrid.*

3.2 Hadron Showers

A typical hadronic shower of 10 EeV has a *size* of 3×10^{10} particles at sea level (atmospheric thickness of 1033 g/cm^2). About 99% of these are photons and electrons/positrons in a ratio of 6 to 1 and they transport 85% of the total energy. The remaining 1% is shared between mostly muons with an average energy of 1 GeV (and carrying about 10% of the total energy), pions of a few GeV (about 4% of the total energy) and, in smaller proportions, neutrinos and baryons. At each step of the cascade the hadronic energy is shared between 70% hadronic and 30% electromagnetic. The shower grows until the charged pions start decaying into muons instead of interacting. The shower development is then at its maximum $X_{\rm max}$ (about 830 g/cm^2 of atmospheric depth) and starts to very slowly decrease. Light nuclei are more penetrating than heavy ones, having therefore a maximum at a larger slant depth: at a given energy there is a difference of about 100 g/cm^2 between the maximum of a shower induced by a proton and one induced by an iron nucleus. Moreover, an iron primary gives 80%

more muons than a proton of the same energy. These two properties - difference in the position of the shower maximum, and in the muon to electron/photon ratio - are the basic parameters used to separate heavy from light nuclei in the primary composition.

The measurement of the direction of the incident cosmic ray is straightforward with a ground array, and mostly model independent. For that, one uses the time of arrival of the particles in the shower front if three or more stations are hit close to the shower axis (Sect. 3.2 of the contribution by P. Billoir). With a FD, a good angular resolution (reconstruction of the shower axis) needs one or several of the following conditions:

- A large number of pixels hit by the produced fluorescence light;
- A stereoscopic view of the EAS by two telescopes set 10-40 km apart;
- A hybrid observation of the EAS where the ground array provides the position of the shower core on the ground.

The angular resolution can reach a fraction of a degree, at high energies and in the hybrid detection mode.

The energy measurement is more direct with a FD than a ground array. The longitudinal development of the shower obeys the rather simple Gaisser-Hillas function (Sect. 2.2 of the contribution by P. Billoir) whose integral is proportional to the total amount of the fluorescence light deposited in the atmosphere by the charged secondaries of the EAS, therefore to the energy of the primary. The difficulty here is to estimate properly the various background sources (including the diffused or direct Čerenkov light), and to monitor as precisely as possible the extinction length of the light due to scattering by air (Rayleigh) or aerosols (Mie, the most delicate). A detailed account on the sources of systematic errors and on the methods used in such measurements is given in Sect. 2 and 3 of the contribution by S. Yoshida. It is shown that with a very good atmospheric monitoring, one can reach systematic errors as low as 10% for UHE showers detected at a distance of about 30 km. To this one should add uncertainties coming from the telescope mirrors, phototubes and associated electronics (Sect. 4 of the contribution by S. Yoshida), the undetected energy (carried away by neutrinos and high energy muons) and the poorly known fluorescence yield due to low energy electrons. The overall energy resolution with a FD is usually of order 30% in the UHE range.

A ground array uses the data recorded by stations up to several kilometers from the shower core to reconstruct the lateral distribution function, namely the density of particles per unit area as a function of the distance to the shower core, $\rho(r)$. The density interpolated for a typical distance of 1 km (where the shower fluctuations are minimized) is a simple (almost linear) function of the initial energy, independent of the nature of the primary. Several analytic functions are used for the lateral distribution function (Sect. 3.2 of the contribution by P. Billoir). At UHE, a resolution of 10% should be reachable, to which one should add some systematic uncertainties especially at large zenith angles for which mainly the high energy muonic halo of the shower will be observed by the ground stations.

As was said above, the main parameters used for the identification of the primaries are the position of the shower maximum $X_{\max}$ and the relative content in muons at ground level. Other shower properties such as the steepness of the lateral distribution function, the signal risetime in the ground array stations, the flatness of the shower front etc derive from these and the geometry of the shower development (zenith angle). Unfortunately, physical fluctuations (Sect. 2.6 of the contribution by P. Billoir) make it almost impossible to identify the primary on a shower-by-shower basis (e.g. the shower maximum for a given particle at a given energy fluctuates by several tens of g/cm^2). One expects to improve the identification methods for showers observed in the hybrid mode where a multi-dimensional analysis becomes possible. However, it is very likely that separation between heavy and light nuclei will be possible only on a statistical basis.

3.3 Photon Showers

Photons with energies above 10 EeV have a large probability of converting into e^+e^- pairs if they cross magnetic fields with non-zero transverse components, as is the case with the geomagnetic field in most of the incident directions. In such a case, a preshower is started before arrival in the atmosphere. However, in some directions the photon arrives in a direction parallel to the geomagnetic field vector and does not convert before entering the atmosphere. Then a second process, the Landau-Pomerantchuk-Migdal (LPM) effect takes over which can be interpreted as a decrease of the electromagnetic interaction cross sections. In short, UHE photons unconverted in the geomagnetic field (i.e. precise directions in the reference frame of the Earth) penetrate deeply in the atmosphere and start their development late. The converted photons, on the other hand, start their development very high, much above the atmosphere and a large part of their electromagnetic component is absorbed when arriving at the ground level. The difference between the two categories can be detected e.g. by measuring the curvature of the shower front: large curvature for unconverted photons, small curvature for converted ones. Such important differences correlated with specific directions related to the geomagnetic fields should be easily detectable. A contamination of about 5-10% of the incident cosmic ray sample by UHE photons is expected to be visible (Sect. 2.7 and 4.2 of the contribution by P. Billoir).

3.4 Neutrino Showers

At a slant depth of 2000 g/cm^2, the electromagnetic component of the EAS is mostly extinguished, leaving only high energy muons to arrive at the ground level. This property is used in the detection of UHE neutrinos. The principle is to look for horizontal air showers (zenith angles larger than 70°) which reach the detector after an atmospheric thickness as large as 36,000 g/cm^2 (for a zenith angle of 90°). A neutrino, unless it has non-standard interaction properties, will interact uniformly in the atmosphere (with a probability close to 10^{-4} for nearly horizontal directions), have various shower front curvatures depending on the depth at which the interaction takes place, and have an electromagnetic component. The background events however (hadronic showers starting far from the detector) will have very flat shower fronts mainly from deeply penetrating muons (and their accompanying bremsstrahlung halo), easily distinguishable from a neutrino initiated hadronic or electromagnetic shower (Sect. 4.1 of the contribution by P. Billoir). An interesting special case, that of the tau neutrino, is also briefly mentioned in this section (see also [11]).

4 UHECR Sources

Today's understanding of the phenomena responsible for the production of UHECR, i.e. the transfer of macroscopic amounts of energy to microscopic particles, is still limited. One distinguishes two classes of processes: the so-called "top-down" and "bottom-up" scenarios. In the former, the cosmic ray is one of the stable decay products of a supermassive particle. Such particles with masses exceeding 1 ZeV can either be metastable relics of some primordial field or highly unstable particles produced by the radiation, interaction or collapse of TDs. In the bottom-up mechanism the energy is transferred to the cosmic rays through their interaction with electromagnetic fields. This classical approach does not require new physics as opposed to the "top-down" scenario, but does not exclude it either since, in some models, the accelerated particle - the cosmic ray - is itself "exotic". The GZK cutoff puts severe constraints on the distance that a cosmic ray can travel before losing most of its energy or being absorbed. The absence of prominent visible astrophysical objects in the direction of the observed highest energy cosmic rays together with this distance cutoff adds even more constraints on the "classical" bottom-up picture.

It is beyond the scope of this summary to describe all the scenarios - they are far too numerous - proposed for the production of the UHECR. Let us simply agree on the fact that the profusion of models shows that none of them is totally satisfactory and that the data are not very constraining. Consequently we will try to present the main features of the various acceleration mechanisms and production models presented in these proceedings by P. Bhattacharjee and G. Sigl, G. Pelletier, B. Rudak, E. Waxman, P. Biermann et al., and P. Biermann and G. Sigl.

4.1 Conventional Acceleration: Bottom-Up Scenarios

The first and most straightforward classical acceleration mechanism one can think of is certainly the direct one-shot acceleration by very large electric fields which can be found in or near very compact objects such as highly magnetized neutron stars or the accretion disks of black holes. However, extensive studies have shown that this idea, although very simple, requires far from elementary MHD modeling. In his lecture B. Rudak reviews the main characteristics of strongly magnetized neutron stars as UHECR accelerators (notably Sect. 4 and 7).

Three fundamental features differentiate the various models: the site of acceleration with respect to the neutron star, the mechanism that transfers the rotational energy to charged particles, the nature (and source) of the accelerated particles. For UHECR the acceleration site must be beyond the light cylinder to make full use of the maximum potential drop and to limit synchrotron losses. Actually, the neutron star could play the role of the final kick for Fermi pre-accelerated particles from the interstellar medium and supernovae remnants. The final energy depends solely on the entry and exit points of the particle inside the pulsar nebula and is maximum when these points are the pole (defined by the rotation axis) and the equator. In such a model the final particle spectra depend very strongly on the spectral and spatial properties of the pre-accelerated particles.

A different approach involves acceleration of proton or iron nuclei from inside the light cylinder i.e. supplied by the neutron star itself. In such models the ratio of the Poynting flux to the particle kinetic energy flux must not remain constant, being larger than one at the light cylinder and much smaller far away from it, reflecting the energy transfer from the spin-down flux to the particles. As B. Rudak puts it *"no consensus about likely mechanisms responsible for the dissipation of the Poynting flux has been reached so far, though several models have been proposed"*. In such models a single iron nucleus could reach a maximum energy of:

$$E_{\mathrm{max}} \simeq Z_{26} B_{13} P_{\mathrm{ms}}^{-2}\,\mathrm{ZeV}$$

where P is the rotation period.

The second mechanism is based on diffusive stochastic shock acceleration in magnetized plasma clouds which generally occurs in all systems where shock waves are present such as supernova remnants or radio galaxy hot spots. This statistical acceleration known as the Fermi mechanism is reviewed by G. Pelletier. Among all the astrophysical objects where strong shocks may occur only a few can possibly accelerate particle up to a ZeV. These are Active Galactic Nuclei (AGN) and Fanaroff-Riley Class II (FRII) radio galaxies and Gamma Ray Bursts (GRB) (see the contributions by P. Biermann *et al.*, G. Pelletier, and E. Waxman).

AGN Cores and Jets

Blast waves in AGN jets have typical sizes of a few percent of a parsec with magnetic fields of the order of 5 gauss. They could in principle lead to a maximum energy of a few tens of EeV. Similarly for AGN cores with a size of a few 10^{-5} pc and a field of order 10^3 G one reaches a few tens of EeV. However those maxima, already marginal, are unlikely to be achieved under realistic conditions. The very high radiation fields in and around the central engine of an AGN will interact with the accelerated protons producing pions and e^+e^- pairs. Additional energy loss due to synchrotron radiation and Compton processes lead to a maximum energy of about 10 PeV, much below the initial value. To get around this problem, the acceleration site must be away from the active center and in a region with a lower radiation density such as in the terminal shock sites of the jets: a requirement possibly fulfilled by FRII radio galaxies.

FRII Radio Galaxies
Radio-loud quasars are characterized by a very powerful central engine ejecting matter along thin extended jets. At the ends of those jets, the so-called hot spots, the relativistic shock wave is believed to be able to accelerate particles up to ZeV energies. This estimate depends strongly on the value assumed for the spot's local magnetic field, a very uncertain parameter. Nevertheless FRII galaxies seem the best potential astrophysical source of UHECR. Unfortunately, no nearby (less than 100 Mpc) object of this type is visible in the direction of the observed highest energy events. The closest FRII source, actually in the direction of the Fly's Eye event at 320 EeV, is at about 2.5 Gpc, way beyond the GZK distance cuts for nuclei, protons or photons. A possible solution would be to admit that all observed UHECR come from M87 (a powerful radio galaxy only 20 Mpc away) and that arrival directions are all randomized by unexpectedly strong magnetic fields (see the contributions by Biermann et al., G. Sigl, and G. Medina Tanco).

γ–Ray Bursts
γ–ray bursters (GRB) are intense sources of γ–rays, sometimes of a few milliseconds duration, with γ energies ranging from about 1 keV to a few GeV. Several hundreds have been observed by satellites. The most favored GRB emission model is the "expanding fireball model" where one assumes that a large fireball, as it expends, becomes optically thin hence emitting a sudden burst of γ–rays. The engine (the power source) of such a fireball remains unknown while the explanation of the non thermal spectra observed needs some additional modeling (such as internal shocks in the expanding fireball).

The observation of afterglow (low energy γ–ray emission of the heated gas in which the fireball expands) allowed to measure the redshift of the GRB from which their cosmological origin was confirmed (and support brought to the fireball model). Under certain conditions, GRB can be shown to accelerate protons up to a ZeV, therefore making them a good candidate site for UHECR production, as explained in detail in the contribution by E. Waxman. However in such a framework the UHECR spectrum should show the GZK cutoff while above 100 EeV the distribution of arrival directions should be anisotropic. Although more data are needed, the most recent results from the AGASA experiment [12] confirm the absence of the GZK cutoff in contradiction with the GRB hypothesis for the acceleration of UHECR. GRB remain however one of the best "classical" candidates for UHECR acceleration (if not the only one). In the future, the detection of high energy neutrinos (from 0.1 PeV up to 1 EeV depending on the GRB environment) in coincidence with a GRB would be a strong evidence for this model.

4.2 "Exotic" Sources: The Top–Down Scenario

One way to overcome the many problems related to the acceleration of UHECR, their flux, the visibility of their sources and so on, is to call upon the decay of super massive relic particles (SMRP) or of TDs. These decays produce, among other things, quarks and leptons. The quarks hadronize, producing jets of hadrons which, together with the decay products of the unstable leptons, result in a large cascade of energetic photons, neutrinos and light leptons with a small fraction of protons and neutrons, part of which become the UHECR (see the contribution by P. Bhattacharjee and [13]).

For this scenario to be observable three conditions must be met:

- The decay must have occurred recently since the decay products must have traveled less than about 100 Mpc because of the attenuation processes discussed above.
- The mass of this new particle must be well above the observed highest energy (100 EeV range).
- The ratio of the volume density of this particle to its decay time must be compatible with the observed flux of UHECR.

According to the current picture on the evolution of the Universe, several symmetry breaking phase transitions from a Grand Unified Theory group (GUT) such as $GUT \Longrightarrow H \dots \Longrightarrow SU(3) \times SU(2)_L \times U(1)_Y \Longrightarrow E(1)_{EM}$ occurred during the cooling. In TD models the defects are leftovers from these GUT symmetry breaking phase transition that occurred in the very early universe. Quantitative predictions of the TD density that survives a possible inflationary phase rely on a large number of theoretical hypotheses. In general TD will be cosmologically distributed and produce GUT scale particles (around $10^{24} - 10^{25}$ eV).

Supermassive particles are relics from some primordial quantum field, produced after the now commonly accepted inflationary stage of our Universe. The ratio of their lifetime to the age of the universe must match their relative abundance to account for the observed rate of UHECR while their mass must exceed 10^{21} eV. It is worth noting that relic particles may also act as non-thermal Dark Matter and cluster in the halo of our galaxy. In such a case the secondary particles will not be affected by the GZK cutoff.

Topological Defects

The very wide variety of TD models together with their large number of parameters makes them difficult to review in detail. Many authors have addressed this field. Among them, let us mention Vilenkin and Shellard [14] and Vachaspati [15] for a review on TD formation and interaction, and Bhattacharjee [16], Bhattacharjee and Sigl [13] and Berezinsky, Blasi and Vilenkin [17] for a review on experimental signatures in the framework of the UHECR and of course the contribution of P. Bhattacharjee in this volume.

As mentionned above, in the current picture on the evolution of the Universe several symmetry breaking took place. For those "spontaneous" symmetry breaking to occur, some scalar field (called the Higgs field) must acquire a non vanishing expectation value in the new vacuum (ground) state. Quanta associated to those fields have energies of the order of the symmetry breaking scale, e.g. $10^{15} - 10^{16}$ GeV for the grand unification scale. Such values are indeed perfectly in the range of interest for the above mentioned $X-$particles.

During the phase transition, regions not causally connected may evolve towards different states - the correlation length is smaller than the horizon - in such a way that at the different domain borders, the Higgs field is forced to keep a vanishing expectation value for topological reasons. Energy is thus trapped at the border called a TD whose properties depend on the topology of the manifold where the Higgs potential reaches its minimum (the vacuum manifold topology).

Possible TDs are classified according to their dimensions: magnetic monopoles (0-dimensional, point-like); cosmic strings (1-dimensional); vortons, a subvariety of the previous which carry current and is superconducting; domain walls (2-dimensional); textures (3-dimensional). Among those, only monopoles and cosmic strings are of interest as possible UHECR sources: textures do not trap energy while domain walls, if they were formed at a scale that could explain EHECR, would overclose the Universe [18]. In GUT theories, magnetic monopoles always exist because the unbroken symmetry group contains at least the electromagnetic $U(1)$ invariance. In fact it is the predicted overabundance of magnetic monopoles in our present universe that led Guth [19] to come up with the now well adopted idea of an inflationary universe. Cosmic strings on the other hand are the only defects that can be relevant for structure formation. It is possible, from the scaling property of the cosmic string network, to relate the cosmic string formation scale η to the cosmic string contribution to the density fluctuations in the Universe. Using the large scale density fluctuation value of $\delta\rho/\rho \sim 10^{-5}$ this gives $\eta \lesssim 10^{16}$ GeV and similar conclusions are drawn if one uses the COBE results on CMB anisotropies [20]. It is striking to see that if cosmic strings were to play a role in large scale structure formation, hence making the Hot Dark Matter scenario viable,

- the proper energy scale is approximately the grand unification scale of GUT theories,
- this scale also corresponds to the one relevant for UHECR production.

When two cosmic strings intercommute, the energy release sometimes leads to the production of small loops that will free more energy when they collapse. These are, among other mechanisms, fundamental dissipation processes that prevent the cosmic string network from dominating the energy density in the Universe. For monopoles, it is the annihilation of monopolonia (monopole-antimonopole bound states) [21,22] that releases energy[1] - although the existence of monopoles of the proper energy scale is very questionable as they are either over abundant or washed out by inflation.

Cosmic strings and monopoles come in various forms according to the scale at which TDs are formed and to the vacuum topology. They may even coexist.

[1] In fact monopolonia are too short lived but monopole-anti-monopole pairs connected by a cosmic string have appropriate lifetimes. This happens when the $U(1)$ symmetry is further broken into Z_2

Nevertheless, the decay rate may, on dimensional grounds, be parameterized in a very general way [23]. Due to the cosmological distribution of the defects the electromagnetic component of the decay will cascade and release its energy into low energy photons (10 MeV - 100 GeV). The density of such photons depends on the time evolution of the decay rate and can be compared to the diffuse extra-galactic γ−ray background as measured by EGRET, putting severe constraints on models with slow time evolution. On the other hand in models with rapid time evolution the large density of γ−rays released in the early Universe impacts on the ^{4}He production and on the uniformity of the CMB. There again severe constraints make it difficult to accommodate the TD scenario for the production of UHECR.

Supermassive Relics
Supermassive relic particles may be another possible source of UHECR [24,25]. Their mass should be larger than 10^{12} GeV and their lifetime of the order of the age of the Universe since these relics must decay now (close by) in order to explain the UHECR flux. Unlike cosmic strings and monopoles, but like monopolonia, relics aggregate under the effect of gravity like ordinary matter and act as a (non thermal) cold dark matter component. This is a strong argument in favor of those models in the UHECR context as their distribution should be biased towards galaxies and galaxy clusters allowing their decay products to evade the GZK cutoff. A high statistics study of the UHECR arrival distributions will be a very powerful tool to distinguish between aggregating and non-aggregating top-down sources.

If one neglects the cosmological effects, a reasonable assumption on the decay rate would simply be, since the decay should occur over the last 100 Mpc/c:

$$\dot{n}_X = \frac{n_X}{\tau}$$

where τ is the relic's lifetime and where the relic density n_X may be given in terms of the critical density of the Universe ρ_c as:

$$n_X = \frac{\rho_c \Omega_X}{m_X} = 10^{-17}(\Omega_X h^2)\left(\frac{m_X}{10^{12}\ \text{GeV}}\right)^{-1}\ \text{cm}^{-3}$$

From which, using the measure of the UHECR flux and a relic mass $m_X = 10^{12}$ GeV, one obtains a lifetime of the order of $10^{21}(\Omega_X h^2)$ years. To obtain such a value, orders of magnitude larger than the age of the Universe, one needs a symmetry (such as R−parity) to be very softly broken unless the fractional abundance Ω_X represents only a tiny part ($\sim 10^{-11}$) of the density of the Universe, in which case the production mechanism of relics must be extraordinarily inefficient.

5 Conclusions

The chemical composition of the cosmic rays, the shape of their energy spectrum and the distribution of their directions of arrival will prove to be powerful tools to distinguish between the different acceleration or decay scenarios.

There is a basic creed on the UHECR problematics which makes them a puzzle, and there are models capable of circumventing the puzzling elements. Basically, if the UHECR are conventional hadrons accelerated by bottom-up mechanisms, they should correlate with their sources, with a quite specific distribution in the sky and a spectrum clearly showing the GZK cutoff (if the sources are cosmologically distributed). If, on the other hand, the accelerated particles are not conventional, they should at least be neutral particles in order not to interact with the CMB and therefore can only be secondary collision products putting even more requirements on the source power. Moreover, they must interact strongly with the atmosphere. Many examples showing how one can put up with those requirements were given in this volume.

For top-down mechanisms and above a ZeV, one should observe a flux of $\gamma-$rays (and neutrinos) as the $\gamma-$ray absorption length increases (up to $\simeq$ 100 Mpc) at extreme energies. Below 100 EeV the spectrum shape will depend on the relative values of a few parameters: the characteristic distance between TD interactions or relic particle decays and Earth, the proton attenuation length, and the $\gamma-$ray absorption length.

For relic particles and TDs like vortons and monopolonia, because of the possible accumulation in the galactic halo, photons will dominate the flux. Some anisotropy should be visible due to Earth's eccentric position in the halo. In this case, the spectrum will not show any GZK cutoff and the EGRET constraint on the injection rate is not crucial as the emitted photons have no time to cascade over the short distances.

Today, we seem to be a long way from untangling the UHECR puzzle. Fortunately, the numerous models will soon have to face the next generation of experiments capable of gaining orders of magnitude in statistics with respect to the present data. For example, a dominant presence of UHE photons and neutrinos in the chemical composition would no doubt be fatal to any bottom-up mechanism, whereas the presence of unambiguously identified heavy nuclei would definitely exclude all top-down models. Finding *the* actual mechanism should then be achievable with a precise reconstruction of the energy spectrum and a detailed study of the (an)isotropy of the source images, probably a few years' work away.

References

1. M. Nagano, A. A. Watson: Rev. Mod. Phys. **72**, 689 (2000)
2. V. Berezinsky, A. Z. Gazizov, S. I. Grigorieva: hep-ph/0107306 (extended version of paper in *Proc. 27th International Cosmic Ray Conference*, Hamburg, Germany 07-15 August, 2001)
3. A. V. Olinto: Proc. *International Workshop on Observing Ultra High Energy Cosmic Rays From Space and Earth*, Metepec, Puebla, Mexico (2000), astro-ph/0011106
   A. V. Olinto: Phys. Rep. **333-334**, 329 (2000)
4. A. Dar, R. Plaga: Astron. Astrophys. **349**, 259 (1999)
   A. Dar, A. De Rujula: astro-ph/0102115

5. P. Sreekumar, et al.: Astrophys. J. **494**, 523 (1998)
6. D. Fargion, B. Mele, A. Salis: Astrophys. J. **517**, 725 (1999)
T. J. Weiler: Astropart. Phys. **11**, 303 (1999)
7. P. Billoir, A. Letessier-Selvon: astro-ph/0001427
8. P.G. Tinyakov, I. I. Tkachev: Pis'ma Zh. Eksp. Teor. Fiz. **74**, 3 (2001) [JETP. Lett. **74**, 1 (2001)]
9. P. W. Gorham, et al.: Phys. Rev. E **62**, 8590 (2000)
10. G. L. Cassiday: Ann. Rev. Nucl. Part. Sci. **35**, 321 (1985)
11. D. Fargion: astro-ph/0002453; astro-ph/0101565 (2001)
X. Bertou, et al.: Astropart. Phys., in press, astro-ph/0104452
12. N. Sakaki, et al.: Proc. *27th International Cosmic Ray Conference*, Copernicus Gesellshaft 333 (2001)
13. P. Bhattacharjee, G. Sigl: Phys. Rep. **327** 109 (2000)
14. A. Vilenkin, E. P. S. Shellard: *Strings and Other Topological Defects*, Cambridge Univ. Press (Cambridge, 1994)
15. T. Vachaspati: *ICTP Summer School 1997 - High energy physics and cosmology*, hep-ph/9710292
T. Vachaspati: *Les Houches 1999 - Topological defects and the non-equilibrium dynamics of symmetry breaking phase transitions* , astro-ph/9903362
16. P. Bhattacharjee: *Observing the Highest Energy Particles (*$> 10^{20}$ *eV) from Space*, College Park, Maryland, 1997, astro-ph/9803029
17. V. Berezinsky, P. Blasi, A. Vilenkin: Phys. Rev. D **58**, 103515 (1998)
18. Ya. Zel'dovich, I. Kobzarev, L. Okun: Zh. Eksp. Teor. Fiz. **67**, 3 (1974)
19. A. H. Guth: Phys. Rev. D **23**, 347 (1981)
20. R. H. Brandenberger: Pramana **51**, 191 (1998)
21. C. T. Hill: Nucl. Phys. **224B**, 469 (1983)
22. D. N. Schramm, C. T. Hill: Proc. *18th International Cosmic Ray Conference*, 393 (1983)
23. P. Bhattacharjee, C. T. Hill, D. N. Schramm: Phys. Rev. Lett. **69**, 567 (1992)
24. V. Berezinsky: Proc. *TAUP97*, Nucl. Phys. B Proc. Suppl. **70B** (1999) 419-430
25. S. Sarkar, R. Toldra: hep-ph/0108098

Index